The Origin and the Evolution of Life
Based on a Novel Philosophy

Yoshihiro NAKATO
Professor Emeritus
Osaka University, Japan

Copyright © 2018 Yoshihiro NAKATO
All rights reserved.
ISBN-13: 978-1979357074

Preface

Science and technology play an overwhelming role in the present societies. The situation produces an atmosphere which leads us to the belief that science correctly catches the truth about nature and the world and a scientific approach is the only way to get at it. In fact, many of scientists have made research under the premise that science is in principle complete. Some of scientists even claim that philosophy is unnecessary in scientific study because it has enough ability to reach the truth. A Nobel laureate, Steven Weinberg, also claims in his book, "Dreams of a Final Theory" (1992), that philosophy has been completely useless in intellectual progress. Well, is science really complete? Our knowledge in the present days solely relies on science. Therefore, it is of fundamental importance to clarify whether science is in principle complete or not. If it is not complete, we have to work out a way to make up for the incompleteness.

In reality, science is not complete. Indeed, science is useful but in essence of an approximate character. In fact, science has now come to show severe difficulty in explaining what life is. For example, all living things including primitive bacteria demonstrate autonomous dynamic self-organizing ability or free independent harmonious spirit and wisdom (creativity). Science has yet given no sufficient explanation to why living things can demonstrate such ability. Science has also not yet succeeded in disclosing how the first living organism emerged on the primitive earth, how it evolved later, and where human ego and free will come from. In addition, modern high-performance computers, humanoid robots and artificial intelligence (AI), which are constructed based on scientific knowledge, are still only machines and entirely lack life, spirit and creativity. Somebody may say that such an unsuccessful situation of the present science about life is simply due to historical limitations. However, such an opinion seems to be too optimistic. Recently, various new sciences such as information science, nonlinear science and complexity science have risen but they also have still been unable to clarify the above problems.

In this way, it is certain that the incompleteness of science has now been clearly embodied in the study of life. Thus, *the purpose of this book is to reveal where the incompleteness of science comes from and construct a new theory of the origin and the evolution of life and related topics, based on it.* Fortunately, I have succeeded in dealing with these issues by investigating them from an absolutely novel standpoint. A fundamental idea is that human consciousness and hence science, which is based on human consciousness, only catch "common" or "repeatedly observed" properties of natural things and fail to catch natural things themselves. Here is why we have to say that science is in essence of an approximate character.

Indeed, it may be well recognized that science only catches things with common or repeatedly observed properties or things with individual unchanging properties of natural things. However, it has in general been believed that things with such properties catch the essence of nature and thus neglected parts, which are in most cases only individual phenomena under particular conditions, can be correctly recovered by combining scientific knowledge (essence) and particular conditions. Unfortunately, this belief is not correct. The autonomous dynamic self-organizing ability or free independent harmonious spirit and wisdom (creativity) of living things can never be recovered based on things with common or repeatedly observed properties or things with individual unchanging properties. Such things can only produce machines or reproduce events or phenomena of a mechanical character. This book shows that free independent harmonious spirit and wisdom (creativity) of living things come from natural things themselves that human consciousness has failed to catch. The truly correct understanding of nature can never be obtained without taking into account natural things themselves.

Strikingly, what is stated above implies that natural things themselves, which provide us with the truly correct understanding of nature, are in *the internal world we cannot be conscious of*. We usually recognize what we look at or what we are conscious of as real things. However, this is a big mistake. In reality, such things are like pictures on a screen in a movie. Namely, they are only images in the external visible world, produced by something internal. Therefore, for gaining the truly correct understanding of nature, it is indispensable not only to obtain scientific knowledge about such external images but also to catch what produce them in the internal invisible world.

Now, this book consists of six chapters, apart from Appendix. Chapter 1 has revealed that nature really has the internal invisible world and life comes from it. These conclusions are obtained from detailed investigations of the basic qualities of living things, the historical origin of human ideas or words, and experiences in our daily life called *Tai-toku*. Chapter 2 and Chapter 3 show that modern science leads to the same conclusions as in Chapter 1 if it is interpreted from a novel standpoint. These chapters demonstrate that the recognition of the internal invisible world leads to the conclusion that the law of increase of entropy is directly deduced from quantum mechanics without using Boltzmann's statistical interpretation. The origin of this law is not a probability but the basic qualities of microscopic particles such as electrons and atomic nuclei. These chapters also demonstrate that the ultimate origin of free independent harmonious spirit and wisdom of living things is the basic qualities of microscopic particles.

Based on these results, Chapters 4 to 6 demonstrate that a new interpretation of modern science by taking into account the internal invisible world leads to completely convincing solutions to many important problems that have been left unsolved to date, such as how the first living organism emerged on the primitive earth, how it evolved later, where human ego and free will come from, and how creation is achieved. Interestingly, this book shows that natural evolution and human creation happen by essentially the same principle. It should be mentioned also that this book provides a key instruction on how we can produce artificial intelligence with free spirit and creativity.

For making discussions in this book easier to understand, let us here explain some novel concepts introduced in this book. Firstly, this book introduces novel concepts of *truly real things* and *the very truth*, apart from traditional concepts of *real things* and *the truth*. It was mentioned earlier that human consciousness only catches "common" or "repeatedly observed" properties of natural things and fails to catch natural things themselves. This means that natural things themselves are *truly real things*. Thus, *the very truth* stands for the way that natural things themselves exist or truly real things exist. An important point is that both truly real things and the very truth, which lead us to the truly correct understanding of nature, are in the internal world we cannot be conscious of. According to this understanding, real things in traditional understanding and philosophies refer to "common" or "repeatedly observed" properties of truly real things or imaginary things with such properties. Thus, the truth in traditional understanding and philosophies stands for the way that such properties or such imaginary things exist. We can be conscious of both real things and the truth. However, they both are of an approximate character.

This book also introduces novel concepts of *a continuous internal connection* and *the internal ability*. A truly real thing exists as *a continuous internal connection* of all other "truly real things" formed in a particular quality level and aspect, though it demonstrates individual unchanging properties in a common or averaged form in the external visible world. We human beings are conscious of such individual unchanging properties. An important property of such a continuous internal connection is that it has *the internal ability* to organize itself so that it realizes a free independent harmonious stable state. Now, according to quantum mechanics in modern science, everything in nature is described by a non-observable continuous wave function and demonstrates individual unchanging properties in the external visible world. Therefore, the above characteristics of a truly real thing are in harmony with the view of nature which quantum mechanics displays. In this way, we can show that a truly real thing has a continuous harmonious and dynamic self-organizing character and acts as the origin of free independent harmonious spirit and wisdom of living things.

Nature outwardly looks as if it consisted of a variety of things with individual unchanging properties but in reality it exists in a fully continuous, dynamic and harmonious state. All individual things we look at or think of are only the external appearance of truly real things in the internal invisible world. The importance of taking into account the internal invisible world is that it makes our understanding extend deep into the area we cannot be conscious of. On the other hand, word-based scientific understanding is always restricted within the area we can be conscious of.

The purpose of this book mentioned at the beginning has been fundamentally achieved by the discussions in Chapter 1 to 6. However, to be strict, there remains an important problem. We cannot be conscious of truly real things and the very truth, as mentioned above. This means that we cannot gain the truly correct understanding of nature and the world by the use of words even if we take into account the internal invisible world. The internal world expressed by words is not the internal world itself. Therefore, some discussions of how we can surmount this problem are added as Appendix at the last part of this book for readers who are interested in this issue. A brief consideration of this problem is also given in Section 1.3 and 1.4. Fortunately, this problem can be successfully handled based on the philosophy of Buddhism, which teaches us that we can transcend word-based understanding and arrive at the very truth. Just arrival at the very truth provides us with the truly correct understanding of nature and the world. Note that only the philosophy of Buddhism deals with this issue because the other philosophies including science are based on what we are conscious of and do not deal with the very truth. The philosophy of Buddhism provides us with a way to make up for the incompleteness of science.

This book deals with a large variety of issues over a wide range of knowledge from quite a novel standpoint. In fact, this book covers not only natural sciences such as life science, complexity science, information science, brain science, physics, chemistry and biology but also philosophy. Therefore, <u>this book focuses on the most basic parts of each issue and does not deal with advanced detailed theories</u>. This is also because the most basic parts have been left uncertain to date, as mentioned above, though they should be the most important parts. Accordingly, this book becomes easy to understand for university and graduate-course students as well as professional researchers.

However, I should like to emphasize that arguments in this book provide many important novel ideas and insights which play key roles in professional study. In general, a change in the basic part of knowledge brings about a drastic alteration in wide areas of knowledge. I do hope that novel ideas and insights proposed in this book will lead to new developments and fruitful results in combination with detailed theories in each field.

Lastly, I would like to express my sincerest thanks to Dr. S. Nagakura, Emeritus Professor of the University of Tokyo, for warm encouragement and kind suggestions. I would also like to express my hearty thanks to Dr. H. Tributsch, Emeritus Professor of Free University and Hahn-Meitner-Institute (Helmholtz-Zentrum Berlin in the present), for reading drafts of initial and later versions and giving a number of valuable comments together with warm encouragement. Thanks are also expressed to Drs. S. Nakanishi and R. Nakamura for offering information about recent scientific studies. I am deeply indebted to my wife for her continuous support and encouragement. Some of discussions in this book are based on considerations in my previous book: Y. Nakato, "A Method of Improving Creativity" (in Japanese), Daigaku Kyo-iku Shuppan, Okayama (2001).

January 2018

Yoshihiro NAKATO

Contents

Japanese technical terms and analogs are highlighted by italics with the first letter written in capital in such a way as *Tai-toku*, *Kuh* and *Satori*.

1.	**Does the Internal Invisible World Really Exist?**	**1**
1.1	A Living Thing Has a Strange Way of Existing	1
1.1.1	Comparison of Living Things and Machines	2
1.1.2	How Are Living Things Understood in Modern Science?	8
1.2	Features and Limits of Scientific Understanding	16
1.2.1	Science Only Catches Common Properties of Natural Things	16
1.2.2	Scientific Knowledge Is Useful as a Technique	26
1.2.3	Various Problems Come from Word-based Scientific Understanding	30
1.3	Experiences of Arriving at the Very Truth in Daily Life	37
1.3.1	Splendid Abilities Are Acquired by Arrival at the Very Truth	38
1.3.2	The Origin of Splendid Abilities Acquired by Arrival at the Very Truth	43
1.3.3	A Human Being Lives with the Internal Invisible World as the Base	51
1.4	A Novel View of Nature and the World	53
1.4.1	How Do Truly Real Things Exist?	54
1.4.2	Nature Has a Dual Structure	66
1.4.3	How Can We Obtain the Truly Correct Understanding?	71
2.	**The Internal Invisible World in Modern Science**	**85**
2.1	The World of Microscopic Particles	85
2.1.1	Fundamental Principles of Quantum Mechanics	85
2.1.2	A Continuous Internal Connection in Quantum Mechanics	93
2.1.3	The Internal Ability in Quantum Mechanics	105
2.2	The Internal Ability in Macroscopic Visible Systems	109
3.	**Dynamic Self-organization in Non-equilibrium Systems**	**124**
3.1	Preceding Studies of Non-equilibrium Systems	124

3.1.1	Fundamental Characteristics of Non-equilibrium Systems	124
3.1.2	A Stationary State and a Dissipative Structure	133
3.1.3	A Reaction Network	145
3.2	A New Approach to Non-equilibrium Systems	150
3.2.1	The Internal Ability in a Stationary State	151
3.2.2	A Reaction Network with the Internal Ability to Organize Itself	156
3.2.3	Can We Construct a Humanoid Robot with Free Spirit and Creativity?	163

4.	**A New Theory of the Origin and the Evolution of Life**	**169**
4.1	A Brief Survey of Previous Studies on Life	169
4.1.1	Previous Studies on the Evolution of Life	169
4.1.2	Previous Studies on the Origin of Life	174
4.2	A New Theory of the Origin of Life	178
4.3	How Have Living Things Evolved?	195
4.4	The Principle of Evolution	201

5.	**The Origin of Human Ego and Consciousness?**	**213**

6.	**How Is Creation Achieved?**	**221**
6.1	How Has Creation Been Understood to Date?	221
6.2	The Principle of Creation	227
6.3	How Does the Internal Ability Work in Processes of Achieving Creation?	237

Appendix

A1.	**The Philosophy of Buddhism (I)** – The Disclosure of the Internal World and the Internal Ability –	**249**
A1.1	Fundamental Characteristics of the Philosophy of Buddhism	250
A1.2	The Birth of Buddhism	255
A1.2.1	Shaka-muni Opened the Door to the Very Truth	255
A1.2.2	Initial Buddhism	260
A1.3	Mahayana Buddhism	274
A1.3.1	A Brief Survey of Mahayana Buddhism	275
A1.3.2	Mahayana Buddhism in the Primary Stage	290
(A)	The Concept of Emptiness (*Kuh*)	291

(B)	The Concepts of Buddha and Buddha's Land	296
A1.3.3	Mahayana Buddhism in the Middle and the Last Stages	302
A1.4	Buddhism in North East Asia	304

A2.	**The Philosophy of Buddhism (II)** – The Disclosure of How the Internal Ability Works in This World –	**310**
A2.1	A Historical Background of Shin-ran's Work	310
A2.2	The *Satori* of Shin-ran	319
A2.3	The Correct Way to Attain *Satori*	329
A2.4	The Way to Live in This World in Accord with the Very Truth	340

A3.	**Human Understanding** – The Past, the Present and the Future –	**353**
A3.1	Original Philosophies in Europe and East Asia	353
A3.1.1	Original Philosophies in Europe	353
A3.1.2	Original Philosophies in East Asia	359
A3.2	How Does Human Understanding Advance?	363
A3.3	The Internal Ability in Human Life and Society	367

A4.	**Explanations of Words**	**373**

About the Author	379
Author Index	380
Subject Index	382

1. Does the Internal Invisible World Really Exist?

This chapter offers an introductory outline of the discussions developed in this book. As mentioned in Preface, science is useful but in essence of an approximate character. Therefore, science has now come to show severe difficulty in explaining what life is. The purpose of this book is to reveal where the incompleteness of science comes from and construct a new theory of the origin and the evolution of life and related topics, based on it.

A unique feature of this book is to introduce novel concepts of *truly real things* (i.e. natural things themselves which human consciousness and hence science fail to catch) and *the very truth* (the way that natural things themselves exist or truly real things exist), both of which are in *the internal world* we cannot be conscious of. Then, this book asserts that the basic qualities of living things such as free independent harmonious spirit and wisdom (creativity) come from truly real things in the internal invisible world. Accordingly, it is of primary importance to clarify whether on earth the internal invisible world really exists. In addition, it is also important to reveal how we can disclose the very truth lying in the internal invisible world. This chapter offers introductory discussions of these issues.

1.1 A Living Thing Has a Strange Way of Existing

In this section we compare living things and word-based scientific understanding of them. As mentioned in Preface, all living things including primitive bacteria demonstrate autonomous dynamic self-organizing ability or free independent harmonious spirit and wisdom (creativity). However, the present science cannot explain why living things can demonstrate such ability. Therefore, the comparison of living things and word-based scientific understanding of them is expected to clarify what intrinsic limits word-based scientific understanding has. The clarification will lead to the revelation of why we need to introduce novel concepts such as truly real things and the very truth. Here, word-based scientific understanding refers to understanding by the use of words including concepts, laws and mathematical equations. Words are regarded as a linguistic expression of things we are conscious of, such as ideas, opinions and sensory images. Therefore, word-based scientific understanding has essentially the same characteristics as things we are conscious of.

1.1.1 Comparisons of Living Things and Machines

Machines are fabricated based on scientific knowledge and thus features and limits of scientific knowledge are clearly embodied in properties of machines. Therefore, we here compare living things and machines for clarifying the difference between living things and word-based scientific understanding of them.

Features and limits of machines compared with living things
Let us begin with a simple example. We look at birds, fishes and persons and acquire scientific knowledge about them and then fabricate airplanes, submarines and humanoid robots, respectively. As can be readily seen, technological products thus fabricated such as airplanes, submarines and humanoid robots are quite different in qualities from original living things.

First of all, technological products have functions much superior to living things. For example, airplanes can fly much faster than birds. Humanoid robots can produce goods much more efficiently than human beings. Computers can perform calculation much faster and more accurately than human beings. In addition, technological products can work under hard conditions such as at a high temperature and in the presence of radioactive rays. These facts indicate that scientific knowledge, based on which technological products are fabricated, has superior characteris-

Figure 1-1 The motion of living things is quite dynamic, smooth and free, in contrast to technological products such as a car and a humanoid robot.

tics.

However, technological products have a severe limit as well. Airplanes, submarines and humanoid robots all are only machines, in sharp contrast to living things. The motion of living things is quite dynamic, smooth and free, while that of technological products is more or less awkward and clumsy. A more important difference is that living things demonstrate free independent harmonious spirit and wisdom (creativity) while technological products display no such ability at all. This fact clearly indicates that the present science fails to correctly catch the basic qualities of living things such as free independent harmo- nious spirit and wisdom and as a result has a mechanical character.

Can we remove the limits of the present science in the future?

The above example clearly indicates what features and limits the present science has. Somebody may say that the above-mentioned limits of the present science simply come from the fact that it is still in a less advanced stage and thus the limits will be removed in a more advanced stage of science in the future. Somebody may also say, "A humanoid robot in the future will behave like a human being". This may be an opinion of the majority of people.

Indeed, science will succeed in producing artificial living things and humanoid robots which behave like a human being in the future. I do believe that such brilliant technological successes will necessarily come in the future. Modern science has revealed that the first living organism on the primitive earth was produced spontaneously from nature and high plants and animals including human beings evolved from it. We human beings should be able to reproduce what happened in nature in the past. There is no reason that we cannot reach this goal.

However, the above opinion includes some fundamental issues to be considered further. In the first place, technological successes in producing artificial living things or intelligence, even if they are achieved in the future, do not guarantee that we correctly understand the basic qualities of living things or intelligence. In other words, we can produce artificial living things or intelligence even if we do not correctly understand the basic qualities of them or even if we have scientific knowledge of a mechanical character. As a simple example, a high-school student can successfully produce a high-temperature superconductor even if the student does not understand the basic qualities of it. A student needs only mix metal oxides and heat them. Thus, *producing artificial living things or intelligence and correctly understanding the basic qualities of them are essentially different matters.*

Another important issue to be considered is that the above opinion says that science in principle has no intrinsic limit and some limits in the present science simply arise from historical limitations. Now, is this opinion correct? Can science in the future really describe the basic qualities of living things correctly? In other words, is there no non-removable gap between living things and scientific knowledge of them? This is an important question because the answer to it determines what view of nature we can have. If there is a non-removable gap between living things and scientific knowledge of them, we have to say that nature has the internal world we cannot be conscious of.

Why does scientific knowledge have a mechanical character?

For investigating whether or not there is a non-removable gap between living things and scientific knowledge of them, let us consider the difference between living things and machines in more detail. At first, let us investigate why scientific knowledge has a mechanical character, as mentioned earlier. We can successfully deal with this problem by comparing "thinking based on scientific knowledge" and a machine's operation.

In thinking based on scientific knowledge, we first clearly define the meanings of words, concepts and laws and then consider various phenomena, keeping the meanings thus defined unchanged. For example, in Newtonian mechanics, we define the mass of substances, the law of gravity, and the laws of motion such as the proportional relation between an external force and acceleration and then investigate a diversity of motion of bodies under various conditions, keeping the first-given definitions unchanged. This is natural because considerations are thrown into confusion if the first-given definitions are changed during consideration. If a student writes such an answer in an examination, all teachers will put an error mark (×) on it.

On the other hand, in machines which are constructed from various devices such as transistors and meters and various materials such as iron metal, glass and plastics, properties and functions of such devices and materials are kept unchanged during operation of a machine. For instance, a transistor displays fixed functions anytime when used in a machine. This is also natural because a machine will go into confusion if properties and functions of devices and materials change during its operation. It is an important requirement for high-quality devices and materials that they exhibit the same unchanging properties and functions anytime, irrespective of how they are used.

Thus, we can see that in both thinking based on scientific knowledge and a machine's operation, constituent elements (words, concepts and laws in thinking and devices and materials in a machine's operation) do not change their qualities (meanings, properties and functions) throughout consideration or operation. Constituent elements always keep the first-given meanings or qualities unchanged. Thinking based on scientific knowledge has the same characteristics as a machine's operation on this point. This fact clearly indicates that thinking based on scientific knowledge or scientific knowledge itself has a mechanical character.

Living things behave in a different way from machines

The above conclusion becomes much clearer if we compare living things and machines. Constituent elements of living things behave in quite a different way from those of machines. For example, modern biology has revealed that protein molecules in a living cell act as a morphological element in a certain time and act as an agent for information transfer in another time and moreover act as an energy source in a third time, depending on the circumstances of a living cell. Protein molecules in a living cell have no fixed structures, properties and functions. They change their qualities so that they are in harmony with the surroundings or so that a whole consisting of "protein molecules and the surroundings" realizes a harmonious state.

There are a large number of similar examples in the behavior of constituent elements of living things. It is known that when a part of the brain is damaged, another part makes up for the function of the damaged part. Recent brain science has also revealed that a visual cortex in the occipital lobe, which becomes active in sighted persons when they see objects, is active in blind persons when they read Braille points [1]. It is interesting to note also that when a person is located in a gravity-free place such as in an artificial satellite, the muscular strength of his or her legs becomes prominently weak in a few weeks. No such change happens in constituent elements of machines.

The difference in the behavior of constituent elements between machines and living things is important. Constituent elements of a living thing incessantly change their structures, properties and functions so that they realize a harmonious state with the surroundings. This means that constituent elements of a living thing always interact with the surroundings and are never separated from them. On the other hand, constituent elements of machines always keep the first-given structures, properties and functions unchanged. This means that constituent elements of machines are independent of or separated from the surroundings.

Furthermore, not only constituent elements but also machines and living things themselves behave in different ways from each other. For example, birds, fishes, and human beings incessantly alter their behavior, depending on the surroundings, so that they realize a harmonious state with the surroundings. On the other hand, air-planes, submarines, and humanoid robots only behave in a fixed manner according to beforehand given rules, irrespective of how the surroundings change. The difference in the behavior of constituent elements between machines and living things is clearly embodied in the difference in the behavior of machines and living things themselves. When I was a university student, I was surprised to see that a professor, who was a dignified gentleman in a classroom, cheerfully helped his wife with family affairs in his home.

The basic qualities of a living thing emerge from its particular way of existing

We can now understand that a living thing and its constituent elements exist in infinite dependence on others and incessantly change their structures, properties and functions so that they realize a dynamic harmonious state with others. An important point is that such a way of existing is responsible for the emergence of the basic qualities of a living thing such as free independent harmonious spirit and wisdom. It is said that a living thing or a living cell has no special individual site, which acts as the controlling center for its physiological processes. Free independent harmonious spirit and wisdom of a living thing thus come from a whole consisting of "a living thing, its constituent elements and the surroundings". A living thing has a strange way of existing.

On the other hand, a machine and its constituent elements keep fixed structures, properties and functions anytime and are independent of one another and also independent of the surroundings. In other words, constituent elements (devices and materials) of a machine have individual unchanging properties and are only logically connected with one another. This is why a machine behaves as a machine. Here, a logical connection refers to a state in which words or things with *individual unchanging* properties are connected with one another according to results of experiences or experiments. For this reason, a machine needs a controlling center for its successful operation. For example, a computer has a CPU (central processing unit) as the controlling center for its successful operation.

Now, constituent elements of scientific knowledge (i.e. words) have individual unchanging meanings and are only logically connected with one another, in the same way as constituent elements of machines. This fact explains why scientific knowledge has a mechanical character. It should be noted also that machines fabricated based on scientific knowledge have functions much superior to living things or human beings, as mentioned at the beginning of this section. Such splendid functions of scientific knowledge also arise from the fact that words in it have individual unchanging meanings and are logically connected with one another. Splendid features and severe limits of scientific knowledge arise from the same origin. *We cannot remove the intrinsic limits of scientific knowledge without losing its features as far as we remain within the realm of word-based scientific understanding.*

The difference in the way of existing between scientific knowledge and living things

The foregoing discussions indicate that there is an about-face difference in the way of existing between scientific knowledge and living things.

> ***Scientific knowledge and its constituent elements*** *(words including concepts, laws and mathematical equations) – having individual unchanging meanings and logically connected with one another – effective for producing machines with superior functions of a mechanical character – demonstrating no free independent harmonious spirit and wisdom (creativity)*

> ***A living thing and its constituent elements*** *– existing in infinite dependence on others and incessantly changing their structures, properties and functions so that they realize a dynamic harmonious state with others – having no individual unchanging quality or having a freely changing dynamic quality – demonstrating free independent harmonious spirit and wisdom (creativity)*

The basic qualities of a living thing such as free independent harmonious spirit and wisdom (creativity) arise from the particular way that a living thing and its constituent elements exist, which is quite different from the way that scientific knowledge exists. This means that the basic qualities of a living thing are beyond scientific understanding. Thus we have to say that there is a non-removable gap between a living thing and scientific knowledge of it.

Somebody may say, "The above conclusion is obtained based on machines fabricated in the present days. Therefore, there remains a possibility that technological products such as humanoid robots or artificial intelligence (AI) demonstrate free independent harmonious spirit and wisdom (creativity) like a human being if they are fabricated based on advanced technologies in the future." Indeed, this opinion deserves careful consideration. It will be actually discussed in detail in Section 3.2.3. The final conclusion is that a system (machine) comprised of constituent elements with *individual unchanging* properties has no effective mechanism for generating free independent harmonious spirit and wisdom. Only a system comprised of constituent elements with a freely changing dynamic quality such as a living thing can demonstrate free independent harmonious spirit and wisdom.

1.1.2 How Are Living Things Understood in Modern Science?

As mentioned at the beginning of this chapter, science has now come to show severe difficulty in explaining what life is. Therefore, we in this section investigate how living things are understood in modern science and consider again whether or not there is a non-removable gap between a living thing and scientific knowledge of it.

Molecular biology and reductionism

Living things have long been studied from the standpoint of reductionism, where reductionism refers to a traditional way of understanding which says that even a thing with complex properties can be readily understood if it is divided into elements, each element is investigated in detail, and the elements thus investigated are combined as they were. The method of dividing a thing into elements and investigating them in detail is called the analytic method. Reductionism and the analytic method were both first clearly proposed by R. Descartes in the 17th century and have been a leading method in scientific research since then.

A typical example of understanding of living things in modern science from the standpoint of reductionism is seen in a book, "The Accidental and the Inevitable" (1971), written by a French biochemist, J. Monod [2]. He was awarded Nobel Prize by the clarification of a mechanism of heredity control for synthesis of proteins and virus. In this book, Monod emphasizes the importance of the analytic method. He also stresses that natural things are objective existents and therefore they should be understood based on objectivism. Here, objectivism refers to the belief that properties of observed things are independent of the mind of an observer (a human being). Monod's assertion is that anything should not be added to the principle of objectivism. Such an opinion also comes from reductionism. Thus, Monod criticizes Bergson's theory of creative evolution for the reason that it includes an artificially added principle of self-creation. Monod also criticizes the philosophy of Hegel (idealistic dialectic) and that of Marx and Engels (materialistic dialectic) for the reason that they include an artificially added principle of self-development. (Note that such criticisms of Monod were not necessarily adequate because his philosophy itself had a limit, as will be explained later.)

Now, Monod discusses how characteristic qualities of living things can be understood based on modern molecular biology. He pays attention to the fact that living things have structures and functions to suit their purposes. Such a characteristic of living things is called finality. With the aim of clarifying why living things can have finality, Monod explains morphogenetic processes of living things in detail, based on advanced molecular biology. Then, he finally shows that the finality of living things comes from the fact that they have the "invariable mechanism for self-replication" achieved by a combination of DNA, RNA and proteins. Namely, codes written on DNA are accurately transmitted by RNA to a site for protein synthesis in a ribosome, just in the same way as signals (information) are accurately transmitted among various parts in a machine. Thus, codes on DNA completely determine primary structures (arrangements of amino acids) of protein molecules and hence completely determine three-dimensional morphological structures of protein molecules, irrespective of however complex they may be. Moreover, three-dimensional structures of protein molecules in turn completely determine their physicochemical properties such as catalytic activity. In this way, living things can have definite functions to suit their purposes, based on codes on DNA. Thus, Monod's final conclusion is that all living things form a wholly unified functional unit which works like a machine. He then declares that living things are machines.

Monod points out, as a supporting evidence for the above conclusion, that an intracellular organism, ribosome, is spontaneously reproduced in a test tube. A ribosome is a small particle of about 20 nanometers in diameter and proteins are synthesized in it. Modern biology has revealed that constituent molecules of a ribosome homogeneously dissolved in an aqueous solution again spontaneously gather together and reproduce a ribosome which has the same composition, the same molecular weight and the same functions as the original ribosome.

Although Monod asserts that living things are machines, as mentioned above, he still acknowledges that a human being has free will. Accordingly, he is finally led to the dualism of body and mind, where the dualism refers to the belief that both a human body and the human mind exist. The dualism involves the difficult problem of why a human body following objective laws of a mechanical character can display free will. This problem was for the first time clearly recognized in the philosophy of R. Descartes (see Section A3.1.1) and has been a central problem in traditional western philosophies since then.

Features and limits of reductionism

Monod's discussions clearly indicate that reductionism leads to the failure to correctly catch the basic qualities of living things such as free independent harmonious spirit and wisdom and leads to a mechanistic view of living things. Certainly, *the analytic method in reductionism has a splendid feature in that it allows us to penetrate complexities of nature and clarify details of it.* Monod emphasizes that plentiful complex functions of living things have never been clarified without analysis. This is true but reductionism also has severe limits.

As mentioned earlier, in reductionism, an object of research is first divided into parts, each part is investigated individually and separately, and then the object of research is understood by a combination of parts thus investigated. Accordingly, *reductionism leads to an incorrect understanding of the basic qualities of an object of research if constituent elements of it change their qualities by their mutual interaction when they are combined with one another.* In other words, *reductionism fails to catch properties coming from a whole about widely interacting things or properties peculiar to such a whole.* This is a serious fault. In fact, a living thing and its constituent elements incessantly change their structures, properties and functions by their mutual interaction or their interaction with the outer world so that they as a whole realize a harmonious state, as mentioned in Section 1.1.1. Free independent harmonious spirit and wisdom of a living thing arise from such a way of existing. Monod's discussion overlooks this point.

Originally, *objectivism itself is an approximate way of understanding.* Natural things, in particular, living things are by no means objective existents because everything in nature exists in infinite dependence on others, as discussed in Section 1.1.1. In fact, a dog will show a friendly attitude by wagging its tail if a person approaches it with a smile. On the other hand, a dog will growl if a person glares at it. The mind of a dog is not independent of the mind of a person as an observer. Quantum mechanics also says that an observed system is not independent of an observing system.

A confrontation between contingency and determinism

Let us next consider how the origin and the evolution of life have been understood to date. According to Luisi's book, "The Emergence of Life – From Chemical Origins to Synthetic Biology" (2006)[3], there has been a confrontation between contingency and determinism about the origin of life. Here, contingency refers to the opinion that the origin and the evolution of life on the primitive earth occurred under favorable accidental conditions while determinism stands for the opinion that they occurred according to inevitable natural laws. Reductionism has the same standpoint as contingency as to the origin of life because it fails to correctly catch the basic qualities of living things and can give no reasonable explanation to their adaptive and evolutional behavior. In fact, Monod explains the emergence of the first living organism on the primitive earth in terms of lucky accidents [2].

At present, it appears that contingency is supported by many modern molecular biologists. Luisi says in his book [3] that many of modern scientists will support contingency in a confrontation between contingency and determinism, though not a few scientists still reject the reasonableness of contingency and uphold determinism regarding the origin of life. Luisi also says that determinism becomes meaningful only when it can show an evidence for the argument that nature has a tendency to go toward the construction of living organisms.

However, contingency has a serious problem in that it cannot give a meaningful probability of the emergence of the first living organism. In fact, it is quite difficult to assume that highly ordered structures of the first living organism were produced by chance at a sufficiently high probability *against* the second law of thermodynamics, which says that nature has a tendency to change toward complete disorder. Contingency has another serious problem. It cannot explain the emergence of the basic qualities of living things such as free independent harmonious spirit and wisdom (creativity). This is because a system in contingency is in an entirely passive state to an attack of the outer world and has no internal ability to organize itself so that it realizes a free independent harmonious state, even though it may have the ability to behave in an organized manner like a machine.

After all, both determinism and contingency can give no sufficient explanation to the origin and the evolution of life at present.

Complexity science and holism

Reductionism fails to correctly catch the basic qualities of living things such as free independent harmonious spirit and wisdom and leads to a mechanistic view of living things, as mentioned earlier. Therefore, scientists who had paid attention to the emergence of such spirit and wisdom in living organisms and disagreed with the standpoint of reductionism started new study based on holism in the 1980's. Here, holism refers to the belief that a whole is more than the sum of its parts. Such new study is called complexity science [4,5] though it appears that no definite definition has yet been given to complexity.

Complexity science had its rise in various preceding approaches [4,5] such as cybernetics by N. Wiener in the 1940's, the general system theory by L. von Bettalanffy in the 1950's, autopoiesis by H. R. Maturana and F. J. Varela in the 1970's, synergetics by H. Haken in the 1970's, and the dissipative structure theory by I. Prigogine in the 1970's. Research activity in complexity science became prominent after Santafe institute in U.S.A. was established as a center for it in 1984.

According to Mitchell's book, "Complexity: A Guided Tour" (2009) [4], complexity science deals with a diversity of complex systems in nature and society, such as living organisms, the human brain, the ecological system and stock markets. Such complex systems show some common prominent characteristics: (1) a complex system is composed of a huge number of components forming a large-scale network, (2) individual components interact with one another according to relatively simple rules, (3) a complex system has no central controlling unit or leader, and nevertheless (4) a complex system exhibits ingeniously organized complex and adaptive dynamic behavior, often leading to evolution. For example, a living cell, which is a typical example of complex systems, is composed of a huge number of various organic compounds which form a large-scale chemical reaction network. It has no central controlling unit but exhibits ingeniously organized complex and adaptive behavior leading to evolution. Such characteristics are difficult to explain by a conventional way of understanding such as reductionism.

Complex systems also show prominent characteristics in that they demonstrate "emergence" and hierarchical structures. Here, "emergence" refers to the appearance of a complex system with essentially new structures, properties and functions, which cannot be explained by existing things and conditions. This book expresses the word of emergence of this meaning by adding quotation marks in such a way as "emergence". "Emergence" is also called spontaneous order or self-organization. The appearance of unicellular living organisms on the primitive earth and their later evolution are typical examples of "emergence". The formation of various hierarchical structures in nature is a result of the occurrence of "emergence".

Because of the above-mentioned characteristics of complex systems, researchers in complexity science pay attention to a whole about widely interacting components rather than individual components, in sharp contrast to researchers in reductionism. Thus, main concerns in complexity science are to clarify how various complex systems are produced and what novel properties they demonstrate. The final goal is to disclose general principles of "emergence" or self-organization.

Researchers in complexity science have another special attitude. They consider that natural processes occur by algorithm-based calculation, where algorithm refers to a set of rules for operation including information treatment. In other words, they consider that natural processes can be reproduced by computer calculation. Therefore, they usually do not directly deal with actual complex systems but treat theoretical models created by idealization.

According to Mitchell[4], a diversity of theoretical models has been constructed, such as logistic equation, fractal geometry, the Turing machine, the self-reproducing automaton, the cellular automaton, genetic algorithm, copycat, and network theory. These models have been successfully utilized in various fields such as physics, chemistry, biology, meteorology, economics, and sociology. For example, the study of logistic equation has been successfully combined with that of nonlinear chemical dynamics dealing with chemical oscillations and spatiotemporal pattern formation [6,7] (see Section 3.1.2 for details).

For clarifying mechanisms for the origin and the evolution of life, it is important to disclose how "emergence" happens. An interesting theory about this issue was proposed by S. Kauffman and his group in Santafe institute [8]. They investigated the behavior of a network of chemical reactions based on the model of cellular automaton, in which components in a network act simultaneously in parallel to one another according to simple rules. Then, they revealed that a reaction network with "collective catalytic actions" was suddenly formed when a reaction network became sufficiently complex. The appearance of such a jump is of much interest because it can be regarded as "emergence". In fact, Kauffman et al. assert that the reaction network with collective catalytic actions has the ability to replicate itself and sustain itself and can be regarded as a precursor of the primitive life (see Section 3.1.3 for details). Based on these results, Kauffman also emphasizes that complexity science has a possibility of explaining the origin and the evolution of life without relying on lucky accidents.

The advancement of complexity science has now allowed us to explain the adaptive and evolutional behavior of complex systems [4]. However, there still remains a serious problem at the most basic level. A computer only operates according to algorithm of a theoretical model, which a human being creates. In actual complex systems such as living organisms, such algorithm is created by nature itself. A serious problem is that it is entirely unknown how nature creates such algorithm. How does nature determine the standards of information treatment in the adaptive and evolutional behavior of a complex system? How does nature recognize the meaning of an actual situation of a complex system and how does nature choose a next suitable action? Mitchell says [4] that these are very difficult fundamental problems in the present complexity science.

In conclusion, complexity science has made remarkable progress but still not clarified the true origin of autonomous dynamic self-organizing ability or free independent harmonious spirit and wisdom of living organisms. Thus it has provided no sufficient explanation to the origin and the evolution of life.

Quantum mechanics and quantum thermodynamics

Finally, let us briefly explain how living things are treated in a new theory proposed in this book (see Section 4.2). This book adopts a completely different approach from traditional ones such as reductionism, nonlinear science and complexity science. A main conclusion is that the origin and the evolution of life can be reasonably explained based on principles of quantum mechanics and quantum thermodynamics if they are appropriately interpreted by taking into account the internal invisible world. Here, "quantum thermodynamics" is a terminology newly introduced in this book. It refers to thermodynamics directly derived from quantum mechanics without using Boltzmann's statistical interpretation (see Section 2.2).

To date, the second law of thermodynamics (or the law of increase of entropy) has been interpreted as expressing the fact that nature has a tendency to change toward complete disorder, as mentioned earlier. Therefore, it has been regarded as a fundamental subject to clarify how highly ordered structures in living organisms are produced *against* the second law of thermodynamics. For this reason, keen attention has been paid to dissipative structures, "emergence" or self-organization arising from nonlinear dynamics or the complexity of organization, as discussed just earlier [4-8].

However, a new interpretation of quantum mechanics and quantum thermodynamics by taking into account the internal invisible world has led to the conclusion that "emergence" or self-organization comes from the second law of thermodynamics itself under a particular condition (see Section 3.2.1). In other words, such a new interpretation has revealed that a non-equilibrium macroscopic system lying in a stationary state has the internal ability to organize itself so that it realizes a free independent harmonious stable state. Autonomous dynamic self-organizing ability of living things can be reasonably explained based on the internal ability. Amazingly, it is not true that living things emerged and evolved against the second law of thermodynamics. The truth is that living things emerged and evolved according to the second law of thermodynamics under a particular condition. Originally, the second law of thermodynamics is the only universal law that describes spontaneous and irreversible processes in nature. Therefore, it is the most reasonable to explain the origin and the evolution of life based on this law[*1].

> [*1]This argument never means that self-organization in complexity science or dissipative structures in nonlinear dynamics are useless for studying living things. Living things use such self-organization or dissipative structures for improving their properties and functions (see Section 4.4 for details).

Where is the essential difference between a system described by quantum mechanics and quantum thermodynamics and one described by complexity science? In either of these systems, not individual constituent elements but a whole about mutually interacting constituent elements is the basic reality. Now, a prominent feature of the former system is that constituent elements (electrons and atomic nuclei) have a wave nature. Therefore, constituent elements and also a system itself is described by a wave function with a freely changing dynamic and continuous character. Moreover, wave functions are superposed one another and as a result a system described by a wave function changes its basic qualities when it interacts with others (see Section 2.1.2 for details). This is an important feature because a living thing and its constituent elements also change their basic qualities by their interaction with others, as mentioned in Section 1.1.1. On the other hand, constituent elements of a system in complexity science are described by words with individual unchanging meanings and thus maintain individual unchanging qualities even when they interact with others[*2,*3]. Therefore, a system in complexity science can indeed behave in an organized manner but can only behave mechanically like a machine, as discussed in Section 1.1.1 (see also Section 3.2.3).

> [*2]This argument means that changes in the basic qualities of living things by their mutual interaction or interaction with others are impossible to correctly describe by the use of words with individual unchanging meanings. This also implies that such changes are difficult for us human beings to notice because our ideas have individual unchanging meanings. Probably for this reason traditional understanding and philosophies have in general overlooked such changes. In fact, complexity science

as well as reductionism has failed to catch changes in the basic qualities of a living thing and its constituent elements by their interaction.

*3This argument indicates that there are two kinds of holism. In complexity science, constituent elements of a system have individual unchanging qualities and thus the difference between a whole and the sum of its parts only arises from the complexity of organization in a system. On the other hand, in quantum mechanics and quantum thermodynamics, the difference between a whole and the sum of its parts also arises from changes in the basic qualities of parts by their mutual interaction.

Even quantum mechanics and quantum thermodynamics do not lead to the truly correct understanding of living things

Comparative examinations of various scientific approaches to living things indicate that scientific understanding comes closer and closer to the correct understanding of living things as it advances. However, it should be noted that even quantum mechanics and quantum thermodynamics provide no truly correct understanding of living things. They only give us an explanation of the basic qualities of a living thing and do not give us the basic qualities themselves. In fact, knowledge of a living thing, which quantum mechanics and quantum thermodynamics give us, exhibits no free independent harmonious spirit and wisdom, in sharp contrast to a real living thing. This means that we cannot correctly understand free independent harmonious spirit and wisdom of a living thing by such knowledge.

Why do quantum mechanics and quantum thermodynamics have such a limit? A main reason is that a wave function still has an unchanging character as a mathematical function, in contrast to a real living thing which has a freely changing dynamic character in every aspect.

1.2 Features and Limits of Scientific Understanding

The preceding section has disclosed that there is a non-removable gap between a living thing and scientific knowledge of it even at the most advanced stage of science. A main reason is that words (including mathematical functions and equations) in scientific knowledge have individual unchanging meanings, in contrast to a living thing and its constituent elements. Now, why do words have individual unchanging meanings? In this section we consider this issue and clarify what features and limits scientific understanding has.

1.2.1 Science Only Catches Common Properties of Nat-

ural Things

How were human ideas born?

For clarifying why words have individual unchanging meanings, it is the most effective to consider how human ideas and words were born. Only human beings have ideas and words and therefore it is certain that they emerged simultaneously with the birth of human beings. The birth of human beings is characterized by the use of tools such as stone implements and fire, the formation of an enlarged brain, two-leg walk, the custom of the burial of a dead body, and so on. The use of tools such as stone implements probably started empirically from necessity in daily life. In the age of the birth of human beings, i.e. in the time from about seven million years to three hundred thousand years ago, a cold age and a warm age came alternately on the earth. Thus, when a cold age came, ancient monkeys (or the ancestor of human beings) will have made desperate efforts to avoid starvation. Such a severe situation will have brought about the leap to the constant use of tools such as stone implements.

Naturally, ancient monkeys will have had no idea of "stone" when they started to use stone implements. This is because "stone" is not a directly seen thing in nature. A directly seen thing is only a variety of "lumps" lying in the ground, rivers or mountains. They are different in size, shape and color and can never be regarded as the same thing as far as they are merely seen. However, through repeated experiences of using such various "lumps" as implements, ancient monkeys will have learned that such "lumps" had *common* properties of "heavy" and "hard" and were useful for implements to the same extent. Thus, they came to have an idea of "stone" as representing an imaginary thing with the common properties of "heavy" and "hard", together with ideas of "heavy" and "hard".

Ancient monkeys may have picked out various common properties and imaginary things with such common properties through experiences of using implements in the same way as the above. Then, they will have noticed that common properties of natural things and imaginary things with such properties were useful for their life. Probably, this discovery led to the evolution (enlargement) of their brain (see Section 4.4). As a result, ancient human beings, who had the ability to pick out a diversity of common properties of natural things, were born. Simultaneously, words were created as marks or signals to express common properties and imaginary things with common properties, because they were impossible to show directly as such and such. The emergence of ideas (consciousness), words (knowledge) and tools (technology) was great creation of ancient monkeys or ancient human beings. This fact made the use of consciousness, knowledge and technology inherent in human beings.

The above brief survey of the history of the birth of human ideas clearly indicates that *human ideas and words emerged as representing common properties of natural things or imaginary things with such properties.* Such a characteristic of human ideas and words is clearly seen in ideas and words we use in the present days. For example, there are actually a great variety of apples such as red apples, green apples, large apples, small apples, round-shaped apples and distorted-shaped apples. There is no apple that is exactly the same as the others. In addition, actual apples change in quality with time. In such a complex situation, the word of "apple" exists as expressing an imaginary substance that has some common properties of actual apples. The same argument applies to all words such as a mountain, a river, a tree, a dog, a human being, life and mind.

Human ideas (consciousness) and words (knowledge) are entirely useful

Now, let us consider characteristics of ideas and words which ancient human beings came to have. Firstly, they were entirely useful for their life. Naturally, this characteristic applies to ideas and words we have in the present days. In fact, common properties of natural things are the kernel of them and exist anytime, anywhere, to any person and with inevitability. Thus, *common properties of natural things and imaginary things with such properties and hence human ideas and words have splendid features of immutability, universality, objectivity and inevitability.* The importance of such features is evident if we consider that everybody can *necessarily* achieve his or her purposes based on these features.

Moreover, common properties and imaginary things with such properties, which have features of immutability, universality, objectivity and inevitability, can be observed *repeatedly*. Therefore, their correctness can be examined by experiences or experiments. In addition, for the same reason, they can be described quantitatively by using mathematics or statistics. Accordingly, human ideas and words can be made more and more accurate and hence more and more useful.

Human ideas and words only catch common or repeatedly observed properties or individual unchanging properties of natural things

Secondly, ideas and words which ancient human beings came to have were indeed useful but only caught common or repeatedly observed properties of natural things or imaginary things with such properties, as mentioned earlier. Now, common or repeatedly observed properties are the kernel of natural things and have individual unchanging qualities. Therefore, we can also say that ideas and words ancient human beings came to have only caught individual unchanging properties of natural things and imaginary things with such properties.

It should be noted here that ancient human beings will have had only sensory images before they came to have ideas and words. Sensory images also only caught common or repeatedly observed properties or individual unchanging properties of natural things, as argued in Chapter 5. Accordingly, the above discussion, after all, implies that ancient human beings were only able to be conscious of common or repeatedly observed properties or individual unchanging properties of natural things and imaginary things with such properties.

Now, ideas and words we have in the present days possess the same characteristics as those which ancient human beings came to have. In fact, we human beings usually think that nature consists of a variety of individual things such as stones, water, iron metal and dogs. We also think that such individual things have inherent unchanging properties and are clearly distinguished from one another. Such a view of nature is entirely familiar to us and constitutes a basis for scientific understanding. The world we are conscious of consists of things with individual unchanging properties. This fact clearly indicates that *we human beings can only be conscious of things with individual unchanging properties.*

Somebody may say, "Science catches various changes by the use of variables, functions, differential calculus or differential equations." Indeed, this is true. However, this does not mean that we human beings can be conscious of changes themselves as they are. Science only catches "common or repeatedly observed parts" of real changes, as will be discussed later (on page 25) in this section.

Human ideas and words are of an approximate mechanical character

The above argument has an important philosophical meaning. In general, it has long been believed that things with common properties or things with individual unchanging properties, which have features of objectivity, immutability, universality and inevitability, express the essence of nature, as already mentioned in Preface. *Human beings have attached the highest importance to usefulness for their life and for this reason they have regarded things with common properties or things with individual unchanging properties as the essence of nature.*

However, this belief has been incorrect. In reality, things with incessantly changing dynamic properties represent the essence of nature. Things with common properties or things with individual unchanging properties are only what human beings have picked out from nature for their own use and are of an approximate character. Human beings, or to be accurate, human consciousness has only caught common or individual unchanging properties of natural things and failed to catch natural things themselves. In fact, neither autonomous dynamic self-organizing ability nor free independent harmonious spirit and wisdom come from things with common properties or things with individual unchanging properties, as discussed in Section 1.1.1 and 1.1.2 (see also Section 3.2.3). They can only produce machines.

Accordingly, we have to correct the way of understanding so that it includes what human consciousness has failed to catch. Namely, we have to work out a way to catch natural things themselves which human consciousness fails to catch. Therefore, this book introduces novel concepts of *truly real things* and *the very truth*, apart from *real things* and *the truth* in traditional understanding and philosophies, as mentioned in Preface. Truly real things refer to natural things themselves which human consciousness fails to catch. The very truth stands for the way that truly real things or natural things themselves exist. Then, this book reveals that truly real things really act as the origin of autonomous dynamic self-organizing ability or free independent harmonious spirit and wisdom of living things (see Section 1.4.1, 2.1.3, 3.2.1 and so on). In this way, this book proposes the complete antithesis of a traditional view of nature in which things with common properties or things with individual unchanging properties are regarded as the essence of nature.

If we use a metaphor, truly real things are like dynamic bending curves while real things in traditional understanding and philosophies are like straight lines touching such dynamic bending curves. Note that this metaphor also indicates that truly real things do not exist separately from real things which human ideas and words express. Therefore, we can have a hope of reaching the very truth based on human ideas and words. In fact, we can arrive at the very truth by transcending individual ideas and words, as will be discussed in Section 1.3 and 1.4 and in Appendix. Discussions in this book are irrelevant to agnosticism.

Human life is in a rather perplexing state

Strikingly, the above arguments indicate that human life is in a rather perplexing state. Human ideas (consciousness) and words (knowledge) are very useful, effective and indispensable for human life. In addition, we human beings can be conscious of only human ideas and words and thus they are the only thing we can rely on, on a conscious level. However, human ideas and words do not represent truly real things or natural things themselves as they are. They only represent common or repeatedly observed properties of truly real things or imaginary things with such properties and are of an approximate character.

Such a perplexing state arises just from the fact that truly real things (or natural things themselves) are by no means useful for human life because they have an incessantly changing dynamic character and have no immutability, universality, objectivity and inevitability, as will be explained later. Therefore, ancient human beings had no way but to pick out common or repeatedly observed properties of truly real things for catching useful things for their life. Thus we obtain the following important conclusion.

Usefulness is gained at the expense of correctness. Human ideas and words are useful because they are of an approximate character.

Note again that we have a possibility of surmounting such a perplexing situation. We can transcend individual ideas and words and arrive at the very truth (the way that truly real things exist). If we have arrived at the very truth, we can live in accord with the very truth lying in the internal invisible world and simultaneously live a practical life based on useful scientific knowledge. This is a strange and ideal life. The philosophy of Buddhism at a completed stage has disclosed that we can live such a life. Details will be explained in Chapter A2.

Science has made progress as an effective method for picking out common or repeatedly observed properties of natural things

We have thus far considered the historical origin of human ideas (consciousness) and words (knowledge) and revealed that they only catch common or repeatedly observed properties of natural things. Naturally, the same argument applies to scientific knowledge which is an advanced form of human ideas and words. Now, scientific knowledge plays a critically important role in the present societies. Therefore, it will be important to investigate the actual history of science and confirm the above characteristics of scientific knowledge.

It is well known that idealization and analysis methods have played a leading role in scientific study. These methods can be regarded as sophisticated methods for picking out common or repeatedly observed properties of natural things or truly real things. Therefore, the fact that these methods have played a leading role in scientific study clearly indicates that science has made progress by picking out common or repeatedly observed properties of truly real things.

The importance of the idealization method in scientific study was first clarified by G. Galilei. It is said that modern science started with the construction of Newtonian mechanics near the end of the Renaissance age. According to a book by A. Einstein and L. Infeld[9], Galilei played a key role in the construction of Newtonian mechanics. In this study, he invented the method of idealization.

Galilei studied various kinds of motion such as a rolling motion of a spherical body, free descents of various substances, and an orbit of a cannon ball. In his age, people believed the Aristotelian idea that velocity was directly related to an externally applied force. Such an idea probably came from intuitive observation of the motion of bodies such as a handcart. In fact, a handcart starts to move when it is pushed by a hand and stops when a pushing force is removed. Namely, velocity is directly related to an externally applied force.

However, a rolling motion of a spherical body does not obey this law. A spherical body continues to roll for a long time after an externally applied force was removed. Probably from such observation, Galilei noticed that not only an externally applied force but also a frictional one exerted an important influence on the motion of a body. Thus, he did various experiments to make the effect of a frictional force clear and finally imagined an *idealized* experiment of "motion with no friction". Based on such idealization, he discovered the law of inertia, which says that a moving body continues to move endlessly at the same speed if no external force is present. He also discovered that it is a change in acceleration that is brought about by an externally applied force.

The "motion with no friction" does not exist in nature. It is only an imaginary model. However, it was by this idealization that Galilei was able to discover the law of inertia. It is often said that scientific laws are discovered by looking at nature as it is. However, this opinion is not correct. We cannot discover scientific laws when we look at nature as it is. Scientific laws do not exist anywhere if we look at nature as it is.

It should be emphasized also that Galilei was the first scientist to use quantitative experimental investigations for discovering scientific laws. Galilei's method of idealization and quantitative experimental investigation has become an important general method in later scientific studies.

Idealization is indispensable for discovering scientific laws

Galilei's work clearly indicates that idealization is indispensable for discovering scientific laws. Interestingly, if we look back on the history of science, idealization was already used for discovering natural laws in the ages before Galilei.

According to the literature [10], the arithmetic law such as "one plus one equals two" was first discovered by Sumer people who dwelt in the basin of the River Tigris-Euphrates. They already had a multiplication table until 2,500 years B.C. The arithmetic law was probably discovered from need to count the number of livestock such as cows and sheep. Sumer people discovered a method of expressing the number of livestock by using stamps of reed sticks on clay [10], allotting one stamp to one cow or sheep. Then, they came to be able to know the number of cows or sheep by a bunch of stamps of reed sticks. The arithmetic law will have been discovered by looking at relations between two or more bunches of stamps of reed sticks. It should be noted here that when Sumer people expressed the number of sheep by a bunch of stamps of reed sticks, they regarded various sheep such as large sheep, small sheep, old sheep and young sheep as sheep of the same kind. Here is clearly idealization because all differences in qualities of actual sheep are disregarded. However, it was by this idealization that Sumer people were able to count the number of sheep. If Sumer people distinguished sheep with different qualities, they were unable to count the number of sheep.

A similar argument applies to the discovery of laws of elementary geometry. Various theorems in elementary geometry were studied in ancient Greece, in particular, in the Pythagoras school in around the 5th century B.C. [10] The theorems were later arranged into a regular form with some axioms by Euclid in the 3rd century B.C. [10] Probably, people in the ancient age paid attention to various geometrical relations in nature and expressed them by *idealized* geometrical shapes such as points, lines, triangles, circles, and so on. In geometry, a point has only a location and no size and a line has only a length and no width. However, such a point and a line do not exist anywhere in the real world. Thus, points and lines in geometry express results of idealization. Again, it is by such idealization that strict geometrical laws can be proved [*1].

[*1] From these examples, we can clearly see that science is accurate, strict and useful because it is of an approximate character, as mentioned earlier.

There is another reason for the importance of the idealization method. A scientific law to be discovered is already tacitly embodied in properties of an idealized object of research. For example, the concepts of numbers and arithmetic laws were already tacitly embodied in properties of stamps of reed sticks on clay. The law of inertia, discovered by Galilei, was already tacitly embodied in properties of "motion with no friction". Accordingly, a great many efforts are made to construct a suitable idealized model in scientific study even in the present days. For example, it was mentioned in Section 1.1.2 that enormous efforts have been made in complexity science to construct fundamental theoretical models by the idealization of actual complex systems.

A Japanese atomic physicist, M. Taketani, investigated the history of physics and proposed the theory that fundamental laws of motion are discovered via three stages of study; (1) phenomenological stage, (2) the clarification of fundamental entities, and (3) the construction of a fundamental law of motion (such as Newton's laws of motion) [11]. In particular, he emphasized the importance of the second stage, in agreement with the above argument.

The analytic method became important in advanced scientific studies

Scientific study in the 19th century is characterized by the establishment of thermodynamics and electromagnetism on the one hand and the discovery of the microscopic world composed of atoms and molecules on the other hand. After the invention of a steam engine in the latter half of the 18th century, many people came to direct their eyes to the driving force for motion in nature and pay attention to power (energy) included in the interior of substances such as heat, electricity, magnetism and vitality. This trend finally led to the establishment of thermodynamics and electromagnetism and the discovery of the microscopic world.

The idealization procedure became more and more important with the progress of science, in particular, in the fields of thermodynamics and electromagnetism, because objects of investigation came to have more and more complex properties. For example, thermodynamics was established by using various idealized models such as the ideal gas, an isolated, closed or open system, the equilibrium state, a quasi-static process, a reversible process, an ideal solution, and so on.

The analytic method, first invented by R. Descartes (see Section 1.1.2), came to play a key role in studies of inner structures of substances. In the analytic method, *"(imaginary) substances with common or repeatedly observed properties"* are caught as *"(imaginary) substances which are kept unchanged in various physicochemical processes"*. Certainly, these two kinds of (imaginary) substances have the same meaning as each other. Firstly, concepts of (chemical) elements, atoms and molecules were discovered as "substances which are kept unchanged in various physicochemical processes". Then, ions were discovered as the first sub-atomic particle by S. Arrhenius in 1887. A little later, electrons were discovered by J. J. Thomson in 1897 and atomic nuclei were discovered by E. Rutherford in 1911. These discoveries led to the clarification of atomic structures and finally led to the establishment of quantum mechanics in 1925-1926. Furthermore, later advanced studies have clarified structures of atomic nuclei. It is now believed that the universe is comprised of various elementary particles such as quarks, leptons, and others.

The idealization and analysis methods were also important for revealing fundamental laws of motion. In fact, *mathematical concepts such as variables, functions, differential calculus and integral calculus represent quantities or relations which are kept unchanged in various natural processes*. Namely, they represent common or repeatedly observed properties of truly real things or truly real changes.

The aim of science is to acquire objective knowledge useful for human life, not to reveal the very truth

It is now certain that science has made progress as an effective way to catch common or repeatedly observed properties of natural things or imaginary things with such properties. This fact clearly indicates that *the original aim of science is to acquire objective knowledge useful for human life, not to reveal the very truth*. Accordingly, it is natural that there is a non-removable gap between scientific knowledge and the very truth (the way that natural things themselves exist).

It seems that there has been a wrong recognition that the objective (scientific) truth represents the very truth in traditional understanding and philosophies except some philosophies such as existentialism and East Asian Philosophies. To be brief, there has been confusion about usefulness and truth. A typical example is seen in pragmatism. In fact, W. James claimed that the truth is a thing that allows us to obtain a meaningful result in our actual practice and hence the criterion for judging the truth is usefulness [12].

In my opinion, scientific knowledge is products of human creative activity, similar to sport and artworks. In fact, science and technology have created the new artificial objective world quite different from nature itself. Cars and airplanes are quite different from horses and birds, respectively. A view of streets in a city is quite different from the scenery of mountains and fields full of trees, birds and insects. Therefore, there is no problem even if there is a non-removable gap between scientific knowledge and the very truth. Nobody will say that paintings and sculptures in art have less important meanings for the reason that they do not represent natural things as they are. *We need only distinguish the scientific (objective) truth and the very truth.* We have the very truth (the way that natural things themselves exist) with profound abilities and qualities behind the scientific truth.

Finally, it is worth noting that it is because scientific knowledge has intrinsic limits that it has made progress to date. Human ideas and words only catch common or repeatedly observed properties of natural things and are of an approximate character, as mentioned earlier. Therefore, clear recognition of a piece of scientific knowledge necessarily leads to the appearance of a discrepancy between it and natural things. The appearance of a discrepancy induces further scientific study and hence further progress of scientific knowledge. In this way, science has penetrated complexities of nature deeper and deeper and constructed a great system of knowledge. Limitless progress of science proves that scientific knowledge has intrinsic limits.

1.2.2 Scientific Knowledge Is Useful as a Technique

The discussions in the preceding section have made it clear that there is really a non-removable gap between scientific knowledge and the very truth. The discussions have also revealed that human life is in a perplexing state. Human ideas (consciousness) and words (knowledge) are very useful for human life. In addition, we human beings can only be conscious of them and thus they are the only thing we can rely on, on a conscious level. However, human ideas and words only catch common or repeatedly observed properties of truly real things and are of an approximate character. Such a perplexing state is an inevitable result of the fact that there is a non-removable gap between scientific knowledge and the very truth.

Now, how can we surmount such a perplexing state? This is a main subject of this book and will be discussed in detail later. In this and the next sections we only consider (1) why human ideas and words are useful though they miss truly real things and are of an approximate character and (2) what problems actually appear in our life when we remain within the realm of word-based understanding of an approximate character. Considerations of these problems are useful for understanding features and limits of scientific knowledge in detail.

Scientific knowledge is useful for fabricating machines

In the first place, let us consider why human ideas and words or word-based scientific understanding is useful though it fails to catch truly real things and is of an approximate character. Now, the usefulness of word-based understanding can be divided into two cases: usefulness in handling artificial technological products and usefulness in handling natural things, in particular, living things.

It is natural that scientific knowledge is useful in handling artificial technological products (machines). This is because logical connections among words with individual unchanging meanings in scientific knowledge can be exactly converted to those among devices and materials with individual unchanging properties in machines, as discussed in Section 1.1.1. Therefore, the progress of scientific knowledge necessarily leads to improvement in properties and functions of machines. Indeed, word-based scientific understanding is of an approximate character but machines are also of an approximate character.

Let us consider in more detail why word-based scientific understanding is useful in handling artificial technological products. Firstly, human ideas and words only represent common or repeatedly observed properties of truly real things (or natural things themselves) and are separated from truly real things. Therefore, human ideas and words can be freely combined with one another on a logical basis so that they produce materials or devices useful for human life. High-quality materials such as iron, silicon and plastics and high-function machines such as TV, cars and robots are fabricated by this principle.

The second reason is concerned with the aforementioned fact that machines have functions much superior to living things or human beings (see Section 1.1.1). For example, airplanes can fly much faster than birds. Robots can produce goods much more efficiently than human beings. Computers can perform calculations much faster and more accurately than human beings. How can we explain these facts? In a machine, constituent elements have individual unchanging properties and are only logically connected with one another. Therefore, various processes in a machine happen independently of one another. In addition, the number of constituent elements or logical connections in a machine is finite and not extremely large. On the other hand, in a living thing, quite a huge number of constituent elements exist in infinite dependence on others and incessantly change their structures, properties and functions so that they are in harmony with one another, as argued in Section 1.1.1. Therefore, connections in a living thing are much more complex than those in a machine. Accordingly, it is likely that a machine can work much faster and much more efficiently than a living thing. This argument implies that *a mechanical function can be achieved effectively by a relatively limited number of mutually-independent logical connections among constituent elements.* Therefore, scientific knowledge is useful for fabricating high-function machines.

Scientific knowledge is useful even in handling living things

Strikingly, scientific knowledge is useful even in handling living things, though it fails to correctly catch their basic qualities such as autonomous dynamic self-organizing ability or free independent harmonious spirit and wisdom, as discussed in Section 1.1.1. For example, J. C. Venter and his group recently reported [13] that they succeeded in creating a new bacterial cell that was controlled by artificially synthesized chromosome. According to their report, artificial bacterial cells had expected phenotypic properties and were capable of continuous self-replication. Such a result is really strange. Here is a great paradox. How can we understand it?

It was mentioned in Section 1.1.1 that we human beings can produce artificial living things even if we do not correctly understand the basic qualities of living things or even if we have scientific knowledge of an approximate mechanical character. In fact, we already succeed in agriculture and medicine though we have not yet clarified what the essence of life is. Here is also a great paradox, similar to the one mentioned above. Evidently, successes in agriculture are achieved by making use of the inherent ability of seeds to spontaneously reproduce the original plants under a suitable condition. Successes in medicine are also attained by making use of the inherent ability of a human body to maintain harmonious physiological processes. Thus, the success of Venter's group in creating artificial bacterial cells can also be understood to be due to the utilization of the inherent ability of a molecular system involving artificially synthesized chromosome to spontaneously reproduce a living bacterial cell.

The above argument can be explained as follows. Indeed, scientific knowledge only catches common or repeatedly observed properties of truly real things and has an approximate mechanical character. However, if we consider this fact in the reverse direction, it means that truly real things have properties and functions much more than those expressed by scientific knowledge. *In actual processes such as agriculture, medicine and experiments, we treat truly real things with such extra properties and functions even though we are only conscious of ideas or words with an approximate mechanical character.* Accordingly, we can produce living things in actual processes. We need only combine truly real things correctly under suitable conditions so that their extra properties and functions work effectively.

The importance of scientific knowledge is that it correctly catches the way that common or repeatedly observed properties of truly real things exist. Namely, scientific knowledge provides us with correct relations among things with common or repeatedly observed properties of truly real things. Accordingly, for example, in the case of producing a new bacterial cell with artificially synthesized chromosome, scientific knowledge teaches us what bio-molecules should be used and under what conditions they should be treated. Therefore, everybody can in principle produce a new bacterial cell. Note here that *we human beings can produce a new bacterial cell without having the correct knowledge of the basic qualities of bio-molecules and a bacterial cell*. This means that *scientific knowledge is useful only as a technique*. Truly real things actually work in the internal world we cannot be conscious of.

The above argument can be generalized to the case of dealing with non-living things. For example, a high-school student can successfully produce a high-temperature superconductor even though the student does not know the basic qualities of it, as mentioned in Section 1.1.1. A student needs only mix appropriate metal oxides at a proper ratio and heat them at a suitable temperature, according to scientific knowledge. Such a success certainly arises from the fact that a high-school student makes use of the inherent qualities and functions of metal oxides, in the same way as the aforementioned successes in agriculture and medicine.

A possible principle of constructing artificial intelligence with free independent spirit and creativity

Interestingly, the above argument indicates that artificial life can be produced if truly real things (acting as constituent elements of a living thing) are correctly combined with one another so that their extra properties and functions work effectively. This conclusion can be generalized to the production of artificial intelligence. Namely, we can produce artificial intelligence with free spirit and creativity by properly combining truly real things (acting as constituent elements of it) so that their extra properties and functions work effectively. A remaining problem is to disclose how truly real things are combined with one another.

In connection with the above argument, it is important to note that we have two kinds of artificial life and intelligence. One is artificial life and intelligence which we produce by using components of already existing living things such as DNA, proteins, neurons or their analogs. The other is artificial life and intelligence which we produce without using such components. The former can be called *pseudo*-artificial life and intelligence, while the latter can be called truly artificial life and intelligence. For example, artificial bacterial cells created by Venter's group can be regarded as pseudo-artificial life. For pseudo-artificial life and intelligence, it is relatively easy to find how truly real things are combined with one another because scientific study of already existing living things and intelligence provides information about it, as mentioned earlier. On the other hand, for truly artificial life and intelligence, it is in general quite difficult to find how truly real things are combined. Further details on this issue will be discussed in Section 3.2.3 after it is clarified what is meant by "extra properties and functions" of truly real things.

1.2.3 Various Problems Come from Word-based Scientific Understanding

Word-based scientific understanding is of an approximate character. In this section we consider systematically what problems appear when we remain within the realm of word-based scientific understanding.

Word-based scientific understanding only catches "common or repeatedly observed parts" of activities of truly real things

Scientific laws catch "relations which are kept unchanged in various natural processes", as discussed in Section 1.2.1. This fact clearly indicates that scientific knowledge only catches common or repeatedly observed parts or individual unchanging parts of activities of truly real things. Science does not catch truly real things as they are. It is for this reason that scientific knowledge can only produce machines, as argued in Section 1.1.1.

Most of people will acknowledge that science comes to lose power as the object of study goes up to a higher quality level. For example, science is powerful in basic areas such as Newtonian mechanics, electromagnetism, quantum mechanics, and so on. Science is also fairly powerful in the area of chemistry. However, it comes to be less powerful in the area of biology because properties of living things largely depend on individual species and bodies. Science comes to be much less powerful in psychology and sociology because mental activities of human beings strongly depend on individual persons and conditions. Most typically, the evaluation of artistic work is very far beyond word-based scientific understanding. Such a limit of science also comes from the fact that it does not catch truly real things as they are.

Moreover, science shows severe difficulty in correctly catching unrepeated or irreversible changes such as spiritual, intentional or historical changes. This problem also comes from the same reason as the above.

Word-based scientific understanding only catches "results" of activities of truly real things

Catching "common or repeatedly observed parts" of activities of truly real things has the same meaning as catching "results" of activities of truly real things. Thus, we can say that word-based scientific understanding can only catch "results" of activities of truly real things or natural things themselves.

As a simple example, let us consider the spatial motion of a particle in vacuum. This event is described as follows: "At an instant, a particle is located here and at a next instant the particle is located there". This is the only way to describe the spatial motion of a particle by words. However, these words describe nothing about motion itself. When we say, "At an instant, a particle is located here", a particle at this instant does not move but stops here. The same holds for a particle at a next instant. A particle moves between "an instant" and "a next instant" but no description is given to the behavior of a particle for this period. Thus, the above words describe nothing about motion itself. They only describe a "result" of motion.

For a moving particle, we can in principle not define the position of a particle because it moves. Therefore, we can also not define velocity. In harmony with this argument, quantum mechanics says that spatial position and velocity obey an uncertainty principle and cannot be determined simultaneously with full accuracy (see Section 2.1.2). This means that *motion is originally beyond word-based understanding.*

The same argument applies to change in quality. For example, a change in leaf color is described as follows: "At an instant, a leaf is green and at a next instant it is orange". In this example, a real change in leaf color occurs between "an instant" and "a next instant" but the above words give no description to it. Thus, they only describe a "result" of change. Somebody may say that a change in leaf color can be explained as due to changes in the amounts of dye molecules in a leaf. However, in such an explanation, only a changing thing is altered. There is no alteration in the way of describing a change.

In the same way as motion, quantum mechanics says that a change in the quality of a system (or a change in the energy of a system) also obeys an uncertainty principle (see Section 2.1.2). This means that *a change in quality is also beyond word-based understanding.*

Such a problem in the description of motion or change by words was already discussed by F. Engels, German philosopher, near the end of the 19th century [14]. He asserts that the spatial motion of a particle should be described as follows: "At an instant, a particle is here and not here". His assertion is based on Hegel's dialectic, which says that everything exists in the internal unification of mutually opposing concepts. Indeed, such a description may correctly express motion but is quite difficult to understand. Such a description rather indicates a breakdown of clear logic. This also indicates that motion is beyond word-based understanding.

It is well known that differential calculus or a differential equation in mathematics provides exact description of motion or change. A variety of great successes in science is entirely due to the use of differential and integral calculus. Mathematics has higher ability to describe motion or change than words. However, this does not mean that differential calculus gives the complete solution to the above problem. This is because differential calculus uses the same way of describing motion or change as the above. The only difference is that the period of time between "an instant" and "a next instant" is made infinitesimally small. Great successes of differential calculus come from the fact that it uses an idealized concept of "infinitesimal", not from an alteration in the way of describing motion or change. Therefore, differential calculus or a differential equation can still only express "results" of motion or change of truly real things [*1, *2].

> [*1]This conclusion is given definite support by the aforementioned argument that mathematical concepts such as differential calculus only catches common or repeatedly observed parts of activities of truly real things. In fact, this means that differential calculus only catches "results" of activities of truly real things.
>
> [*2]Differential calculus in mathematics is defined based on the concept of continuity for a valuable and a function. In mathematics, continuity is defined by a so-called ε - δ method. This method also only catches a "result" of continuity and does not catch continuity itself. We can rather say that the ε - δ method indicates how difficult it is to define continuity by words. The concept of continuity is also beyond word-based understanding.

It is important to note here that the above argument leads to the following significant conclusion.

Generation and growth and related events such as "emergence", evolution and creation can be regarded as a kind of motion and change. Therefore, word-based scientific understanding cannot correctly describe such an event.

Word-based scientific understanding only catches "the external appearance" of activities of truly real things

Why can word-based scientific understanding only catch "common or repeatedly observed parts" of activities of truly real things or "results" of them? This is because it only catches the external appearance of activities of truly real things.

In general, when we human beings obtain scientific knowledge, we stand face to face to unknown nature. Then we act on nature and look at a response to it from nature. This is a way to obtain scientific knowledge. Importantly, here is clear separation between the subject and the object. Under such a situation, we can only look at the external appearance of nature and hence we can only obtain knowledge of an approximate character. It is impossible to gain the truly correct understanding without the fusion of the subject and the object.

Science fails to catch the origin of the basic qualities of natural things

The foregoing discussions of the characteristics of word-based scientific understanding indicate that it fails to catch the origin of the basic qualities of natural things. Thus, we can say as follows.

> *Science only describes relations among observed properties, by regarding them as given a priori, without providing the origins of them.*

In fact, properties of an apple are regarded as given a priori. The law of inertia, discovered experimentally by Galilei, is also regarded as given a priori.

Somebody may say that properties of an apple can be explained by physicochemical properties of constituent molecules. This is true. However, such an explanation only gives relations among properties observed on a macroscopic level and those observed on a microscopic level. Thus, the origin of physicochemical properties of constituent molecules is left unknown. A similar argument applies to every explanation in science. Originally, scientific knowledge only represents common or repeatedly observed properties of natural things and fails to catch natural things themselves. Therefore, it can never provide the origin of the basic qualities of natural things. The origin of the basic qualities of natural things is beyond word-based scientific understanding.

Somebody may also say that scientific laws provide the origin of motion or change in the form of a driving force. Indeed, for example, Newton's laws of motion involve a physical force such as the universal gravity as a driving force for motion. However, such a driving force does not correctly represent the origin of motion or change of truly real things. It only represents common or repeatedly observed parts of the origin of motion or change. For example, a human being indeed falls over a precipice in the same way as a stone. A human body strictly follows Newton's laws of motion in this respect. However, this is only one aspect of the way that a human being moves or changes. For example, free independent harmonious spirit and wisdom of a human being cannot be explained by the universal gravity in Newtonian mechanics. Everything in nature exists in infinite dependence on others, as mentioned in Section 1.1.1. Therefore, a driving force in a scientific law only partially or fragmentarily represents the origin of motion or change.

Problems arising from mutually conflicting concepts are impossible to overcome within the realm of word-based understanding

Let us consider intrinsic limits of word-based understanding from a different point of view. We have many pairs of mutually conflicting concepts such as freedom and equality, selfishness and altruism, independence and cooperation, conservation and innovation, safety and adventure, and so on. In general, problems arising from such mutually conflicting concepts are impossible to overcome based on word-based understanding.

For example, when we have met a problem arising from mutually conflicting concepts of freedom and equality, if we emphasize freedom, able persons behave as they wish and equality is broken. On the other hand, if we emphasize equality, the freedom of persons, especially, that of able persons is severely restricted. Therefore, freedom and equality cannot stand together. Nevertheless, both freedom and equality are concepts of fundamental importance. Therefore, for successfully solving the problem, it is necessary to find a way to realize both freedom and equality simultaneously in a harmonious manner under a given condition. However, such a thing is impossible to achieve when we remain within the realm of word-based understanding because freedom and equality are mutually conflicting concepts. Thus, we are usually forced to make an appropriate compromise. Such a way of solving a problem is naturally not a complete one. It leaves us a more or less unconvincing feeling. In addition, such a way of solving a problem often leads to a miserable failure in actual life. The same argument applies to all mutually conflicting concepts.

A fatal fault of word-based understanding arises from the fact that words have individual unchanging meanings. *Words have individual unchanging meanings and thus cause various sorts of distinction and separation in our ideas, which in turn impose severe restriction on our thinking.* If mutually conflicting concepts are caught individually and separately, they come to have completely opposing meanings to each other. Thus, we can no longer solve a problem arising from mutually conflicting concepts in a harmonious manner.

In general, it may be believed that problems arising from mutually conflicting concepts are in principle impossible to overcome on a reasonable basis. However, this idea is not correct. A living thing and its constituent elements exist in infinite dependence on others and incessantly change their structures, properties and functions so that they realize a harmonious state with others, as discussed in Section 1.1.1. Thus we can expect from this fact that *in the very truth even mutually conflict- ing concepts can exist in a harmonious manner.* Further details will be discussed in Section 1.4.1. Difficulty in overcoming mutually conflicting concepts solely arises from an approximate mechanical character of word-based understanding.

The dualism of body and mind cannot be overcome under word-based scientific understanding

Philosophically, we have many important "mutually conflicting concepts" such as life and death, subject and object, existence and non-existence, freedom and inevitability, independence and cooperation, finiteness and infiniteness, part and whole, phenomenon and essence, result and cause, exterior and interior, the known and the unknown, and so on. In the same way as the aforementioned mutually conflicting concepts, these mutually conflicting concepts also cause many difficult problems. Let us look at some examples of such problems.

The dualism of body and mind, mentioned in Section 1.1.2, involves the serious problem of why a human body, which follows inevitable natural laws of a mechanical character, can display free will. It is evident that this problem arises from mutually conflicting concepts of freedom and inevitability. For successfully solving the problem, it is necessary to find a way to realize both freedom and inevitability simultaneously in a harmonious manner. However, such a thing is impossible to achieve when we remain within the realm of word-based understanding. In fact, the above problem has been a central problem in traditional western philosophies since Descartes but has still not been solved to date. For example, Monod finally fell into the dualism of body and mind, as mentioned in Section 1.1.2. He may have felt the lack of clarity but had to put up with such a conclusion.

True freedom cannot be achieved if we remain within the realm of word-based scientific understanding

How we can achieve true freedom is also one of difficult problems arising from mutually conflicting concepts. In general, it is believed that knowledge provides us with freedom. Hegel says, "Freedom is insight into inevitability". Indeed, knowledge provides us with freedom in that it teaches us how we can solve a problem. However, freedom given by knowledge is not true freedom because in this case we cannot be free from restrictions arising from knowledge itself. In true freedom, we have to be free from restrictions arising from knowledge itself, for example, restrictions arising from inevitable natural laws. Thus, here is also a problem arising from mutually conflicting concepts of freedom and inevitability.

This problem has long since been pointed out by existentialists. In fact, they show strong resistance to deterministic logical thinking based on knowledge and stress the primary importance of free independent personal decisions of individuals. For example, J. P. Sartre says [15], "Existence precedes essence" [*3]. F. M. Dostoevskii declares [16], "I do have freedom not to obey (mathematical) natural laws even if they have iron inevitability".

> [*3] In many of traditional philosophies, things we are conscious of have been divided into two groups, existence and essence [16]. Existence refers to an individual particular existent, while essence refers to an individual but general existent. For example, a red apple on a table is existence while a thing expressed by the word of "apple" is essence. When we say, "Mr. A is a human being", Mr. A is existence while a human being is essence.

Now, an important problem is how we can achieve true freedom. It does not seem that existentialists have succeeded in solving this problem. In fact, it appears that they have not yet clarified where the free will of a human being comes from and how we can improve the quality of it. In my opinion, it is because existentialism or other traditional philosophies have failed to clarify where human free will comes from that they have not yet overcome the dualism of body and mind.

Scientific knowledge is not fully effective in fields of ethics, morality, art, and value

For making the correct judgment about ethics, morality, art or value, we need to grasp the essence of a huge variety of things as a whole in an instant. Scientific understanding is strong in explaining a thing in detail by using a large number of words but is weak in catching the essence of a huge variety of things as a whole in an instant. Therefore, scientific understanding is not fully effective in fields of ethics, morality, art, and value.

Here I would like to tell my past personal experience as an example. When I was a university student, I encountered a problem of "How should I live?" or "What is the ultimate purpose of my life?" and thought it over for a long time. The learned persons usually said, "Work for societies" or "Work for other people". However, I was at that time unable to be convinced of such an opinion because I had been unable to solve a problem of "What meaning does it have to live for other persons if all of them live for their own interests or if all of them have no definite meaning in their lives?" In the meanwhile I noticed that it was necessary to find the absolute value of human life and then learned various ideas and opinions in books. However I was unable to reach any convincing answer to the above problem. I only reached the following tentative conclusion: "There is no absolute value in the world. We can only live under the present conditions with past experiences as the base" and "As to the purpose of my life I have to decide it for myself. Any existing knowledge can only provide me with materials for consideration." At that time I was unable to find the absolute basis for my life.

The question of "How should I live?" or "What is the ultimate purpose of my life?" usually becomes important when a person tries to create his or her own ideas of good, molarity, beauty and value. Everybody has his or her own absolute (irreplaceable) particular life and thus everybody has to create his or her own ideas of good, molarity, beauty and value. For creating such ideas, it is necessary to grasp the essence of the whole affairs in the world, including the unknown as well as affairs which may happen in the future. Therefore, such a question as mentioned above is originally impossible to answer based on knowledge in books. *Grasping the essence of the whole affairs in the world and having much knowledge about them are essentially different matters.*

1.3 Experiences of Arriving at the Very Truth in Daily Life

We have thus far considered features and limits of word-based scientific understanding and revealed that it is useful but in essence of an approximate character. The very truth (the way that natural things themselves exist or truly real things exist) is in the internal world we cannot be conscious of. We can never catch the very truth by the use of words. Now, how can we reveal the very truth we cannot be conscious of?

Fortunately, I have become aware that we human beings have unawares noticed that word-based understanding has intrinsic limits and have unconsciously overcome such limits. In fact, if we carefully look at our daily life, we can see that we have various experiences of jumping over intrinsic limits of word-based understanding and arriving at the very truth. Thus, based on such daily experiences, we can clarify how we can arrive at the very truth and how natural things themselves or truly real things exist.

Of various daily experiences of arriving at the very truth, experiences called *Tai-toku* in Japanese are of the most interest because abilities gained by arrival at the very truth are clearly embodied in the mind and the behavior of a person who has attained *Tai-toku*. Therefore, in this section we consider *Tai-toku* in detail. Note that subjects of discussion in this section may look too familiar and common but the meaning of it is very deep. Readers will be surprised to know how novel a view of nature is revealed by considering such a simple daily experience. Readers will also be surprised to know how great differences are between word-based understanding and arrival at the very truth.

1.3.1 Splendid Abilities Are Acquired by Arrival at the Very Truth

What is Tai-toku?

In some particular fields of human activity such as sport, art and technical professions, sufficient abilities cannot be gained by word-based understanding alone. Therefore, keen attention has been paid to the acquisition of "ability beyond word-based understanding" by training or exercise though no theoretical study has been made on it. The word of *Tai-toku*, which means "understanding with a body", is mainly used in such fields. Now, the same problem exists in our daily life. For example, we cannot gain sufficient abilities without training or exercise when we learn how to ride a bicycle, how to drive a car, how to swim, how to play the piano, and so on. To be strict, "ability beyond word-based understanding" is necessary in all fields of human activity.

Let us consider a simple example of learning how to ride a bicycle. A beginner will at first learn techniques by words and acquires knowledge such as "to mount a bicycle", "to take the balance of a body", "to exert an appropriate force on right and left legs alternately", "to keep shoulders relaxed", "to see ahead", and so on. However, a beginner who only learns by words can actually not ride a bicycle. When a beginner is about to take the balance of a body, he or she forgets to exert an appropriate force on left and right legs alternately. When a beginner is about to exert an appropriate force on left and right legs alternately, he or she forgets to take the balance of a body. A beginner cannot do both things simultaneously in a harmonious manner. Thus, a beginner's behavior becomes awkward and clumsy and his or her mind gets full of worries and agonies. Besides, for the same reason, a beginner is strongly attached to learned words and makes his or her head full of ideas such as "I have to take the balance of a body", "I have to exert an appropriate force on right and left legs alternately", and so on.

On the other hand, if a beginner continued exercise and has gained the ability to ride a bicycle, he or she can ride a bicycle freely, skillfully and unconsciously with whistling, in harmony with the surroundings. At this stage, worries and agonies as well as strong attachments to learned words, which have been heavily covering a beginner's mind till then, completely disappear and a wholly clear state of mind full of pleasure appears instead.

Nearly the same processes are experienced in the case of learning how to drive a car, how to ski, how to swim, how to make artworks, how to play the piano, and so on. The initial stage at which a person only learns by words is called the stage of "understanding by words" or "word-based understanding". On the other hand, the stage at which a person can behave freely, skillfully and unconsciously is called the stage of *Tai-toku*.

The attainment of Tai-toku guarantees arrival at the very truth

Now, let us consider characteristics and meanings of the attainment of *Tai-toku*. The most important thing is that a person arrives at the very truth when he or she has attained *Tai-toku*.

When a person is at the stage of "understanding by words", his or her behavior is awkward and clumsy and his or her mind is full of worries and agonies, as mentioned above. Understanding by words does not work effectively in the real world[*1]. This fact clearly indicates that there are some discrepancies between a person's understanding at this stage and the way that things exist in the real world, i.e. a person has not yet arrived at the very truth. The situation is hardly changed even if a person increases the amount of learned knowledge. Examples of *Tai-toku* clearly indicate that understanding by words has severe intrinsic limits.

[*1]This statement is not in contradiction to the arguments given in Section 1.1.1 and 1.2.2. Understanding by words or scientific understanding works effectively for fabricating high-function technological products (machines) but does not work

effectively for achieving free independent harmonious spirit and wisdom.

On the other hand, when a person continued exercise and has reached the stage of *Tai-toku*, he or she can behave freely, skillfully and unconsciously with whistling, in harmony with the surroundings. In general, at the stage of *Tai-toku*, everything goes well in the real world and free independent harmonious spirit and wisdom is achieved together with a wholly clear state of mind full of pleasure. These facts clearly indicate that a person's understanding at the stage of *Tai-toku* and the way that things exist in the real world have completely agreed with each other, i.e. a person has arrived at the very truth.

The above conclusion about arrival at the very truth is important. Therefore, it will be discussed again at the end of Section 1.3.2.

The ability to do a variety of things simultaneously in harmony with one another

What essential difference is between word-based understanding and the attainment of *Tai-toku*? At the stage of word-based understanding, a person can merely do things *one by one*. On the other hand, a person at the stage of *Tai-toku can do a variety of things simultaneously in harmony with one another* so that they go well as a whole.

Let us again consider an example of learning how to ride a bicycle. For a person who is at the stage of "understanding by words", when he or she is about to exert an appropriate force on left and right legs alternately, he or she forgets to take the balance of a body, as mentioned earlier. On the other hand, when he or she is about to take the balance of a body, he or she forgets to exert an appropriate force on left and right legs alternately. Such a person cannot do two or more things simultaneously in a harmonious manner. Such a person can only do things one by one. For this reason, his or her behavior becomes awkward and clumsy and his or her mind gets full of worries and agonies. Examples of *Tai-toku* clearly indicate where intrinsic limits of "understanding by words" are. Words have individual unchanging meanings and exist individually and separately. They are only logically connected with one another. Therefore, they only work one by one.

On the other hand, a person who has attained *Tai-toku* can do a variety of things simultaneously in harmony with one another so that they go well as a whole. For example, a person who has attained *Tai-toku* about riding a bicycle can do a variety of things such as "to take the balance of a body", "to exert an appropriate force on left and right legs alternately", "to keep shoulders relaxed", and so on simultaneously in harmony with one another. For this reason, a person at the stage of

Tai-toku can ride a bicycle freely, skillfully and unconsciously, in harmony with the surroundings. The same argument applies to other examples of *Tai-toku* such as driving a car, skiing, swimming, and playing the piano.

The achievement of true freedom

Let us consider abilities gained by the attainment of *Tai-toku* in more detail. As mentioned above, a person at the stage of *Tai-toku* can behave freely, skillfully and unconsciously, in harmony with the surroundings. Such a person does not use words and needs not use words and therefore is completely free from dependence on learned words and hence completely free from the limits of them. Thus, we can say as follows.

A person who has attained Tai-toku achieves true freedom [*2].

On the other hand, a person who is at the stage of "understanding by words" strongly relies on learned words and cannot be free from them. Such a person is only free in that he or she is aware of how he or she should behave.

> [*2]Note, however, that a person who has attained *Tai-toku* about riding a bicycle achieves true freedom only in the quality level and the aspect of riding a bicycle. See Section 1.4.1 about "quality level and aspect".

A person who has attained *Tai-toku* achieves "understanding without using words". Such a phrase may sound strange because it is in general believed that understanding is achieved by the use of words. However, the achievement of "understanding without using words" is indispensable for arriving at the very truth. In fact, awkward and clumsy behavior and worries and agonies at the stage of "understanding by words" all come from dependence on learned words with individual unchanging meanings. Anybody cannot escape from awkward and clumsy behavior when he or she depends on learned words.

It is important to note here that the achievement of "understanding without using words" never means that a person throws learned words away. It only means that a person has achieved *high-level ability to freely control learned words* [*3] and no longer needs to rely on them. Accordingly, a person at the stage of *Tai-toku* can still use words in the same way as at the stage of understanding by words. For example, a person who has attained *Tai-toku* about riding a bicycle can explain how to ride a bicycle by using words.

We should rather say that a person at the stage of *Tai-toku* can use words much more effectively than one at the stage of "understanding by words" because the former person has achieved high-level ability to freely control learned words, as mentioned above. These arguments lead to the following important conclusion.

> *The attainment of Tai-toku (or arrival at the very truth) allows a person to overcome the intrinsic limits of word-based scientific understanding without losing its feature of usefulness.*

Such a thing is impossible to achieve when a person remains at the stage of understanding by words, as mentioned on page 7 in Section 1.1.1.

[3] Theoretically, human ideas and words merely represent common or repeatedly observed properties of truly real things or imaginary things with such properties. Therefore, we can freely control them when we arrive at the very truth and grasp truly real things themselves.

The emergence of free independent harmonious spirit and wisdom (creativity)

The attainment of *Tai-toku* has another prominent meaning. It allows a person to have free independent harmonious spirit and wisdom (creativity). As mentioned earlier, a person at the stage of *Tai-toku* can do a variety of things simultaneously in harmony with one another so that they go well as a whole. Now, doing a variety of things simultaneously in harmony with one another is equivalent to catching a whole about a variety of things and doing a variety of things so that it goes well. *The attainment of Tai-toku allows a person to catch a whole about a variety of things.* Free independent harmonious spirit and wisdom (creativity) arise from here.

For example, a person who has attained *Tai-toku* about riding a bicycle can ride a bicycle with an umbrella in a hand in the rain. This is because such a person can catch a whole about riding a bicycle and do a variety of things so that such a whole goes well.

The emergence of free independent harmonious spirit and wisdom (creativity) by the attainment of *Tai-toku* can be understood more clearly in an example of playing the piano. A person who has attained *Tai-toku* about playing the piano can look at a whole about a music tune and play individual notes so that it becomes as elegant as possible. Thus, the performance of such a person sounds as if it were guided by "a whole about a music tune" itself. Here is clearly free independent harmonious spirit and wisdom (creativity) coming from catching a whole.

A person who is at the stage of "understanding by words" can never behave in the above way. This is because such a person can only look at individual notes of a music tune one by one and cannot catch a whole about a music tune. Accordingly, such a person entirely relies on individual notes and plays just as individual notes indicate. In general, a person who is at the stage of understanding by words can only do what he or she has learned as he or she has learned. Here is only mechanical behavior and no free independent harmonious spirit and wisdom (creativity).

The attainment of Tai-toku brings about a big revolution in human understanding

Let us summarize the foregoing discussions of the characteristics and meanings of the attainment of *Tai-toku*. It brings about a big revolution in human understanding. A mechanical awkward state subordinated to individual things is converted into an internally ingeniously organized harmonious stable state full of free independent spirit and wisdom (creativity).

> ***Understanding by words (or word-based understanding)*** *– looking at individual things individually – one-by-one thinking and action – strong dependence on learned words – awkward mechanical behavior – full of worries and agonies – only doing what a person has learned as a person has learned.*
>
> ***The attainment of Tai-toku*** *– catching a whole about a variety of things – simultaneous and harmonious thinking and action – completely free from dependence on learned words – free, skillful and unconscious behavior in harmony with the surroundings – achievement of true freedom – achievement of free independent harmonious spirit and wisdom (creativity) – achievement of a wholly clear state of mind full of delight.*

We can clearly see that the attainment of *Tai-toku* provides us with abilities far beyond word-based understanding.

1.3.2 The Origin of Splendid Abilities Acquired by Arrival at the Very Truth

Based on the discussions in the preceding section, we in this section consider why we can achieve various splendid abilities by the attainment of *Tai-toku*. The attainment of *Tai-toku* is a well-known event and its properties have been widely used to improve skill in sport, art, and technical professions. However, no theoretical study has been made on it to date.

The formation of a continuous connection of various things (constituent elements) to one another

It is evident that the big jump happens when a person has attained *Tai-toku* because the impossible becomes possible at this moment. In fact, it is entirely impossible for a person who is at the stage of "understanding by words" to do a variety of things simultaneously in harmony with one another. The occurrence of the big jump can also be understood from the fact that severe worries and agonies that have been heavily covering the mind till then suddenly disappear and a clear state of mind full of delight appears instead. Here is an about-turn change in the state of mind. Furthermore, it is known that while a person continues exercise, the ability acquired by it does not go up in proportion to the amount of it. The ability remains on a low level for a while and steeply goes up at a stage of a certain amount of exercise, as schematically illustrated in Figure 1-2. Such non-linear improvement of ability also suggests that the jump happens during exercise.

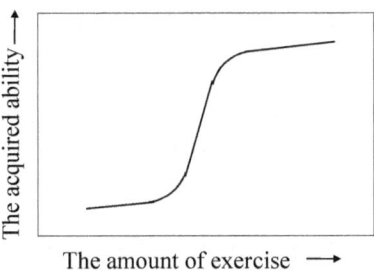

Figure 1-2 The ability goes up non-linearly with the amount of exercise.

Now, what jump happens upon the attainment of *Tai-toku*? As stated earlier, a person who is at the stage of "understanding by words" can merely do things one by one. This indicates that constituent elements of "the ability gained by the attainment of *Tai-toku*" exist individually and separately at this stage and are only logically connected with one another, in a similar way to constituent elements of a machine.

In general, constituent elements of "the ability gained by the attainment of *Tai-toku*" consist of a huge variety of things. For example, constituent elements of the ability to ride a bicycle consist of a human body, eyes, hands, legs, a bicycle, the ground, the air (wind), the gravitation, friction, and so on, together with ideas (nervous patterns in the brain) of these substances and forces. At the stage of "understanding by words", such constituent elements exist individually and separately and are only logically connected with one another. Thus, a person at this stage can do things only one by one.

On the other hand, a person at the stage of *Tai-toku* can behave freely, skillfully and unconsciously, in harmony with the surroundings. Also, a person at this stage can do a variety of things simultaneously in harmony with one another so that a whole about them goes well. Besides, the attainment of *Tai-toku* provides a person with free independent harmonious spirit and wisdom (creativity) and a wholly clear state of mind full of delight. Thus, we have to say that at the stage of *Tai-toku* constituent elements of "the ability gained by the attainment of *Tai-toku*" are fully continuously connected with one another and *fused into unity*[*1]. In other words, ideas (nervous patterns) in the human brain and things in the real world are fully continuously connected with one another and have completely agreed with each other. Here is clearly the fusion of the subject and the object. If no such continuous connection were formed, i.e. if any discrepancy, separation or distinction remained anywhere, a person at the stage of *Tai-toku* could neither behave freely, skillfully and unconsciously nor do a variety of things simultaneously in harmony with one another. This conclusion is also in agreement with the aforementioned argument that the attainment of *Tai-toku* allows a person to catch a whole about a variety of things.

[*1]To be strict, not only constituent elements of "the ability gained by the attainment of *Tai-toku*" but also related actions are continuously connected. Thus, both constituent elements and related actions work simultaneously in harmony with one another. For example, for a person who has attained *Tai-toku* about riding a bicycle, not only constituent elements such as a body, hands, legs, a bicycle, etc. but also various actions such as "to take the balance of a body", "to exert an appropriate force on left and right legs alternately" and so on are continuously connected with one another and work simultaneously in harmony with one another.

It should be noted also that constituent elements of "the ability gained by the attainment of *Tai-toku*" cover *all* things in nature and the world. This is because otherwise no fully continuous connection would be formed and a person who has attained *Tai-toku* could never achieve splendid abilities such as mentioned above. Thus, for example, when a person has attained *Tai-toku* about riding a bicycle, *everything* in nature and the world is continuously connected with one another in the quality level and the aspect of riding a bicycle. The constituent elements of the ability to ride a bicycle, raised above, only express some examples.

The attainment of Tai-toku gives direct evidence for the existence of the internal invisible world

There is another important point to be noted in the attainment of *Tai-toku*. Strangely, no clear change happens in the external visible world when a person has attained *Tai-toku*, i.e. when constituent elements are fully continuously connected with one another. For example, when a person has attained *Tai-toku* about riding bicycle, constituent elements of the ability to ride a bicycle such as a human body, hands, legs, a bicycle, the air, the ground, the gravitation, and so on exist individually and separately in the external visible world, in the same way as at the stage of understanding by words.

How can we understand this fact? Note first that we cannot deny the formation of a continuous connection such as mentioned earlier even though we cannot look at it in the external visible world. If we deny it, we cannot explain the attainment of *Tai-toku*, i.e. the attainment of various splendid abilities such as true freedom, free independent harmonious spirit and wisdom, a wholly clear state of mind full of delight, and so on. Therefore, we have to say that such a continuous connection is formed *internally* in the region we cannot be conscious of. This argument is reasonable if we take into account that a continuous connection has no individual unchanging quality and thus we cannot be conscious of it even if it is formed. We human beings can only be conscious of things with individual unchanging properties, as discussed in Section 1.2.1. The attainment of *Tai-toku* cannot be understood without taking into account the internal invisible world.

Somebody may say that the attainment of *Tai-toku* or the attainment of the above-mentioned splendid abilities can be explained in terms of a "jump" in nonlinear dynamics without assuming the formation of a continuous connection. However, I cannot agree with such an opinion because nonlinear dynamics, which is based on words with individual unchanging meanings, is of a mechanical character, as discussed in Section 1.1.2, and thus cannot explain the emergence of free independent harmonious spirit and wisdom (creativity) upon the attainment of *Tai-toku*. This issue will be discussed again in Section 3.2.3.

Based on the foregoing arguments, we can now say that the attainment of *Tai-toku* is to form a state in which constituent elements of "the ability gained by the attainment of *Tai-toku*" are fully continuously connected with one another in a quality level and aspect of this ability. Namely, the attainment of *Tai-toku* is to form an infinitely-spreading fully-continuous internal connection of constituent elements of "the ability gained by the attainment of *Tai-toku*" to one another in a quality level and aspect of this ability. It is very likely that strong interaction between various things (constituent elements) during thorough exercise leads to the formation of such a continuous connection. On the other hand, a simple increase in the amount of learned knowledge does not lead to the attainment of *Tai-toku*, as mentioned earlier. There is a great difference in meaning between carrying out thorough exercise and simply increasing the amount of learned knowledge.

The formation of a continuous internal connection is responsible for the emergence of free independent harmonious spirit and wisdom (creativity)

Now, for getting better understanding of the foregoing argument, let us consider again the difference between a person who remains at the stage of "understanding by words" and one who has attained *Tai-toku*. A main purpose is to clarify how free independent harmonious spirit and wisdom (creativity) emerge by the attainment of *Tai-toku*.

For a person who is at the stage of "understanding by words", constituent elements of "the ability gained by the attainment of *Tai-toku*" (i.e. various things in the real world and ideas of them in the brain) exist individually and separately and are only logically connected with one another, as schematically illustrated in Figure 1-3(A). Namely, constituent elements in this case *always* have individual unchanging properties and thus exist as the basic governing reality, in the same way as constituent elements of a machine. Accordingly, a person at this stage can only behave one by one like a machine. Thus, such a person can demonstrate no free independent harmonious spirit and wisdom (cre-

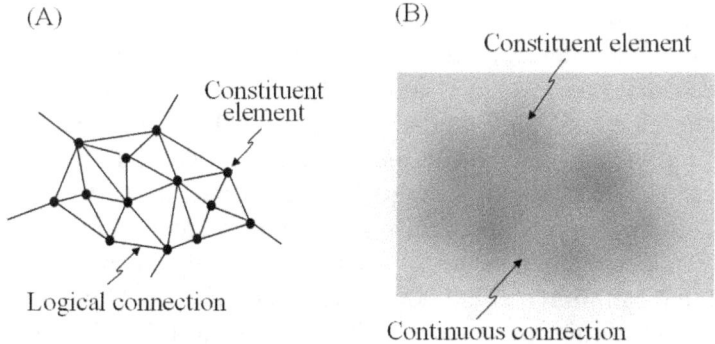

Figure 1-3 Schematic illustrations of (A) a logical connection of constituent elements with individual unchanging properties to one another and (B) a continuous internal connection of constituent elements with freely changing dynamic qualities to one another.

ativity).

On the other hand, for a person who has attained *Tai-toku*, constituent elements of "the ability gained by the attainment of *Tai-toku*" are fully continuously connected with one another and fused into unity, as illustrated in Figure 1-3(B). Therefore, individual constituent elements no longer have unchanging properties though they still demonstrate individual unchanging properties in the external visible world. They incessantly change their structures, properties and functions so that they as a whole realize a free independent harmonious state. For this reason, a person who has attained *Tai-toku* can behave freely, skillfully and unconsciously, in harmony with the surroundings, as mentioned earlier. Also, such a person can do a variety of things simultaneously in harmony with one another so that a whole goes well. After all, such a person realizes a free independent harmonious state both in the interior and in relation to the outer world.

From these considerations, we can say that for a person who has attained *Tai-toku*, fully continuously connected constituent elements as a whole constitute the basic governing reality and controls individual constituent elements so that they as a whole realize a free independent harmonious state. Interestingly, *the formation of a continuous internal connection leads to the transfer of the basic governing reality from individual constituent elements to a whole about continuously connected constituent elements.* Free independent harmonious spirit and wisdom come from such a whole.

The understanding of a person who has attained Tai-toku extends deep into unknown areas

There is another important aspect about the emergence of free independent harmonious spirit and wisdom. As mentioned above, when constituent elements are internally continuously connected with one another, individual constituent elements no longer have unchanging properties. They incessantly change their structures, properties and functions so that they as a whole realize a free independent harmonious state. This fact means that it is only things with freely changing dynamic qualities that can form a continuous internal connection. Now, what thing has freely changing dynamic qualities? It is a thing existing as a continuous internal connection, such as a person who has attained *Tai-toku*. In fact, such a thing (or such a person) realizes a harmonious state both in the interior and in relation to the outer world, as mentioned earlier.

These considerations indicate that in a continuous internal connection of constituent elements to one another, such as illustrated in Figure 1-3(B), the constituent elements each again exist as a continuous internal connection of other constituent elements. Thus, here are only continuous internal connections. Namely, here is nothing with individual unchanging qualities and hence here is no discrepancy, distinction and separation. Importantly, this means that in such a state there is no separation between the finite and the infinite or no separation between the known and the unknown.

Amazingly, a person who has attained Tai-toku has understanding that is not restricted within the known area. It extends deep into vast unknown areas and has no limit. This is natural, in a sense, if we consider that such a person attains "ability beyond word-based understanding" or the fusion of the subject and the object, as mentioned earlier. It is for this reason that such a person can achieve free independent harmonious spirit and wisdom (creativity).

The attainment of Tai-toku really guarantees arrival at the very truth

Finally, let us consider again that the attainment of *Tai-toku* really guarantees arrival at the very truth. As argued earlier, for a person who has attained *Tai-toku*, constituent elements of "the ability gained by the attainment of *Tai-toku*" (i.e. various things in the real world and ideas of them in the brain) are fully continuously connected with one another and fused into unity. In other words, ideas (nervous patterns) in a person's brain and things in the real world are fully continuously connected with one another and have completely agreed with each other. Accordingly, we can safely say that a person who has attained *Tai-toku* has arrived at the very truth.

It should be emphasized here that only the above condition can be adopted as the criterion for arrival at the very truth. The importance of the attainment of *Tai-toku*, i.e. the formation of a fully continuous connection of ideas in the human brain and things in the real world is that it guarantees the complete agreement between human understanding and the real world *in the whole area including the unknown and what we human beings cannot be conscious of.* Here is the fusion of the subject and the object. This is a critically important point because human consciousness only catches common or repeatedly observed properties of natural things and fails to catch natural things themselves, as argued in Section 1.2.1. The attainment of true freedom, free independent harmonious spirit and wisdom (creativity), a wholly clear state of mind full of delight, etc. guarantees that we have recovered what human consciousness has failed to catch[*1].

*¹This does not mean that we can be conscious of what human consciousness has failed to catch when we have recovered it. We cannot be conscious of what human consciousness has failed to catch even when we have recovered it. We accept it as the internal wisdom (see Section 1.4.3 and A1.1).

In science, the correctness of knowledge is in general guaranteed by experimental proof and logical consistency. Such criteria are suitable for acquiring the objective (scientific) truth but unsuitable for judging arrival at the very truth. This is because such criteria can only be applied to what we can be conscious of, i.e. things with common or repeatedly observed properties of natural things. Here is only agreement between human understanding and the real world in the area we can be conscious of. Here is definite separation between the known and the unknown and between the subject and the object. Originally, science has made progress with the aim of acquiring the objective truth useful for human life, not the very truth, as discussed in Section 1.2.1. Therefore, criteria in science are not suitable for judging arrival at the very truth.

We sometimes have an experience of feeling a doubt or suspicion about a conclusion that looks completely correct on a logical basis. This means that a human mind to achieve a wholly clear state has power to find a fault in a logical conclusion. Logical thinking is based only on the known (what has been disclosed) while the human mind to achieve a wholly clear state works in areas including the unknown and what we cannot be conscious of.

Verification by a clear fact does not necessarily guarantee truth

In relation to the above argument, there is an important issue to be noted. In general, it is believed that the correct knowledge is obtained based on clear facts. Science has also been constructed based on clear facts. However, this belief is not completely correct. This is because

Even "a clear fact" is of an approximate character.

Namely, a clearly observed thing only represents common or repeatedly observed properties of truly real things (see Section 1.2.1 and 1.2.3). Therefore, verification by a clear fact does not necessarily guarantee truth. We should rather say as follows.

Relying on a clear fact alone often leads to an incorrect conclusion.

In fact, it will be argued in Section 2.2 that the atmosphere of attaching the highest importance to "a clear fact" in science has led to an unsuitable interpretation of the second law of thermodynamics.

As another example, let us here consider the philosophy of Descartes, who is known as the originator of modern philosophy. It is said that Descartes started from the statement of "Any philosophy should be constructed based on a clear fact that nobody can doubt any further" (see Section A3.1.1). Indeed, this attitude of Descartes led to the establishment of the concept of human ego and the construction of modern philosophy. However, this attitude also had a severe problem in that it gave the absolute value to human intellect or reason, thus leading to the exclusion of the internal world lying beyond the bounds of human consciousness. As a result, Descartes failed to catch the free will of a human being and fell into the dualism of body and mind (see Section 1.1.2 and 1.2.3). Taking a clear fact as the basis for judgment is a kind of self-centered attitude of a human being and results in the failure to become aware of the intrinsic limits of human consciousness.

We should not forget that we human beings can only be conscious of common or repeatedly observed properties of natural things. Therefore, *we cannot deny the existence of the internal invisible world for the reason that we cannot be conscious of it.* Behind the objective or scientific truth, there is the very truth with profound abilities and qualities. Therefore, *denying the existence of the internal invisible world is equivalent to lowering human understanding to the level of human word-based understanding.* It is evident that this is not a right way of understanding.

1.3.3 A Human Being Lives with the Internal Invisible World as the Base

The origin of abilities gained by the attainment of Tai-toku is in the internal world we cannot be conscious of

We have thus far argued that the attainment of *Tai-toku* allows us to transcend word-based understanding and arrive at the very truth. The attainment of *Tai-toku* provides us with splendid abilities that are far beyond word-based understanding. An important point to be noted here is that we cannot be conscious of the origin of such splendid abilities. The development of this argument leads to another interesting conclusion.

At first, let us confirm that we can never be conscious of the origin of splendid abilities gained by the attainment of *Tai-toku*. We can easily see this fact by considering how a person who has attained *Tai-toku* explains the origin of his abilities. For example, when a person who has attained *Tai-toku* about riding a bicycle is asked how to ride a bicycle, the person will answer as follows: Carry out a variety of things such as to take the balance of a body, to exert an appropriate force on left and right legs alternately, and so on simultaneously in harmony with one another. Indeed, this is the correct answer. A person who has attained *Tai-toku* about riding a bicycle can say nothing more than this. However, this answer is trite to even persons who cannot ride a bicycle because they already know such an answer by learning by words. Regretfully, a person who has attained *Tai-toku* cannot be conscious of the origin of his or her splendid abilities and hence cannot express it by words.

Understanding by words and the attainment of *Tai-toku* have a great difference in abilities of actual behavior. However, a person who is at the stage of understanding by words and one who has attained *Tai-toku* speak nearly the same words. This fact clearly indicates that we can never be conscious of the origin of abilities gained by the attainment of *Tai-toku*. The origin of such abilities is in the internal world we cannot be conscious of and works spontaneously and unconsciously when it is necessary as if it were instinctive ability. Accordingly, it is probable that a good golf player is not necessarily a good teacher in a golf lesson.

We human beings can live a free independent harmonious life with the internal invisible world as the base

Now, a glance of our daily life indicates that we human beings have a great diversity of splendid abilities without being conscious of why we can have such abilities. For example, we can live a healthy life though we have no enough knowledge of biology, physiology and psychology. We can consider various things in various ways though we have no enough knowledge of logic. Also, we can judge various matters about ethics, morality, beauty and value appropriately though we have no enough knowledge about them. In general, it is not true that we know everything and then we live. The truth is that we live and then we know something. We live intentionally only about some chosen issues. Our life is in essence supported by a great many abilities, the origin of which we cannot be conscious of.

Traditional understanding and philosophies have explained such abilities as due to the instinctive ability of a human being. However, such an explanation only describes observed things as they are, by regarding them as given a priori, without providing any origin of them. Such a way of understanding is characteristic of word-based scientific understanding, as emphasized in Section 1.2.3. Now, if we take into account the characteristics of the attainment of *Tai-toku* argued in Section 1.3.1 and 1.3.2, we can easily explain why we human beings can have such various splendid abilities.

At first, let us look back on the characteristics of the attainment of *Tai-toku*. A person at the stage of *Tai-toku* has formed a continuous internal connection of constituent elements of "the ability gained by the attainment of *Tai-toku*" (a variety of things in the real world and ideas of them in the brain). Such continuously connected constituent elements spontaneously and simultaneously work in harmony with one another so that they as a whole realize a free independent harmonious state both in the interior and in relation to the outer world. Accordingly, a person who has attained *Tai-toku* can stand on such continuously connected constituent elements, which he or she cannot be conscious of, and can live a free independent harmonious life without relying on anything in the external visible world. All splendid abilities come from continuously connected constituent elements, which work spontaneously and unconsciously when they are necessary.

Thus, by the same principle, we human beings can have a diversity of splendid abilities if a human being itself exists as a continuous internal connection of constituent elements of it. Fortunately, we can show that a human being or generally a living thing exists as a continuous internal connection of constituent elements of it. Details on this conclusion will be discussed in Section 4.2 to 4.4. Here, let us consider only a simple example. It was mentioned in Section 1.1.1 that a living thing and its constituent elements exist in infinite dependence on others and incessantly change their structures, properties and functions so that they realize a dynamic harmonious state with others. On the other hand, a person who has attained *Tai-toku* also does a variety of things simultaneously in harmony with one another so that they as a whole realize a free independent harmonious state. Thus, we can see that a living thing behaves in the same way as a person who has attained *Tai-toku*. This indicates that a living thing or a human being is produced by the same principle as a person who has attained *Tai-toku*.

Accordingly, based on the characteristics of the attainment of *Tai-toku*, we can easily see that we human beings can have a diversity of splendid abilities without being conscious of the origin of them. The origin of such abilities is in continuous internal connections of constituent elements that a human being has formed. Human life cannot be correctly understood without taking into account the internal invisible world.

1.4 A Novel View of Nature and the World

The arguments in the preceding sections have made it clear that the internal world we cannot be conscious of really exists and the very truth is in it. In general, traditional understanding and philosophies have not taken into account the internal invisible world. Therefore, the next important step is to clarify what novel view of nature and the world we have when we acknowledge the existence of the internal invisible world. Thus we in this section deal with this issue. A novel view of nature and the world we discuss here offers a model for constructing a novel view of nature based on modern science in Chapter 2 and later. In fact, this book shows that modern science demonstrates the same view of nature as discussed below.

1.4.1 How Do Truly Real Things Exist?

We can only look at common or repeatedly observed properties of natural things and cannot look at natural things themselves, as discussed earlier. Therefore, the first important question is how natural things themselves or truly real things exist. Then, let us here consider this problem.

The way that truly real things exist, argued here, may look quite strange. Originally, there is a non-removable gap between word-based understanding and the very truth (the way that truly real things exist), as argued in Section 1.2.1. Therefore, it is natural that truly real things have a strange way of existing, which is impossible to correctly understand within the realm of word-based understanding. A feeling of strangeness solely comes from the fact that we human beings are now completely accustomed to the way that things we are conscious of exist or the way that things with common or repeatedly observed properties exist. We now need to become accustomed to the way that truly real things exist.

A truly real thing exists as a fully continuous internal connection of all other things

Now, how can we clarify the structures and properties of truly real things we cannot be conscious of? As argued in Section 1.3.2, the attainment of *Tai-toku* guarantees arrival at the very truth (i.e. the way that truly real things exist). Therefore, the way that a person who has attained *Tai-toku* exists represents the way that a person who has arrived at the very truth exists and hence represents the way that truly real things exist. Namely, the way that truly real things exist is embodied in the way that a person who has attained *Tai-toku* exists.

It was argued in Section 1.3.2 that for a person who has attained *Tai-toku*, constituent elements of "the ability gained by the attainment of *Tai-toku*" (a variety of things in the real world and ideas of them in the brain) are fully continuously connected with one another and fused into unity. Now, constituent elements of "the ability gained by the attainment of *Tai-toku*" actually cover *all* things in nature and the world. Therefore, when a person has attained *Tai-toku*, for example, about riding a bicycle, *all* things in nature and the world are continuously connected with one another in the quality level and the aspect of riding a bicycle. This means that in the very truth the ability to ride a bicycle (or a person who has attained *Tai-toku* about riding a bicycle) exists as an infinitely-spreading fully-continuous internal connection of all other things in nature and the world to one another, formed in the quality level and the aspect of riding a bicycle.

The above way that a thing exists can be generalized to all things in nature and the world. In fact, a human being or generally a living thing exists in the same way as a person who has attained *Tai-toku*, as mentioned in Section 1.3.3. In addition, non-living things exist in the same way as living things though qualities of the way of existing are different. Details of this issue will be discussed in later chapters. Thus, in the very truth all things in nature and the world exist in essentially the same way as the ability to ride a bicycle or a person who has attained *Tai-toku*. Accordingly we can say as follows.

In the very truth, everything in nature and the world (or every "truly real thing") including a living thing and a human idea (a nervous pattern in the brain) exists as an infinitely-spreading fully-continuous internal connection of all other things (acting as constituent elements) to one another, formed in each particular quality level and aspect.

We cannot be conscious of a continuous internal connection, as discussed in Section 1.3.2. Therefore, the above statement indicates that in the very truth everything (or every "truly real thing") is in the internal world we cannot be conscious of. We can only be conscious of common or repeatedly observed properties of truly real things.

Nature is composed of a huge diversity of truly real things (natural things themselves). Therefore, the above statement can be rewritten as follows.

In the very truth, nature and the world consist of a complexly-entangled, multi-fold and multi-dimensional dynamic network of a

> *huge (or an infinite) number of infinitely-spreading fully-continuous internal connections. A thing existing as a fully-continuous internal connection formed in each particular quality level and aspect corresponds to an individual "truly real thing".*

This statement represents a fundamental aspect of a novel view of nature and the world in this book. Though there are only internal continuous connections in the very truth about nature and the world, they can be distinguished from one another because they are formed each in a particular quality level and aspect.

It is interesting to note here that the above statement indicates that nature demonstrates a variety of hierarchical structures in the external visible world. This is because a truly real thing displays individual unchanging properties in a common or averaged form in the external visible world, as discussed in Section 1.2.1 and 1.3.2 (see also Section 1.4.2). In fact, science has revealed that nature has a variety of hierarchical structures such as elementary particles – atoms and molecules – colloids – macroscopic visible things or non-living things – unicellular living organisms – multicellular living organisms – high plants and animals – human beings.

A truly real thing realizes a free independent harmonious state and besides has the internal ability to organize itself so that it realizes a free independent harmonious state

Let us consider properties of a truly real thing in more detail. As stated in Section 1.3.1 and 1.3.2, when a person has attained *Tai-toku*, he or she forms a continuous internal connection of constituent elements of "the ability gained by the attainment of *Tai-toku*". Thus, individual constituent elements no longer have unchanging qualities. They incessantly change their structures, properties and functions so that they as a whole realize a free independent harmonious state both in the interior and in relation to the outer world. Thus, we can say as follows.

> *A truly real thing, i.e. a thing existing as a continuous internal connection of all other things (constituent elements) realizes a free independent harmonious state both in the interior and in relation to the outer world.*

As mentioned above, when constituent elements are continuously connected with one another, they incessantly change their structures, properties and functions so that they as a whole realize a free independent harmonious state. This means that a whole about continuously connected constituent elements or a continuous internal connection of constituent elements controls individual constituent elements so that it realizes a free independent harmonious state. Thus, we can also say as follows.

A continuous internal connection of constituent elements and hence a truly real thing existing as such a continuous internal connection has the internal ability to organize itself so that it realizes a free independent harmonious state.

These considerations clearly indicate that a truly real thing can work as the origin of autonomous dynamic self-organizing ability or free independent harmonious spirit and wisdom of living things.

In the very truth, not individual things in the external visible world but a continuous internal connection of them exists as the basic governing reality

Let us consider properties of a truly real thing from a different point of view. As mentioned above, when constituent elements are continuously connected with one another, they incessantly change their structures, properties and functions so that they as a whole realize a free independent harmonious state. This means that such constituent elements can no longer be regarded as fundamental existents in nature even though they still demonstrate individual unchanging properties in the external visible world. Instead, a whole about continuously connected constituent elements or a continuous internal connection of constituent elements, which controls individual constituent elements and realizes a free independent harmonious state, is regarded as the fundamental existent.

In the very truth, not individual things in the external visible world but a continuous internal connection of them, which we cannot be conscious of, exists as the basic governing reality. The basic qualities of individual things are completely controlled by underlying continuous internal connections of them or a whole about them.

If the basic qualities of individual things are completely controlled by underlying continuous internal connections of them, individual things can no longer be separated from such continuous internal connections. Now, in the very truth, all things in nature and the world are continuously connected with one another, as mentioned earlier. Therefore, we can obtain the following important conclusion.

In the very truth, anything cannot be separated from nature and the world. If something is separated from nature and the world, it loses the true way of existing and hence loses the true qualities.

For example, if something is expressed by words with individual unchanging meanings, it is separated from nature and the world. Therefore, such a thing loses the true way of existing and hence loses the true qualities. This argument explains why word-based scientific understanding is of an approximate character.

Strange dynamic harmony between mutually opposing powers

There is another remarkable feature in a truly real thing. It realizes strange dynamic harmony between mutually opposing powers of "infinite dependence on others" and "independence of others". Here, strange dynamic harmony means that mutually opposing powers are simultaneously and harmoniously realized in a real situation.

The above fact is clearly embodied in the way that a truly real thing exists, discussed earlier. In fact, every "truly real thing" exists as a continuous internal connection of all other things and simultaneously realizes a free independent harmonious state.

The presence of strange dynamic harmony can be shown theoretically as well. An infinitely-spreading fully-continuous internal connection has no discrepancy, distinction and separation. Therefore, it realizes infinite continuous connection to others and simultaneously realizes a free independent harmonious stable state. Moreover, the realization of infinite continuous connection to others leads to the achievement of a free independent harmonious stable state and the achievement of a free independent harmonious state leads to the realization of infinite continuous connection to others. In this way, "infinite dependence on others" and "independence of others" have opposing meanings to each other but cooperate with each other.

Accordingly, from the above cosiderations, we can obtain the following interesting conclusion.

Every "truly real thing" realizes strange dynamic harmony between mutually opposing powers of "infinite dependence on others" and "independence of others" in each particular quality level and aspect.

If a face of "independence" is emphasized, we can also say as follows.

Every "truly real thing" exists as a quasi-independent or quasi-self-completed system though it still exists in infinite dependence on others.

The presence of strange dynamic harmony is very important in that it explains the basic structure of nature. For example, it explains why nature can have a variety of hierarchical structures.

The presence of strange dynamic harmony is also important for correctly understanding mutually conflicting concepts. As argued in Section 1.2.3, we have many mutually conflicting concepts such as freedom and equality, selfishness and altruism, independence and cooperation, and so on. Such mutually conflicting concepts can be understood as an appearance of the above strange dynamic harmony between mutually opposing powers in a particular area. For example, the idea of freedom arises from emphasizing "independence" while that of equality arises from stressing "infinite dependence". Similarly, the idea of selfishness arises from emphasizing "independence" while that of altruism arises from stressing "infinite dependence". Therefore, we can say as follows.

In the very truth, mutually conflicting concepts realize strange dynamic harmony in the same way as the mutually opposing powers of "infinite dependence on others" and "independence of others".

We can reach the same conclusion by taking into account that there is in essence no discrepancy, distinction and separation in the very truth.

The concept of strange dynamic harmony is impossible to correctly understand within the realm of word-based understanding. For correctly understanding it, we have to gain "ability beyond word-based understanding" such as discussed in Section 1.3.1.

In the very truth, all things are controlled by the grand internal ability

We have thus far considered how truly real things exist. Next, let us consider how they are controlled. It was argued in Section 1.3.2 that strong interaction between constituent elements of "the ability gained by the attainment of *Tai-toku*" during exercise leads to the formation of an infinitely-spreading fully-continuous internal connection of them. Thus, based on this argument together with the foregoing arguments, we can say as follows.

In the very truth, all things in nature and the world, including living things and human ideas, are controlled by the grand internal ability to make various things interact with one another according to their inherent properties and motion, form as

> *widely-spreading fully-continuous internal connections of them as possible under given conditions in diverse quality levels and aspects, and complete the formation of such continuous internal connections. In this way, the grand internal ability produces a new thing existing as an infinitely-spreading fully-continuous internal connection of other things.*

This statement represents another fundamental aspect of a novel view of nature and the world in this book. The correctness of this statement will be confirmed later, based on quantum mechanics in modern science in Section 2.1.3 and based on the philosophy of Buddhism in Section A2.3 and A2.4.

Importantly, the grand internal ability explains why a truly real thing exists as a continuous internal connection of all other things, why nature and the world consist of a complex network of a huge number of continuous internal connections, why interaction among a variety of things (constituent elements) leads to the formation of a continuous internal connection, and why a truly real thing realize strange dynamic harmony.

The above statement can be rewritten into a form familiar to our daily experiences.

> *A diversity of motion and behavior of various things and their interaction in the external visible world lead to the formation of as widely-spreading fully-continuous internal connections of them as possible in diverse quality levels and aspects under given conditions and finally lead to the production of new things existing as infinitely-spreading fully-continuous internal connections of them.*

This statement indicates that the internal invisible world is the world of causes while the external visible world is the world of results. Thus we cannot understand the origin of the basic qualities of things as far as we only look at the external visible world, as discussed in Section 1.2.3. It is important to pay attention to the internal invisible world.

There are some comments on the above statement. Firstly, the grand internal ability is immanent in nature and the world and we can pursue the way that it works through investigations of natural phenomena as well as our experiences. Therefore, the grand internal ability should be clearly distinguished from supernatural power of a god or God in traditional understanding or philosophies.

Secondly, an infinitely-spreading fully-continuous internal connection, formed by the grand internal ability, represents a "result" of activities of the grand internal ability. Therefore, such a continuous internal connection catches the grand internal ability within it, namely, such a continuous internal connection has the internal ability of the same character as the grand internal ability. In fact, a continuous internal connection of constituent elements has the internal ability to organize itself so that it realizes a free independent harmonious state, as mentioned earlier.

Thirdly, the grand internal ability works in multiple ways. Namely, it produces various things, makes them interact with others in diverse quality levels and aspects, and produces further various things. Such processes happen everywhere limitlessly in nature and the world. Dynamic aspects of nature and the world such as the evolution of nature, the development of society, the growth of human intellectual ability, etc. come from here, as will be discussed later.

Fourthly, nevertheless, nature and the world are in essence kept in a peaceful harmonious state. This is because the grand internal ability always controls various things in nature and the world so that they as a whole realize a peaceful harmonious state. Note that such a favorable view of nature and the world does not come from theoretical consideration. It comes from *the fact* that the very truth about nature and the world is embodied in a wholly clear state of mind we achieve upon the attainment of *Tai-toku*. In fact, in a wholly clear state of mind, our understanding and the real world completely agree with each other and besides a peaceful harmonious state is realized there.

We can arrive at the very truth only in a particular quality level and aspect

We have thus far considered how truly real things exist and how they are controlled, based on our experiences of the attainment of *Tai-toku*. Indeed, the attainment of *Tai-toku* guarantees arrival at the very truth but we have to note that the very truth we arrive at is not quite the same as the whole "very truth" about nature and the world. Thus, let us next consider how they are related to each other.

We can arrive at the very truth by other ways than the attainment of *Tai-toku*. For example, when we met a problem and have overcome it by thorough consideration and pursuit, we create a new idea, gain free thinking and achieve a wholly clear state of mind full of delight. The achievement of free thinking and a wholly clear state of mind clearly indicates that we have arrived at the very truth (see the argument in Section 1.3.2). Thus, we can say as follows.

> *We reach the very truth whenever we have overcome a problem by thorough consideration and pursuit and achieved a wholly clear state of mind full of delight, irrespective of what problem we have overcome.*

Interestingly, it is not difficult to arrive at the very truth.

However, we have to note also that we can arrive at the very truth only in a particular quality level and aspect, depending on what problem we have overcome. Thus, we also have to say as follows.

> *When we have overcome a problem and achieved a wholly clear state of mind full of delight, we only arrive at the very truth in a particular quality level and aspect, depending on what problem we have overcome.*

Let us look at some examples. When a person has attained *Tai-toku* about riding a bicycle, he or she arrives at the very truth about riding a bicycle. Similarly, when a person has attained *Tai-toku* about skiing, swimming, and playing the piano, he or she arrives at the very truth about skiing, swimming, and playing the piano, respectively. When H. Yukawa, Japanese physicist, achieved a clear state of mind by overcoming the problem of what force acts between a proton and a neutron in an atomic nucleus (see Section 6.2 for details), he arrived at the very truth about how small particles such as protons and neutrons interact with each other. Shaka-muni, originator of Buddhism, felt intense mental pain to look at the death of a person or to become aware of a temporary character of human life. When he overcame his mental pain and achieved a wholly clear state of mind (see Section A1.2.1 for details), he arrived at the very truth about what human life is.

Now, we can arrive at the very truth only in a particular quality level and aspect, as mentioned above. This means that we can achieve a wholly clear state of mind only in a particular quality level and aspect and hence we can form a continuous internal connection only in a particular quality level and aspect.

> *When we have overcome a problem, we form a continuous internal connection in a particular quality level and aspect, depending on what problem we have overcome.*

When we have attained *Tai-toku* about riding a bicycle, we form a continuous internal connection in the quality level and the aspect of riding a bicycle. When we have attained *Tai-toku* about playing the piano, we form a continuous internal connection in the quality level and the aspect of playing the piano. Similarly, when Shaka-muni overcame his mental pain and achieved a wholly clear state of mind, he formed a continuous internal connection in the quality level and the aspect of what human life is.

There is another important point to be noted. It was argued in Section 1.3.2 that a person who has attained *Tai-toku*, i.e. who has formed a continuous internal connection of constituent elements achieves free independent harmonious spirit and wisdom (creativity). Now, if we form a continuous internal connection in a particular quality level and aspect, we also achieve free independent harmonious spirit and wisdom in a particular quality level and aspect.

> *When we have overcome a problem, we achieve free independent harmonious spirit and wisdom (creativity) in a particular quality level and aspect, depending on what problem we have overcome.*

In the very truth, everything in nature and the world exists as a continuous internal connection of all other things, formed in a particular quality level and aspect. Therefore, everything in nature and the world achieves its own free independent harmonious spirit and wisdom (creativity) in each particular quality level and aspect.

How can we approach the very truth about nature and the world?

When we have overcome a problem, we arrive at the very truth only in a particular quality level and aspect, as mentioned above. This means that in such a case we only arrive at a part of the very truth about nature and the world. Now, how can we approach the whole "very truth" about nature and the world?

As discussed in Section 1.3.1, when a person has attained *Tai-toku*, the person acquires new ability. Such new ability comes from a new continuous internal connection the person has formed. Thus, if a person has overcome a large number of problems of various kinds, the person forms new continuous internal connections in various quality levels and aspects and hence acquires new ability in various quality levels and aspects. This means that such a person has reached the very truth on a high level.

> *The very truth on a high level is embodied in the behavior and the mind of a person who has arrived at the very truth over wide and deep areas of knowledge in a diversity of quality levels and aspects.*

It is relatively easy to reach the very truth, as mentioned earlier. However, it is not easy to reach the very truth on a high level. From this consideration, we can say as follows.

Reaching the very truth on a higher level is equivalent to coming closer to the very truth about nature and the world.

Note that this conclusion indicates that even the very truth on a high level still deviates from the very truth about nature and the world and is of an approximate character.

There exist various kinds of hidden separation in the very truth about nature and the world

Let us return to a discussion of the very truth about nature and the world. There remains an important aspect in it. It was mentioned at the beginning of this section that in the very truth everything in nature and the world exists as a continuous internal connection of all other things, formed in each particular quality level and aspect. This means that in the very truth everything in nature and the world exists individually and separately when it is viewed from a *new* quality level and aspect. Thus, we can say as follows.

There are various kinds of hidden separation in the very truth about nature and the world when it is viewed from new quality levels and aspects.

Indeed, there is no discrepancy, distinction and separation in a continuous internal connection, as discussed thus far. However, there are various kinds of discrepancy, distinction and separation between continuous internal connections formed in different quality levels or aspects if no continuous internal connection is formed between them.

Let us look at some examples of hidden separation. When Mr. A has achieved *Tai-toku* about riding a bicycle, Mr. A's ability to ride a bicycle exists as a continuous internal connection formed in this quality level and aspect. Therefore, Mr. B's ability to ride a bicycle, which exists as a continuous internal connection formed in another quality level and aspect, is separated from Mr. A's ability. When Shaka-muni achieved a wholly clear state of mind by overcoming his mental pain, his mind existed as a continuous internal connection formed by his particular experience. Therefore, other people's mind, which exists as a continuous internal connection formed in another quality level and aspect, is separated from Shaka-muni's mind.

There is another type of hidden separation. Let us consider a person who can freely ride a bicycle but cannot ride a monocycle. For such a person, constituent elements of the ability to ride a bicycle are fully continuously connected with one another while those of the ability to ride a monocycle are left individual and separate. Strikingly, however, constituent elements of the ability to ride a bicycle and those of the ability to ride a monocycle are the same, except that a bicycle is replaced by a monocycle. They both consist of a human body, hands, legs, a bicycle or monocycle, the ground, the gravitation, and so on. Therefore, the fact that a person can freely ride a bicycle but cannot ride a monocycle means that these constituent elements are fully continuously connected with one another in the quality level and the aspect of riding a bicycle but are left individual and separate in the quality level and the aspect of riding a monocycle.

In general, everything in nature and the world is fully continuously connected with one another in some quality levels and aspects but exists individually and separately in other quality levels and aspects.

Nature and the world have an evolutional character

The presence of hidden separation has an important philosophical meaning. It brings about various changes in nature and the world such as the evolution of nature, the development of society, the improvement of human intellect, and so on.

Let us look at a simple example. Even a person who only has "understanding by words" about riding a bicycle, i.e. even a person who cannot ride a bicycle can achieve a peaceful harmonious state of mind if he or she does not want to ride a bicycle. Therefore, even such a person can be regarded as lying at the stage of *Tai-toku* in quality levels and aspects other than riding a bicycle. In this case, a person has hidden separation in the quality level and the aspect of riding a bicycle. Such hidden separation becomes apparent when a person has come to want to ride a bicycle. Furthermore, if such a person has achieved the ability to ride a bicycle by exercise, the person achieves self-creation and the hidden separation disappears.

Theoretically, the above example can be explained as follows. In the very truth, all things are controlled by the grand internal ability to make them interact with one another, form a new continuous internal connection of them, and produce a new thing which realizes a free independent harmonious state, as mentioned earlier. Now, when a new continuous internal connection is formed in a certain quality level and aspect, hidden separation in this quality level and aspect becomes apparent. This means that a limit of existing ideas or things in this quality level and aspect becomes apparent. Then, if such a limit is overcome by effort, a new idea or thing is produced and hidden separation in this quality level and aspect disappears. Further details of this process will be discussed in Section 6.2.

1.4.2 Nature Has a Dual Structure

In the preceding section we revealed that truly real things have an infinitely-spreading continuous and incessantly-changing dynamic character. On the other hand, it was stated in Section 1.3.2 that constituent elements of "the ability gained by the attainment of *Tai-toku*" exist individually and separately in the external visible world even after a person has attained *Tai-toku* and formed a fully continuous connection of them. It was also stated in Section 1.2.1 that we human beings can only be conscious of individual unchanging properties of truly real things. From these statements, we have to say that truly real things have a dual nature. In this section we consider this issue in detail and clarify how truly real things actually exist.

A truly real thing demonstrates individual unchanging shapes and properties when it is viewed in a common or averaged form

Indeed, a truly real thing has no individual unchanging quality if it is viewed *as it is*, as argued in Section 1.4.1. However, even such a truly real thing demonstrates individual unchanging shapes and properties if it is viewed in a common or averaged form, i.e. if it is viewed by using the procedure of picking out common or repeatedly observed properties. Human consciousness catches such common or repeatedly observed properties of truly real things.

Let us consider constituent elements of a living thing as an example. Indeed, they exist in infinite dependence on others and incessantly change their structures, properties and functions so that they as a whole realize a free independent harmonious state, as mentioned in Section 1.1.1 and 1.4.1. Therefore, they have no individual unchanging quality when they are viewed as they are. Originally, a moving or changing thing cannot be correctly expressed by words, indicating that it has no individual unchanging quality, as discussed in Section 1.2.3. However, even such constituent elements of a living thing are controlled so that they as a whole realize a free independent harmonious state or so that they as a whole produce a living thing. For this reason, they demonstrate certain individual unchanging shapes and properties when they are viewed in a common or averaged form.

Interestingly, the realization of a free independent harmonious state, which is accomplished by incessant changes in constituent elements, is responsible for the demonstration of individual unchanging shapes and properties in an averaged or common form. The realization of a free independent harmonious state and the demonstration of individual unchanging shapes and properties have entirely opposing characters to each other but they are two faces of one thing. They coexist in a harmonious manner.

A truly real thing has a dual nature: a continuous changing nature and an individual unchanging nature

In this way, we can say that a truly real thing has a dual nature: a continuous changing nature and an individual unchanging nature. Naturally, the two natures are not independent of each other. A continuous changing nature and an individual unchanging nature have opposing meanings to each other but cooperate with each other, in a similar way to mutually opposing powers of "infinite dependence on others" and "independence of others" discussed in Section 1.4.1.

Let us recall that in the very truth all things in nature and the world are controlled by the grand internal ability to make them interact with one another, form a new continuous internal connection of them, and complete the formation of such a new continuous internal connection. Here, it is because truly real things have an individual unchanging nature that they can interact with one another. On the other hand, it is because truly real things have a continuous changing nature, i.e. they each exist as a continuous internal connection that interaction between them leads to the formation of a new continuous internal connection (see Section 6.2 for details). Moreover, the completion of the formation of a new continuous internal connection again leads to the production of a new truly real thing with a dual nature.

However, we should note that a continuous changing nature is the basic character of a truly real thing and an individual unchanging nature emerges from it, as argued just earlier. It is a continuous changing nature that explains the emergence of autonomous dynamic self-organizing ability or free independent harmonious spirit and wisdom in living things. It is also a continuous changing nature that explains a change in the basic qualities of things by their interaction, as discussed in Section 1.4.1.

Human beings have made use of individual unchanging shapes and properties of truly real things extensively

Historically, it was human beings that discovered that truly real things (natural things themselves) have individual unchanging shapes and properties and have made use of them extensively [*1], as discussed in Section 1.2.1. As a result, human ideas (consciousness), words (knowledge) and tools (technology) were created and the new huge extra artificial objective world, i.e. human society and civilization have been constructed, apart from the natural world.

> [*1]To be accurate, it was living things that first discovered that truly real things have individual unchanging shapes and properties and made use of them. Living things emerged on the primitive earth as a system with the internal ability to organize it itself so that it realized a harmonious stable state (see Section 4.2). Thus, it is probable that living things had the self and hence had the ability to "observe" other things in the outer world. Then, living things came to pick out individual unchanging shapes and properties and make use of them. For example, unicellular living organisms make use of things with individual unchanging shapes and properties such as proteins, RNA or DNA and cell membranes. Also, bees form a honeycomb and birds form a nest for holding their eggs. Human ideas, words and technology can be regarded as an advanced form of such things (see Section 4.4 for details).

The emergence of human ideas, words and tools was a big event in the course of natural evolution. This was because they had a unique way of existing, quite different from the way that natural things or truly real things exist. As discussed in Section 1.4.1, natural things including living things each exist as a continuous internal connection of constituent elements and have an incessantly changing dynamic quality and thus demonstrate autonomous dynamic self-organizing ability or free independent harmonious spirit and wisdom. Indeed, natural things demonstrate individual unchanging shapes and properties in a common or averaged form, as argued above. However, for natural things, such shapes and properties are included in properties of continuous internal connections (see discussions of information-based connections in Section 4.4) and play almost no extra role in their behavior.

On the other hand, human ideas (consciousness), words (knowledge) and tools (technology) are produced by picking out individual unchanging shapes and properties of natural things (truly real things) and connecting them logically. Human beings have made use of individual unchanging shapes and properties of natural things for making useful things for their life, irrespective of what continuous internal connections original natural things have. Human beings have regarded "things with individual unchanging properties picked out from nature" as "independent existents" by taking experimental verification or reproducibility as the grounds for the validity of such a way of understanding. Accordingly, human ideas (consciousness), words (knowledge) and tools (technology) in essence demonstrate no autonomous dynamic self-organizing ability (see Section 3.2.3).

Nature has a dual structure comprised of the very truth and the external appearance of it

Now, the foregoing arguments can be summarized, as shown in Figure 1-4. A truly real thing has a dual nature, as mentioned earlier. Accordingly, nature has a dual structure, comprised of the very truth and the external appearance of it. In the very truth, truly real things (or natural things themselves) exist each as a continuous internal connection of all other truly real things, formed in a particular quality-level and aspect, and are controlled by the grand internal ability. We cannot be conscious of both the very truth and truly real things. However, we can reach the very truth by achieving a wholly clear state of mind full of delight through thorough consideration and pursuit or thorough exercise. On the other hand, the external appearance of the very truth consists of

```
┌─────────────────────────────────────────────────────┐
│              The objective world                     │
│   consisting of science, technology, civilization, etc. │
│   The world that human consciousness has constructed.│
└─────────────────────────────────────────────────────┘
```

```
┌─────────────────────────────────────────────────────┐
│        The external appearance of the very truth     │
│              The external visible world              │
│   consisting of common or repeatedly observed properties of │
│   truly real things and imaginary things with such properties. │
│  ─────────────────────────────────────────────────  │
│     The very truth lying in the internal invisible world │
│            in which truly real things each exist as  │
│   an infinitely-spreading fully-continuous internal connection │
│         under the control of the grand internal ability. │
│    A continuous internal connection has the internal ability to │
│         organize itself so that it realizes a free independent │
│                  harmonious stable state.            │
│   The origin of free independent harmonious spirit and wisdom. │
│   Accessible by the achievement of a wholly clear state of mind. │
└─────────────────────────────────────────────────────┘
```

Figure 1-4 The dual structure of nature and its relation to the objective world.

common or repeatedly observed properties of truly real things and imaginary things with such properties. We can in principle be conscious of such properties or such things. Thus, the external appearance of the very truth constitutes the external visible world. The objective world consisting of human ideas, science, technology, etc. or human society and civilization is the world that human consciousness has constructed. The external appearance of the very truth is inseparably connected with the very truth but there is a definite gap between the objective world and the very truth.

Nevertheless, even the objective world is *within* the dual structure of nature. Indeed, there is a definite gap between the objective world and the very truth, as mentioned above, but the objective world is by no means completely separated from the very truth. We human beings, who construct and control science, technology, civilization, etc., are in turn controlled unconsciously by the grand internal ability immanent in nature and the world. Therefore, problems arising from a gap between the objective world and the very truth are reflected in our mind and thus removed by our activities.

The dual structure of nature provides us with the correct view of nature

The dual nature of a truly real thing and hence the dual structure of nature are impossible to correctly understand within the realm of word-based scientific understanding. However, they provide us with the correct view of nature. Many important problems that have been left unsolved to date can be reasonably overcome based on the dual structure of nature.

Let us look at an example. It has long been a central problem in traditional western philosophies how we can overcome *the dualism of body and mind* (see Section 1.1.2 and 1.2.3). We can easily and completely overcome it if we take into account the dual structure of nature. The key point is that natural things, including living things and human beings, each exist as a continuous internal connection of constituent elements and realize a free independent harmonious state. Therefore, a human being originally has free independent harmonious spirit and wisdom by nature. However, even such a human being looks as if it were a machine if it is viewed in a common or averaged form by using the procedure of picking out common or repeatedly observed properties. Thus, the dualism of body and mind simply originates from the intrinsic limit of traditional word-based scientific understanding.

The importance of the dual structure of nature has not been correctly recognized

Unfortunately, the importance of the dual structure of nature has not been correctly recognized. For this reason, the dualism of body and mind has not been overcome up to now. In the present days, it appears that great progress of science is rather leading us in a worse direction. In fact, it is said that great progress of science, in particular, great progress of brain science and computer science has come to show a tendency to deny the existence of the free will of a human being. Thus, we human beings have now come to completely miss the very truth lying in the internal world and live based on the objective world alone.

Why has the importance of the dual structure of nature not been correctly recognized? First of all, this is because we human beings have attached the highest importance to usefulness, not to truth, as mentioned in Section 1.2.1. In other words, this is because we solely rely on words or word-based understanding. Words are useful but only catch common or repeatedly observed properties of truly real things or natural things. Namely, words only catch the external appearance of truly real things. Therefore, when we rely on words, we easily overlook the existence of the very truth behind them (see a footnote #2 of Section 1.1.2).

1.4.3 How Can We Obtain the Truly Correct Understanding?

It is evident that the correct opinion and the correct judgment come from the truly correct understanding of nature and the world. Now, how can we obtain the truly correct understanding if nature and the world have a dual structure, as argued above? We need to understand everything from a deep level by taking into account the internal invisible world, as discussed in Section 1.4.1. However, at this stage, we have still gained no truly correct understanding because the very truth expressed by words is not the very truth itself. Thus, it is necessary to jump over the intrinsic limits of word-based understanding and arrive at the very truth. In this section we consider briefly a basic part of how we can arrive at the very truth.

The fundamental method for arriving at the very truth about learned knowledge

We already discussed how we can arrive at the very truth in Section 1.3.1 and 1.4.1. We arrive at the very truth when we have attained *Taitoku* by thorough exercise or when we have overcome a problem by thorough consideration and pursuit. Strong interaction among truly real ideas in our brain and truly real things in the real world during thorough exercise or thorough consideration and pursuit leads to the formation of a continuous internal connection of them and leads us to the very truth. Thus, we can say that it is not difficult to arrive at the very truth, as already mentioned in Section 1.4.1. However, it is not easy to reach the very truth on a high level or on a meaningful level, as also argued in Section 1.4.1.

In fact, historically, the revelation of how we can arrive at the very truth has been an extremely difficult problem in Buddhism, which for the first time disclosed the existence of the very truth (see Section A2.1). Accordingly, it is natural that this problem has been left unsolved in the history of human beings, in which the existence of the very truth itself has not been clearly recognized. In fact, for example, the attainment of *Tai-toku* is a well-known event and its properties have been widely used to improve skill in sport, art, and technical professions. However, the essence of *Tai-toku* has been left unclarified to date, as mentioned in Section 1.3.2. Similarly, many important problems about life, evolution and creation have been left unsolved, as mentioned in Preface. This is also solely because it has been difficult to reveal how we can arrive at the very truth.

Now, this book provides the definite answer to the problem of how we can arrive at the very truth, as mentioned above. Thorough exercise or thorough consideration and pursuit give a method of arriving at the very truth. Such a definite answer is deduced from the way to attain *Satori* (or the way to arrive at the very truth) which a Japanese Buddhist monk, Shin-ran, disclosed in the early 13th century (see Chapter A2). He discovered that a person can arrive at the very truth by first reaching the utmost limit of his or her word-based understanding and then jumping over it. This is a reasonable way to arrive at the very truth if we take into account that there is a non-removable gap between word-based understanding and the very truth (see Section 1.2.1). Unfortunately, the importance of Shin-ran's way to attain *Satori* has not been clearly recognized to date. It is no exaggeration to say that great successes in this book are solely due to the discovery of the importance of Shin-ran's way to attain *Satori*.

There are various kinds of ideas

For considering how we can arrive at the very truth in further detail, it is important to clarify what ideas we have. In general, we have a large number of ideas we acquired by learning by words or acquired by the sense organs. In addition, after we acquired such ideas, we often have an experience of attaining *Tai-toku* or overcoming a problem by thorough consideration and pursuit and arriving at the very truth (see Section 1.3.1 and 1.4.1). As a result, we have (1) truly real ideas (nervous patterns in the brain) we acquire when we have arrived at the very truth, (2) truly real ideas (nervous patterns in the brain) we acquire when we learn knowledge by words, (3) truly real ideas (nervous patterns in the brain) we gain by the sense organs, (4) ideas we are conscious of, and (5) ideas expressed by words.

Only truly real ideas we acquire when we have arrived at the very truth correctly represent the way that truly real things exist, though in a particular quality level and aspect. Namely, when we have such ideas, we have arrived at the very truth in a particular quality level and aspect and hence achieve true freedom, free independent harmonious spirit and wisdom, etc., as discussed in Section 1.3.1. On the other hand, ideas we acquire by learning by words are at the stage of "understanding by words" (see Section 1.3.1). They exist individually and separately and only work one by one, thus leading to awkward thinking and behavior. Ideas we acquire by the sense organs are of the same character as ones we acquire by learning by words. Ideas we are conscious of are essentially the same as ideas expressed by words.

The relation between (1) truly real ideas we acquire when we have arrived at the very truth and (4) ideas we are conscious of can be explained as follows. As argued in Section 1.4.1, when we met a problem and have overcome it by thorough consideration and pursuit, we create a new idea and achieve free thinking and a wholly clear state of mind full of delight. The achievement of free thinking and a wholly clear state of mind means that truly real ideas in our brain and truly real things in the real world have been fully continuously connected with one another, i.e. we have arrived at the very truth, as discussed in Section 1.3.2. This also means that we have produced "a new truly real idea" given by such a newly formed continuous (internal) connection. This "new truly real idea" is just a truly real idea we acquire when we have arrived at the very truth. However, we cannot be conscious of such a "new truly real idea" itself. We can only be conscious of common or individual unchanging parts of it. Thus, *ideas we are conscious of represent common or individual unchanging parts of truly real ideas we acquire when we have arrived at the very truth*.

Words are a linguistic expression of ideas we are conscious of and thus only represent common or individual unchanging parts of truly real ideas we acquire when we have arrived at the very truth. As a consequence, (2) truly real ideas we acquire by learning by words also only represent common or individual unchanging parts of truly real ideas we acquire when we have arrived at the very truth. Note, however, that (2) truly real ideas we acquire by learning by words are not the same as (4) ideas we are conscious of. The former exist in the brain each as a continuous internal connection of nervous patterns while the latter has no such continuous internal connection. To be brief, the former are truly real things in the brain while the latter are merely a signal or mark.

Now, what difference exists between (1) truly real ideas we acquire when we have arrived at the very truth and (2) those we acquire by learning by words? In learning by words, we receive various ideas one by one. This means that various continuous connections of nervous patterns are formed in the brain one by one *in different aspects*. For this reason, truly real ideas we acquire by learning by words exist individually and separately. On the other hand, truly real ideas we acquire when we have arrived at the very truth exist each as a continuous internal connection of various ideas [*1].

[*1] Even truly real ideas we acquire when we have arrived at the very truth have hidden separation when they are viewed from a new quality level and aspect, as discussed in Section 1.4.1. Therefore, the difference between truly real ideas (1) and (2) is not absolute.

There is a definite difference in the physiological state between truly real ideas (1) and (2)

For getting better understanding of the differences between various ideas, let us consider a simple example. It was argued in Section 1.3.3 that a person who has attained *Tai-toku* and one who is at the stage of "understanding by words" speak nearly the same words though their abilities in actual behavior are entirely different. This fact can be explained by using Figure 1-5, which metaphorically illustrates how ideas exist in the brain. Ideas at the stage of "understanding by words" are given in Figure 1-5(A) while those at the stage of *Tai-toku* are given in Figure 1-5(B). Only two ideas are depicted for simplicity. Patterned gray areas in (A) refer to main areas of continuous internal connections for truly real ideas we acquire by learning by words, the difference in pattern indicating that they are formed in different aspects. Broken circles refer to ideas we are conscious of, which represent common or repeatedly observed parts of truly real ideas. The broken line refers to a logical connection of ideas we are conscious of. The continuous gray area with no pattern in (B) refers to a main area of a continuous internal connection of two truly real ideas, formed upon the attainment of *Tai-toku*. This internal continuous connection is formed in a single aspect.

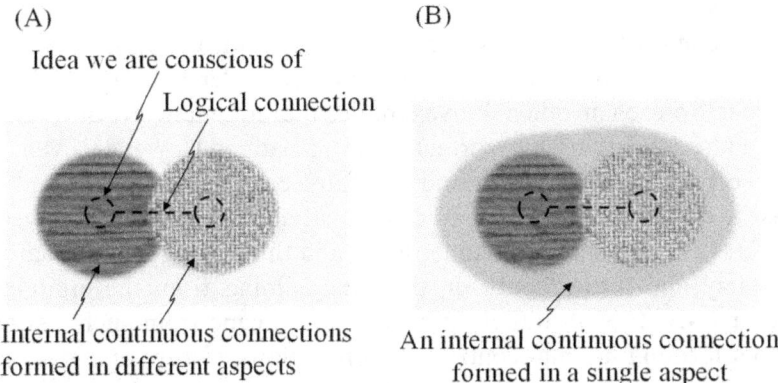

Figure 1-5 Metaphoric illustrations of truly real ideas in the brain (gray areas with or without a pattern) and ideas we are conscious of (broken circle) in (A) the stage of understanding by words and (B) the stage of *Tai-toku*.

When a person is at the stage of "understanding by words", truly real ideas exist individually and separately because continuous internal connections for them are formed in different aspects (see Figure 1-5(A)). Therefore, they only work one by one. On the other hand, when a person has attained *Tai-toku*, the new continuous internal connection of a single aspect is formed (see Figure 1-5(B)). Accordingly, they work simultaneously in harmony with one another so that they as a whole realize a free independent harmonious stable state [*2]. However, nevertheless, ideas we are conscious of (broken circles) and a logical connection of them (the broken line) are the same between Figure 1-5(A) and (B), indicating that a person who has attained *Tai-toku* and one who is at the stage of "understanding by words" speak nearly the same words. A change in continuous internal connections for truly real ideas is not correctly reflected in ideas we are conscious of.

[*2] It is interesting to note that constituent elements of machines exist in a similar way to ideas in Figure 1-5(A) while those of a living thing exist in a similar way to ideas in Figure 1-5(B).

It should be emphasized here that the above argument indicates that there is a definite difference in the *physiological* state of nervous patterns between truly real ideas we acquire when we have arrived at the very truth and ones we acquire by learning by words. Therefore, we have a high barrier for arriving at the very truth. Also we cannot freely control the difference intentionally.

In relation to the above argument, it should be noted also that the correct understanding and the truly correct understanding are entirely different from each other. For example, the argument given in Section 1.4.1 indeed gives us the correct view of nature. However, even if we have correctly understood it and are fully convinced of it, if truly real ideas in our brain are left in an individual and separated state, we shall unawares rely on such individual ideas and think things individually and separately and throw doubt on opinions coming from the argument in Section 1.4.1. It is absolutely impossible to escape from such a situation without forming internal continuous connections of truly real ideas in our brain. On the contrary, even if we know nothing about the argument in Section 1.4.1, if truly real ideas in our brain are continuously connected with one another, we shall unawares have opinions similar to the argument in Section 1.4.1.

In general we have arrived at the very truth in the area of elementary ideas

There is another kind of difference in our ideas. Namely, there are elementary ideas and advanced ideas. As mentioned earlier, we have a

Elementary	Knowledge (ideas or words) ⟶	Advanced
Ideas we acquired by learning by words or by the sense organs		Unknown area
Ideas at the stage of arrival at the very truth (Ideas internally continuously connected with one another)	Ideas at the stage of understanding by words (Ideas existing individually and separately)	
(The acquisition of the internal wisdom) The attainment of true freedom, free independent harmonious spirit and wisdom (creativity), etc.	Ideas working one by one	
The area of ideas in which we can feel interest, unclearness or doubt and carry out thorough consideration and pursuit	The area of ideas in which we feel everything natural and reasonable	

Figure 1-6 A schematic illustration of various kinds of ideas in our brain. The words in the lowest row will be explained later.

large number of ideas we acquired by learning by words or by the sense organs. In addition, we often have an experience of attaining *Tai-toku* or overcoming a problem by thorough consideration and pursuit and arriving at the very truth. An important point is that it is mainly in the area of elementary ideas that we have an experience of attaining *Tai-toku* or overcoming a problem. Therefore, we in general have "ideas lying at the stage of arrival at the very truth" in the area of elementary ideas. On the other hand, we have usually remained at the stage of "understanding by words" in the area of advanced ideas. Such a state of ideas in our brain can be expressed as illustrated in Figure 1-6.

We can see from this figure that it is of primary importance to convert ideas which remain at the stage of "understanding by words" to ideas at the stage of arrival at the very truth.

Re-creation is an effective way to arrive at the very truth about learned knowledge

Word-based understanding is intrinsic ability to human beings and besides is the only thing we can rely on, on a conscious level, as mentioned in Section 1.2.1. Therefore, we are in general apt to want to accumulate much knowledge and understand it as completely as possible. However, accumulated knowledge only represents common or repeatedly observed properties of truly real things, as mentioned earlier. Therefore, even if we have understood it completely, we only understand the skeleton of nature and fail to catch the flesh and blood of it.

Naturally, it is important to accumulate much knowledge but this is only the first step toward arrival at the very truth. Interestingly, we should not accept accumulated knowledge as it is, if we want to understand it correctly. It is necessary to feel interest, unclearness, doubt, etc. and carry out thorough consideration and pursuit until achieving a wholly clear state of mind full of delight, as discussed earlier.

In relation to this argument, it is worth noting that re-creation is an effective way to arrive at the very truth about learned knowledge. Any knowledge was created by somebody else in the past. The re-creation of learned knowledge means that we first feel doubt about it, carry out thorough consideration and pursuit about the origin of such a feeling, and re-create learned knowledge again by our own ability and effort, in the same way as somebody else did in the past. If we have achieved re-creation, we recover a continuous internal connection behind learned knowledge and grasp the real entity of it. In other words, when we have re-created learned knowledge, we accept the spirit and the wisdom of somebody else who first created it in the past.

Re-creation is also important when we understand phenomena in nature and the world. We can only look at the external appearance of nature and the world. Therefore, when we only look at nature or the world, we cannot grasp the substance of it. It is important to feel doubt or interest about a phenomenon or event in nature and the world and re-create it by our own ability and effort. Namely, we need to catch an underlying continuous internal connection in an observed phenomenon or event. When we have achieved re-creation, we can grasp "the spirit and the wisdom" of nature or the world [*3].

[*3] As argued in Section 1.4.1, everything in nature and the world exists as a continuous internal connection of all other things and therefore everything achieves its own free independent harmonious spirit and wisdom (creativity) in each particular quality level and aspect.

We cannot intentionally and freely arrive at the very truth

It should be emphasized here that there remains a serious problem in the foregoing arguments. It was mentioned earlier that it is of primary importance to convert ideas which remain at the stage of "understanding by words" to ideas at the stage of arrival at the very truth. Actually, however, we cannot be aware of which ideas remain at the stage of "understanding by words" and which ideas are at the stage of arrival at the very truth because we cannot be conscious of continuous internal connections formed.

The above problem is usually unawares surmounted in the following way. When we have learnt a variety of knowledge, we in general come to feel interest, unclearness, unease, doubt, etc. about it. Then, based on such a feeling, we can start thorough consideration and pursuit or thorough exercise and achieve a wholly clear state of mind. Namely, we can arrive at the very truth. However, here is again a serious problem. It is not easy to feel interest, unclearness, unease, doubt, etc. about learned knowledge.

Knowledge is constructed so that it includes no contradiction and inconsistency and thus it usually looks natural and reasonable (see Figure 1-6). If we cannot feel interest, doubt, etc. about learned knowledge, we cannot start thorough consideration and pursuit or thorough exercise [*4]. Somebody may say that we can intentionally feel interest, doubt, etc. about learned knowledge. However, such an intentional feeling usually does not work effectively. Namely, it does not lead us to thorough consideration and pursuit or thorough exercise and hence does not lead us to a wholly clear state of mind [*5].

[*4] This argument mainly applies to knowledge in science and philosophy because it

is not difficult to feel unclearness, unease, doubt, etc. about learned knowledge in the fields of sport, art and technical professions, as discussed in Section 1.3.1. In such fields we cannot behave freely, skillfully and unconsciously when we are at the stage of "understanding by words". On the other hand, in science and philosophy, the difference between understanding by words and arrival at the very truth is not clear even in actual behavior.

*5This does not mean that intentional effort is always ineffective. For example, it will be argued in Section A2.2 and A2.3 that the continuation of sincere and strong desires leads to the emergence of a feeling of doubt.

It is important to have the internal wisdom to feel interest, unclearness, unease, doubt, etc. about learned knowledge

From these considerations, we are led to a conclusion that it is important to have *the internal wisdom* to feel interest, unclearness, unease, doubt, etc. about learned knowledge, i.e. *the internal wisdom* to notice a limit of learned knowledge. In general, we cannot arrive at the very truth without having such internal wisdom. Now, how can we have it? Let us consider this problem in some detail.

Note first that we can notice a contradiction (a problem or a limit) in learned knowledge when we can look at ideas or words in it *in true connection*. In other words, we cannot notice a contradiction in learned knowledge when we look at ideas or words in it individually and separately. For example, freedom and equality are mutually conflicting concepts, as discussed in Section 1.2.3. However, if we consider them individually and separately, they both look natural and reasonable. Thus, we overlook a contradiction included between them. The same holds for all mutually conflicting concepts.

There is another interesting example. In the Aristotelian idea in ancient Greece, earthly bodies were understood to have vulgar properties of falling down while heavenly bodies were understood to have noble properties of moving in circles. In the present days, we know that this idea is wrong. Newtonian mechanics established in the Renaissance age disclosed that both earthly and heavenly bodies have essentially the same properties. However, ancient people had believed that the Aristotelian idea was correct for a long time of more than 2,000 years. Why? This was because ancient people looked at the earthly bodies and heavenly ones individually and separately. In fact, the Aristotelian idea is not wrong if we look at earthly bodies and heavenly ones individually and separately. Certainly, almost all earthly bodies fall down while almost all heavenly bodies move in circles.

Thus we can see that it is important to look at ideas or words in learned knowledge in true connection. Now, how can we look at ideas or words in learned knowledge in true connection? When we are at the stage of "understanding by words", we look at things individually and separately, as discussed in Section 1.3.1. Therefore, at this stage, we can never look at ideas or words in learned knowledge in true connection. On the other hand, when we have attained *Tai-toku*, i.e. when we have formed a continuous internal connection of ideas, we look at various things simultaneously in harmony with one another (see Section 1.3.1 and 1.3.2). Here, looking at various things *simultaneously in harmony with one another* has the same meaning as looking at various things in *continuous* connection, i.e. in *true* connection. Accordingly, we can say that we can look at ideas or words in learned knowledge in true connection when we have formed a continuous internal connection of them.

These considerations, after all, indicate that it is important to form a *new* (partial, non-completed) continuous internal connection of ideas or words in learned knowledge. The formation of such a *new* continuous internal connection allows us to look at ideas or words in learned knowledge in true connection and to feel interest, unclearness, unease, doubt, etc. about it. Now, we form a *new* continuous internal connection of ideas when we carry out thorough consideration and pursuit or thorough exercise, as discussed in Section 1.3.1 and 1.4.1. This means that we can *efficiently* form a new (partial, non-completed) continuous internal connection in an area of ideas in which we can carry out smooth thinking, i.e. we have formed a large number of continuous internal connections of ideas.

Here we reach the final conclusion. We can have the internal wisdom to feel interest, unclearness, unease, doubt, etc. about learned knowledge or to notice a limit of it *in or near an area of ideas in which we can carry out smooth thinking, i.e. we have formed a large number of continuous internal connections of ideas*. This conclusion implies that it is important to have an experience of arriving at the very truth repeatedly and to spread the area of ideas in which we can carry out smooth thinking in the direction of advanced ideas (see Figure 1-6). If we have formed continuous internal connections over wide and deep areas of knowledge in a diversity of quality levels and aspects, we come to be able to easily notice limits of learned knowledge and gain the truly correct understanding of nature and the world.

The ability to look at various things as they are

Let us consider the above-mentioned ability to look at various things simultaneously in harmony with one another in more detail. This ability has the same meaning as looking at various things as they are.

In general, it is believed that everybody can look at various things as they are. However, this is a big mistake. It is not easy to look at various things as they are. This is because in the very truth all things in nature and the world are in infinite dependence on others and incessantly change their structures, properties and functions so that they are in harmony with others (see Section 1.4.1). Therefore, we can look at various things as they are only when we can look at all things *simultaneously* in harmony with one another. The ability to look at various things as they are is the ability far beyond word-based understanding.

The difficulty in looking at various things as they are can be understood in the following way as well. Nature and the world consist of not only the known but also the unknown. Therefore, we can look at various things in nature and the world as they are only when we can look at the unknown as well as the known. Similarly, we can look at various things in nature and the world as they are only when we can look at what we cannot be conscious of as well as what we can be conscious of. Surprisingly, we need to look at the unknown and what we cannot be conscious of in order to look at various things in nature and the world as they are.

These considerations explain why arrival at the very truth or the acquisition of the ability to look at various things simultaneously in harmony with one another leads to the acquisition of the internal wisdom to notice a limit of learned knowledge. When we have arrived at the very truth, our understanding jumps over the limit of the known and extends deep into the unknown.

We need not endure all unpleasant conclusions deduced from scientific knowledge

To have the internal wisdom to notice a limit of learned knowledge has another important meaning. As mentioned earlier, word-based understanding is intrinsic ability to human beings and the only thing we can rely on, on a conscious level. Therefore, we are apt to regard many things or matters as impossible or unavoidable, based on scientific understanding, and give up thinking further. However, this is a big mistake. Scientific understanding is not the final goal of human understanding. Therefore, *we have a possibility of overcoming things or matters that look impossible or unavoidable based on scientific understanding*. The internal wisdom to notice a limit of learned knowledge is also the internal wisdom to notice such a possibility.

The attainment of *Tai-toku* indicates that the basic qualities of individual ideas and things are largely changed during thorough exercise or thorough consideration and pursuit even though they look unchanged in the external visible world (see also Figure 1-5). This means that *thorough exercise or thorough consideration and pursuit lead to the formation of a new continuous internal connection or a new internal possibility* (see Section 6.2 for details). Therefore, we have a possibility of overcoming things or matters that look impossible or unavoidable based on the present scientific knowledge through thorough exercise or thorough consideration and pursuit. Traditional understanding and philosophies have overlooked the existence of the internal invisible world and hence have overlooked the fact that the basic qualities of things change internally. Hegel says, "Freedom is insight into inevitability". However, we should rather say, "Freedom is insight into new internal possibilities".

In relation to the above argument, we have to note also that the external visible world does not correctly express the internal invisible world (or the very truth). For example, a person who has attained *Tai-toku* has much more splendid abilities than one who has learned only by words, but they speak nearly the same words, as discussed in Section 1.3.3 (see also Figure 1-5).

There are similar examples in natural things as well. For example, protein molecules in a living cell and ones in an aqueous solution in a test tube have the same chemical structures and look as if they had the same basic qualities. However, in reality, they have quite different basic qualities. As mentioned in Section 1.1.1, protein molecules in a living cell incessantly change their structures, properties and functions so that they are in harmony with the surroundings. All molecules in a living cell are internally continuously connected with one another and are controlled so that they as a whole realize a harmonious stable state (see Section 4.2). Protein molecules in an aqueous solution have no such properties. The moment a protein molecule is separated from a living cell, it changes the basic qualities. Monod's discussions in Section 1.1.2 overlook this fact.

It is of critical importance to catch the grand internal ability within our body

There remains another important issue for obtaining the truly correct understanding of nature and the world. Indeed we can reach the very truth on a high level and come close to the very truth about nature and the world if we have an experience of arriving at the very truth repeatedly (see Section 1.4.1). However, even if we reach the very truth on a high level, it is still of an approximate character because nature and the world are infinite in size and quality and always change.

How can we surmount this problem? It was argued in Section 1.4.1 that all "truly real things" in nature and the world including living things and human ideas are controlled by the grand internal ability. Thus, if we grasp the grand internal ability itself within our body, this allows us to live in accord with the very truth about nature and the world, which is infinite in size and quality and always changes. This is because it is the grand internal ability that governs nature and the world at the present time. Thus, the final goal of our understanding is to grasp the grand internal ability within our body. Details on this issue will be discussed in Chapter A2.

References

(1) K. Sakai, *Brain Science of the Mind* (in Japanese), Chuko-shinsho, Chuo-koron-shinsha, Tokyo, 2008.
(2) J. L. Monod, *Le Hasard et la Nécessité, Essai sur la philosophie naturelle de la biologie moderne,* Alfred A. Knopf, Inc., Paris, 1971: The Japanese edition translated by I. Watanabe and M. Murakami, Misuzu Shobo, Tokyo, 1972.
(3) P. L. Luisi, *The Emergence of Life – From Chemical Origins to Synthetic Biology*, Cambridge University Press, Cambridge, 2006: The Japanese edition translated by T. Shirakawa, P. Y. Gunji, NTT Shuppan, Tokyo, 2009.
(4) M. Mitchell, *"Complexity: A Guided Tour"*, Oxford University Press, Inc., Oxford, 2009: The Japanese edition translated by H. Takahashi, Kinokuniya Shoten, Tokyo, 2011.
(5) H. Tanaka, *Life and Complex Systems* (in Japanese), Baifukan, Tokyo, 2002.
(6) K. Yoshikawa, *Nonlinear Science – Rhythms and Shapes of Molecular Ensembles* (in Japanese), Gakkai-shuppan-sentah, Tokyo, 1992.
(7) H. Mi-ike, Y. Mori, and T. Yamaguchi, *Science of Non-equilibrium Systems – Dynamics of Reaction Diffusion Systems* (in Japanese), Kodansha, Tokyo, 1997.
(8) S. Kauffman, *At Home in the Universe: The Search for Laws of Self-organization and Complexity*, Oxford University Press, Inc., Oxford, 1995: The Japanese edition translated by F. Yonezawa, Nihon-keizai-shinbun-sha, Tokyo, 1999.
(9) A. Einstein and L. Infeld, *The Evolution of Physics,* 1938: The

Japanese edition translated by J. Ishihara, Iwanami-shinsho, Iwanami-shoten, Tokyo, 1939.

(10) For example, S. Mason, *A History of the Sciences – Main Currents of Scientific Thought*, Lawrence & Wishart Ltd., London, 1953: The Japanese edition translated by S. Yajima, Iwanami-shoten, Tokyo, 1955.

(11) M. Taketani, *Some Problems of Dialectic* (in Japanese), Riron-sha, Tokyo, 1962.

(12) W. Hiromatsu, et al. (editors), *Encyclopedia of Philosophy and Thought* (in Japanese), Iwanami-shoten, Tokyo, 1998.

(13) D. G. Gibson, et al. *Science,* **329,** 52 (2010).

(14) F. Engels, *Naturdialektik,* 1873-83: The Japanese edition translated by S. Tanabe, Iwanami-bunko, Iwanami-shoten, Tokyo, 1956/7.

(15) J. –P. Sartre, *Existentialism Is Humanism,* Nagel, Paris, 1946: The Japanese edition translated by T. Ibuki, Jinmon-shoin, Kyoto, 1955.

(16) S. Matsunami, *Existentialism* (in Japanese), Iwanami-shinsho, Iwanami-shoten, Tokyo, 1962.

2. The Internal Invisible World in Modern Science

In the preceding chapter we investigated the behavior of living things, the historical origin of human ideas and words, and our daily experiences called *Tai-toku*, and revealed that the internal world we cannot be conscious of really exists in nature and the world. In this chapter we again examine whether or not the internal invisible world really exists, based on modern science. The purpose of this chapter is to confirm the conclusion of the preceding chapter and clarify how natural things themselves or truly real things exist, on the scientific standard.

2.1 The World of Microscopic Particles

How is the internal invisible world reflected in science? Newtonian mechanics, electromagnetism, thermodynamics, and so on, which are in general called classical mechanics, deal with the macroscopic visible world we can be conscious of. Therefore, we cannot expect that the internal invisible world is reflected in such fields of science. On the other hand, modern science has deeply penetrated complexities of nature and clarified various strange facts that are impossible to understand by traditional concepts. Therefore, it is probable that the internal invisible world is reflected in such areas of modern science. In fact, this book shows that quantum mechanics established in the early 20th century, which deals with microscopic particles such as electrons and atomic nuclei, clearly demonstrates the existence of the internal invisible world. This result has an important meaning because classical mechanics such as Newtonian mechanics is now regarded only as an approximate form of quantum mechanics.

2.1.1 Fundamental Principles of Quantum Mechanics

Microscopic particles with small weights such as electrons and atomic nuclei have a dual nature: a particle nature and a wave nature

Before proceeding to considerations of philosophical meanings of quantum mechanics, let us briefly explain some fundamental principles of it. The construction of quantum mechanics started from the discovery of the strange fact that microscopic particles with small weights such as electrons and atomic nuclei have a dual nature: a particle nature and a wave nature. Radiation and light also have a similar dual nature. To be strict, it is now believed that macroscopic visible substances with large weights also have a similar dual nature though its effect is nearly imperceptible.

The disclosure of the dual nature of microscopic particles began with Planck's discovery. He theoretically studied thermal radiation and discovered in 1900 that radiation energy has a particle nature. A little later, Einstein studied the photoelectric effect and discovered that light energy also has a particle nature. These discoveries showed that radiation and light, which had been regarded as an electromagnetic wave, behaved as a particle. Light which behaves as a particle is called a photon. Such strange properties of radiation or light were further confirmed in 1922 by an experiment called the Compton scattering. On the basis of these findings, L. de Broglie proposed in 1923 the revolutionary opinion that an electron, which had been regarded as a particle, had a wave nature. This proposal was experimentally verified in 1927 by a discovery that an electron beam shows an interference pattern in the same way as light. In this way, it was made clear that both a photon (light) and an electron have a dual nature.

The disclosure of the dual nature of microscopic particles caused a serious problem in theoretical physics because a particle nature and a wave nature were absolutely incompatible properties with each other. An electron has the mass, the electric charge, and the magnetic moment of definite amounts. A divided piece of these quantities is never observed. In addition, the bombardment of a weak electron beam to a fluorescent plate gives intermittent local spots of light emission. These facts clearly indicate that an electron exists as a particle. However, an electron beam shows an interference pattern in the same way as light, as mentioned above. This fact definitely demonstrates that an electron exists as a continuous wave. The observation of an interference pattern undoubtedly demonstrates that *an electron simultaneously passes two or more slits*, which is a phenomenon absolutely impossible to understand based on a particle nature.

Quantum mechanics was created as new mechanics for overcoming the dual nature of microscopic particles. The construction of quantum mechanics in the early 20th century brought about a big revolution in theoretical physics. It is now well known that quantum mechanics gives the correct explanation to almost all natural phenomena on the earth and constitutes a fundamental basis for recent high technologies. However, strikingly, principles of quantum mechanics have still been regarded as strange and rationally incomprehensible, in sharp contrast to their highly successful technological applications. This fact strongly suggests that quantum mechanics involves principles that are impossible to understand within the realm of word-based understanding[*1].

[*1] It was argued in Section 1.4.2 that the consideration of the attainment of *Tai-toku* leads to the conclusion that truly real things or natural things themselves have a dual nature: a continuous changing nature and an individual unchanging nature. Thus, the dual nature of microscopic particles can be regarded as an example of the dual nature of truly real things.

The fundamental law of motion in quantum mechanics – the Schrödinger Equation [1-3)]

Quantum mechanics was created as new mechanics for explaining the dual nature of microscopic particles such as electrons and atomic nuclei. An outstanding feature of quantum mechanics is that a system consisting of a number of microscopic particles (electrons and atomic nuclei) such as an atom, a molecule, a crystal or their aggregate is caught by the concept of a "state" and described by a mathematical function, Ψ, called a state function or a wave function. The function Ψ is a smooth continuous and finite function of the spatial coordinates of all constituent particles and time, spreading to infinite distance and time, and obeys the following differential equation, called the Schrödinger equation,

$$i (h/2\pi) \partial \Psi(q,t)/\partial t = \hat{H}(q,t) \Psi(q,t) \tag{2-1}$$

where the letter, i, refers to the unit of imaginary numbers and h is Planck's constant. The variable q in $\Psi(q,t)$ and $\hat{H}(q,t)$ refers to the spatial coordinates of all constituent particles in a generic form. The variable t refers to time. $\hat{H}(q,t)$ is a mathematical operator expressing the energy of a system and called the Hamiltonian operator. It is described as follows.

$$\hat{H}(q,t) = -\Sigma_k \{(h/2\pi)^2/2m_k\}(\partial^2/\partial x_k^2 + \partial^2/\partial y_k^2 + \partial^2/\partial z_k^2) + V(q,t) \tag{2-1a}$$

The first term of the right-hand side describes the kinetic energy of all constituent particles in the form of a mathematical operator, where each particle is designated by a positive integer k ($k = 1, 2, 3, \cdots, N$, N: the number of constituent particles in a system). Variables x_k, y_k and z_k refer to the spatial coordinates of the k-th particle and the quantity m_k is the mass of the k-th particle. The second term expresses the potential energy in a system at a time t. It usually consists of electromagnetic (and gravitational) interactions among constituent particles in a system but sometimes includes a perturbation term expressing an influence of the outer world.

Equation (2-1) represents the fundamental law of motion for a system consisting of a number of microscopic particles in quantum mechanics. Modern theoretical physics has developed more advanced forms of the fundamental law of motion but equation (2-1) is enough for discussions in this book. Note that the way of describing a system in quantum mechanics is entirely different from that in Newtonian mechanics. In quantum mechanics, a system is described by a wave function $\Psi(q,t)$, as mentioned above. In Newtonian mechanics, a system is described by using a set of the spatial coordinates and the velocities or momentums of individual constituent particles. Another remarkable feature of quantum mechanics is that a wave function $\Psi(q,t)$ follows *the principle of superposition*, as can be seen from equation (2-1). This means that states in quantum mechanics can overlap one another, or in other words, a state in quantum mechanics can be described by a linear combination of various wave functions. Interestingly, a state in quantum mechanics can represent itself in diverse ways.

The explanation of the dual nature of an electron

Now, the dual nature of an electron is explained as follows. A wave function for a free electron in vacuum, moving in the x direction, is given from equation (2-1) as follows.

$$\Psi(x,t) = \{A_n \exp(i\, k_n\, x) + B_n \exp(-i\, k_n\, x)\}\exp(-i\, 2\pi\, \nu_n\, t) \quad (2\text{-}2)$$

where k_n is the wave number of an electron wave, ν_n is the frequency of it, and A_n and B_n are the amplitudes of it. The subscript n is an integer called the *quantum number* and specifies a state of electron motion. From equations (2-1) and (2-2), we obtain the relation of $(h/2\pi)^2 k_n^2/2m = h\nu_n$ (m: the mass of an electron) between k_n and ν_n. Equation (2-2) indicates that an electron wave is characterized by the amplitude and the phase factor and obeys the principle of superposition.

Therefore, the wave nature of an electron is accurately described by the wave function, i.e. equation (2-2). On the other hand, the particle nature of an electron is understood by Born's interpretation, which says that the square of the absolute value of $\Psi(x,t)$, $|\Psi(x,t)|^2 = \Psi(x,t)^* \Psi(x,t)$, where $\Psi(x,t)^*$ is the complex conjugate of $\Psi(x,t)$, gives the probability of observing an electron at a spatial position x and time t. The formation of an interference pattern in an electron beam can be exactly explained by considering the superposition of wave functions and Born's interpretation.

A wave function of a system consisting of a number of electrons and atomic nuclei

For a general system consisting of a number of microscopic particles (electrons and atomic nuclei), a wave function is described in a similar manner to equation (2-2). A state described by a wave function is called a *microscopic state* or a *quantum state*. If the potential energy $V(q,t)$ in equation (2-1a) does not include time t as a variable and is expressed as $V(q)$, equation (2-1) can be rewritten in the following form.

$$\Psi(q,t) = \Phi_m(q) \exp\{-i\,(2\pi/h)\,E_m\,t\} \qquad (2\text{-}3)$$

$$\hat{H}(q)\Phi_m(q) = E_m \Phi_m(q) \qquad (2\text{-}4).$$

Equation (2-4) is independent of time and called *the Schrödinger equation in a stationary state*. This equation is a fundamental equation widely used in treating an electronic system in a stationary state such as an atom, a molecule and a crystal or their aggregate. The subscript m is an integer called the *quantum number* and specifies one of *microscopic (or quantum) states* which a system can take. Thus, $\Phi_m(q)$ and E_m represent a wave function and an observed energy value for a microscopic (or quantum) state designated by m, respectively. E_m can take only a discrete value when constituent particles in a system are in a bound state.

When $V(q,t)$ includes time t as a variable, equations (2-3) and (2-4) no longer hold. However, in case $V(q,t)$ only slightly changes with t, i.e. $V(q,t)$ is expressed in such a way as $V(q,t) = V(q) + V'(t)$ and $V(q) \gg V'(t)$ holds, we can assume mathematically that a wave function $\Psi(q,t)$ for a system can be expressed by a linear combination of $\Phi_m(q)$ for equation (2-4).

$$\Psi(q,t) = \Sigma_m\, C_m(t)\, \Phi_m(q) \qquad (2\text{-}5)$$

$$\Sigma_m\, C_m(t)^* C_m(t) = 1 \qquad (2\text{-}5\text{a})$$

The wave function (2-5) represents a state in which the probability of observing a quantum state $\Phi_m(q)$ with energy E_m at a time t is given by the square of the absolute value of the coefficient $C_m(t)$, $|C_m(t)|^2 = C_m(t)^* C_m(t)$, where $C_m(t)^*$ is the complex conjugate of $C_m(t)$.

For example, the above argument applies to an atom or a molecule irradiated by light, light acting as a perturbation giving $V'(t)$. In this case, equation (2-5) is interpreted as follows: A system (an atom or a molecule) lies in a state described by $\Phi_m(q)$ for a certain period of time and then suddenly changes to another state described by $\Phi_n(q)$. Such a sudden change from a quantum state to another one, accompanied with the absorption or the emission of a photon with energy ($E_m - E_m$), is called a (quantum-mechanical) transition or jump.

Physical quantities of a system consisting of a number of electrons and atomic nuclei

Equation (2-3) or (2-5) expresses how a microscopic state of a many-body system is described. Next, let us consider how physical quantities of such a system, such as momentum, angular momentum and energy, are described. As mentioned above, a system in quantum mechanics is described by a wave function, in contrast to Newtonian mechanics in which a system is described by a set of the spatial coordinates and the velocities or momentums of individual constituent particles. Therefore, physical quantities of a system in quantum mechanics are also described by a wave function, not by a set of the spatial coordinates and the velocities of individual constituent particles.

In quantum mechanics, an observed value for a physical quantity of a system is calculated as follows. We first define a mathematical operator, Ω, which represents a physical quantity, and then obtain the following Eigen (proper) equation for it.

$$\Omega \phi_l(q) = \omega_l \phi_l(q) \tag{2-6}$$

where $\phi_l(q)$ is the Eigen (proper) function for an operator Ω and ω_l the Eigen (proper) value for Ω. An observed value for a physical quantity Ω is given by ω_l. When a system is in a stationary state described by a wave function $\Phi_m(q)$ of equation (2-4), the probability of observing ω_l is calculated by an equation

$$\Phi_m(q) = \Sigma_l C_l \phi_l(q) \tag{2-7a}$$

$$\Sigma_l C_l^* C_l = 1 \tag{2-7b}$$

The square of the absolute value of the coefficient C_l, $|C_l|^2 = C_l^* C_l$, gives the probability of observing ω_l. Similarly, when a system is described by a time-dependent function, $\Psi(q,t)$, such as equation (2-5), the probability of observing ω_l is calculated by an equation

$$\Psi(q,t) = \Sigma_l C_l(t) \phi_l(q) \tag{2-7c}$$

$$\Sigma_l C_l(t)^* C_l(t) = 1 \tag{2-7d}$$

The square of the absolute value of the coefficient $C_l(t)$, $|C_l(t)|^2 = C_l(t)^* C_l(t)$, gives the probability of observing ω_l at a time t.

When a system consists of a number of partial systems, the energy of a partial system is calculated as follows. For example, water vapor in a box consists of a huge number of water molecules weakly interacting with one another. In this case, an individual water molecule constitutes a partial system. The Hamiltonian operator of a system consisting of a number of partial systems is approximately expressed by the sum of Hamiltonian operators of partial systems plus small interaction terms. Accordingly, properties of a partial system are fundamentally determined by the Hamiltonian operator of a partial system but an Eigen (proper) function of it is not the correct wave function for a partial system because of the presence of small interaction terms. Thus, a partial system is described by a linear combination of Eigen (proper) functions of the Hamiltonian operator of a partial system, in a similar way to equation (2-5), though in this case the coefficients C_m is independent of time. This means that a partial system takes various energy values because it interacts with other partial systems. In fact, for water vapor in a box, the energy of an individual water molecule takes a variety of values with a certain probability of observation, even though water vapor as a whole is in a stationary state with constant energy.

In this way, an observed value for a physical quantity in quantum mechanics in general scatters from experiment to experiment even though a system is situated under the same conditions. Accordingly, only the probability of observing a particular value or the average value for a physical quantity obeys a strict inevitable law. An observed value for a physical quantity in quantum mechanics obeys the probability law.

We can only observe physical quantities of a partial system

Finally, let us consider the meaning of the observation of a physical quantity in quantum mechanics. The first point to be noted is that we can only observe physical quantities of a partial system. This is because the observation of physical quantities of a system needs the connection of an observing system to it. Thus, a combination of an observed system and an observing one forms the whole system, an observed system being a partial system.

For example, the Hamiltonian operator $\hat{H}(q)$ for the Schrödinger equation in a stationary state, equation (2-4), is an operator representing the energy of a system, as mentioned earlier. Therefore, equation (2-4) gives the Eigen (proper) equation for the energy of a system. Accordingly, when a system is described by $\Phi_m(q)$, an observed value for the energy of a system is always given by E_m, namely, the probability of observing E_m is always unity. However, this is only a theoretical argument. Actually, we cannot observe such a state. Even such a system is changed to a system described by a wave function such as equation (2-5) when it is connected, e.g. to a light source for measuring an absorption spectrum. This is because illumination gives a perturbation $V'(t)$ to an observed system and the Hamiltonian operator for an observed system is changed from $\hat{H}(q)$ to $\hat{H}(q) + V'(t)$.

Note that in the above example we cannot observe the energy of the whole system consisting of an observed system and an observing one. We can only observe the energy of an observed system. This is just as we cannot look at our own eyes by our own eyes.

Observation has a limit in that it is possible only when an observed system and an observing one weakly interact with each other

The connection of an observing system has another important effect on an observed system. A perturbation by an observing system causes an "uncertainty" for the energy of a microscopic state of an observed system. This is because an observed system is changed to one described by a wave function such as equation (2-5), as mentioned above. Equation (2-5) means that $\Phi_m(q)$ is no longer the correct wave function for an observed system and hence E_m is not the correct energy value. We can say in the following way as well. Equation (2-5) indicates that quantum-mechanical transitions happen between microscopic states $\Phi_m(q)$ and $\Phi_n(q)$, as mentioned earlier. This means that such a microscopic state has a finite lifetime, depending on the transition probability. Such a finite lifetime leads to an energy uncertainty according to Heisenberg's uncertainty principle (see equation (2-10b)).

In relation to the above statement, it should be noted also that an energy uncertainty for a microscopic state of a partial system interacting with other partial systems comes not only from its interaction with an observing system but also from its interaction with other partial systems. In general, an energy uncertainty due to the latter interaction is larger than one due to the former. Therefore, we can in general observe an increase in energy uncertainty by an increase in interaction between partial systems.

Now, the above considerations indicate that there is another important point to be noted about observation. Namely, observation has a limit in that it is possible only when an observed system and an observing one *weakly* interact with each other. Naturally, observation is impossible when there is no interaction between such two systems. In addition, observation is also impossible when interaction between such two systems is too strong. This is because in such a case an observed system and an observing one are fused into unity and the concept of observation loses its basis. We can also say as follows. In such a case, an uncertainty of the energy of a microscopic state in an observed system becomes very large and exceeds an energy difference between microscopic states in an observed system. Therefore, nothing with individual unchanging properties can be observed [*2].

> [*2] In relation to the above argument, it is interesting to consider the difference between the stage of the attainment of *Tai-toku* and that of "understanding by words". We can say that the former stage corresponds to a case in which interaction between an observed system and an observing one is very strong while the latter stage corresponds to a case in which such interaction is weak. In fact, for a person who has attained *Tai-toku*, ideas in his or her brain and things in the real world are fully continuously connected with one another and fused into unity. Here is the fusion between the subject (an observing system) and the object (an observed system). On the other hand, for a person who is at the stage of "understanding by words", ideas in his or her brain are only logically connected with one another and are separated from things in the real world.

2.1.2 *A Continuous Internal Connection in Quantum Mechanics*

A wave function should be regarded as a real existent

Now, let us consider philosophical meanings of principles of quantum mechanics. The important issue to be investigated first is whether or not the wave function, Ψ, can be regarded as a real existent. If the reality of the wave function is denied and only observed physical quantities are taken as real existents, we look at natural phenomena individually and separately through observed physical quantities, in a similar way to in Newtonian mechanics. In this case, quantum mechanics can be wholly understood within the realm of word-based understanding, the wave function being only regarded as a mathematical tool for solving a problem. On the other hand, if the wave function is regarded as a real existent, quantum mechanics gives a novel view of nature involving the existence of the internal invisible world.

It is said that the above issue was really a big problem in the process of constructing quantum mechanics in the early 20th century. In general, there was intense belief that science had to be constructed based on a clear fact. Therefore, orthodox scientists had not liked to acknowledge the reality of an unobserved quantity such as a wave function. A similar problem appeared in the process of constructing electromagnetism. In this case, H. Hertz directly proved by experiments that an electromagnetic wave really exists. However, it is in principle impossible to directly prove the reality of a wave function by experiments because only the square of the absolute value of a wave function is related to observed quantities, as mentioned earlier. Originally, the Schrödinger equation and the wave function are described mathematically by the use of a complex function including the unit of imaginary numbers, in contrast to an electromagnetic wave. This fact clearly indicates that the wave function is in essence of a non-observable character.

In this way, it appears that the principles of quantum mechanics have been interpreted by regarding only observed physical quantities as real existents. In fact, for example, Newtonian mechanics-based statistical thermodynamics and quantum mechanics-based statistical thermodynamics have been constructed in the same mode [4,5]. This indicates that quantum mechanics has been interpreted in essentially the same way as Newtonian mechanics.

Contrary to such a traditional interpretation, we in this book regard the wave function as a real existent. Namely, we think that *the wave function represents the way that truly real things exist.* Interestingly, in this interpretation, it is natural that the wave function is not observed because truly real things are in the internal world we cannot be conscious of, as argued in Section 1.3.2 and 1.4.1. The validity and the importance of acknowledging the reality of the wave function will be discussed in detail later from various points of view.

It should be noted also that we cannot deny the reality of the wave function for the reason that it is not observed. This is because we can in principle not observe a thing such as a wave function, i.e. a thing with an infinitely-spreading continuous and incessantly changing dynamic quality. We can only observe a thing with individual unchanging qualities. Originally, observation has a limit in that it is possible only when an observed system and an observing one weakly interact with each other, as argued in Section 2.1.1. Somebody may say that we can observe an electromagnetic wave which has an infinitely-spreading continuous and incessantly-changing dynamic quality. However, an electromagnetic wave is different from a wave function in that it starts from electric charges or magnetic moments and ends in them and thus has individual unchanging qualities in this respect. In fact, we observe an electromagnetic wave by using electric charges or magnetic moments. The wave function has no individual unchanging quality in any respect.

A wave function has the same meaning as a continuous internal connection introduced in Section 1.3.2

If we acknowledge the reality of the wave function, quantum mechanics demonstrates quite a novel view of nature involving the existence of the internal invisible world, as mentioned earlier. Namely, it demonstrates that nature has a dual structure consisting of the very truth lying in the internal invisible world and the external appearance of it, in the same way as discussed in Section 1.4.2. In quantum mechanics, the very truth (the way that truly real things exist) is described by the Schrödinger equation and the wave function while the external appearance of it is embodied in observed physical quantities. Quantum mechanics is really in an advanced stage of scientific understanding compared with Newtonian mechanics in that it directly catches the internal invisible world lying behind the external visible world.

In the first place, let us consider characteristics of the internal invisible world in quantum mechanics, expressed by the Schrödinger equation and the wave function. Characteristics of the external appearance, embodied in observed physical quantities, will be discussed later. An important point to be noted first is that a wave function in quantum mechanics has the same meaning as a continuous internal connection introduced in Section 1.3.2 and 1.4.1. In quantum mechanics, every substance or system including a macroscopic visible thing such as a living thing and a human being is described by a wave function. Therefore, if a wave function has the same meaning as a continuous internal connection, this means that every substance or system exists as a continuous internal connection, in agreement with the argument in Section 1.4.1. If we consider this agreement in the reverse direction, it indicates that *a wave function certainly represents the way that truly real things exist,* as mentioned earlier.

Now, the above statement that a wave function has the same meaning as a continuous internal connection can be explained as follows. Firstly, a wave function $\Psi(q,t)$ is of a non-observable character, in the same way as a continuous internal connection, and exists in the internal invisible world. Secondly, a wave function $\Psi(q,t)$ spreads continuously to infinite distance both in space and time with no separation, in the same way as a continuous internal connection.

Thirdly, both a continuous internal connection and a wave function allow simultaneous and harmonious realization of all possible constituent elements and related actions or all possible states and processes under given conditions. As argued in Section 1.3.2, a person who has attained *Tai-toku* and formed a continuous internal connection of constituent elements can freely control all constituent elements and related actions so that they work simultaneously in harmony with one another. This means that a continuous internal connection of constituent elements allows simultaneous and harmonious realization of all possible constituent elements and related actions or all possible states and processes under given conditions.

The same holds for a wave function in quantum mechanics. As mentioned in Section 2.1.1, a wave function can be expressed by a linear combination or superposition of various wave functions. In such a case, superposed wave functions are fused into unity and corresponding states are simultaneously and harmoniously realized. For example, a wave function in equation (2-7a) is expressed by a linear combination of various wave functions $\phi_l(q)$. Thus, when a system is described by equation (2-7a), various states expressed by wave functions $\phi_l(q)$ are simultaneously and harmoniously realized. It is for this reason that the square of the absolute value of the coefficient C_l in equation (2-7a), $|C_l|^2$, can give the probability of observing a quantum state $\phi_l(q)$. In this way, a wave function in quantum mechanics also allows simultaneous and harmonious realization of all possible constituent elements and related actions or all possible states and processes under given conditions.

An electron exists simultaneously in all possible spatiotemporal positions under given conditions

The above arguments, in particular, the third item may look strange. However, it represents the way that truly real things exist. For getting better understanding of this point, let us consider again the meaning of Born's interpretation, which says that the square of the absolute value of a wave function gives the probability of observing an electron at a certain spatial position and time.

At first, let us assume that we deny the reality of the wave function and regard it only as a mathematical tool for solving a problem, as was done in a traditional interpretation of quantum mechanics. In this case, Born's interpretation only describes "results" of observation, irrespective of where an electron actually exists. Thus, it becomes uncertain where an electron exists or why the square of the absolute value of a wave function gives the probability of observing an electron at a certain spatial position and time. Such a way of understanding is characteristic of word-based scientific understanding, as emphasized in Section 1.2.3. Namely, in this way of understanding, only observed properties are described, by regarding them as given a priori, without providing anything about the origin of them.

On the other hand, if we acknowledge the reality of the wave function, it follows that an electron exists as a wave. Therefore, we have to say that *an electron exists simultaneously in all possible spatiotemporal positions under given conditions with an appropriate weight distribution*. According to this understanding, it is quite reasonable to say that the square of the absolute value of a wave function $\Psi(q,t)$ provides the probability of observing an electron at q and t. Namely, the acknowledgement of the reality of the wave function leads to the clarification of the origin of observed events. Thus, Born's interpretation offers an example of the above-mentioned argument that a wave function allows simultaneous and harmonious realization of all possible states and processes under given conditions.

Note here that the acknowledgment of the reality of the wave function leads to the emergence of a strange situation in which an electron exists simultaneously in all possible spatiotemporal positions under given conditions, as mentioned above. We have to acknowledge such a strange situation as expressing the very truth (i.e. the way that truly real things exist). Originally, there is a non-removable gap between word-based understanding and the very truth, as discussed in Section 1.2.1. Therefore, it is natural that truly real things have a strange way of existing. We should not forget that we human beings can only be conscious of "common or repeatedly observed parts" or "results" of activities of truly real things (see Section 1.2.3).

An electron really exists simultaneously in all possible spatial positions

Furthermore, we can show definite support to the argument that an electron exists simultaneously in all possible spatial positions under given conditions. As mentioned earlier, the formation of an interference pattern in an electron beam can only be explained by assuming that an electron exists simultaneously in all possible spatial positions under given conditions. Similarly, the formation of atoms, molecules and crystals, which is a universal phenomenon in nature, can also only be explained by making the same assumption as the above.

At first, no atom can be formed if an electron cannot exist simultaneously in all possible spatial positions under given conditions. This is because in such a case an electron is localized at one site and thus sticks to an atomic nucleus. As an electron is a charged particle, it cannot keep a stationary circular or elliptical motion because an accelerated motion (such a circular or elliptical motion) of a charged particle loses energy by the emission of radiation.

The formation of molecules and crystals, in particular, the formation of a chemical bond of a covalent type can also never be explained if an electron cannot exist simultaneously in all possible spatial positions under given conditions. As a simple example, let us consider the formation of a covalent bond for a hydrogen molecule. The bond energy D(H-H) for a hydrogen molecule is defined as the energy difference between two hydrogen atoms lying far apart and a hydrogen molecule lying at the most stable state. W. Heitler and F. London showed in 1927 that the following wave function,

$$\phi_{HL}(H\text{-}H) = \{\chi_{A1s}(1)\chi_{B1s}(2) + \chi_{A1s}(2)\chi_{B1s}(1)\}/(2 + 2S^2)^{1/2} \quad (2\text{-}8)$$

gives the bond energy $D(H\text{-}H)$ which is in good agreement with the observed value. Here, χ_{A1s} and χ_{B1s} refer to the 1s atomic orbital (1s-AO) for two hydrogen atoms named H_A and H_B, respectively. The numerals in parentheses stand for two electrons, named 1 and 2. Thus, $\chi_{A1s}(1)$ means that electron 1 exists in the 1s-AO of H_A while $\chi_{B1s}(2)$ means that electron 2 exists in the 1s-AO of H_B. S is an overlap integral between χ_{A1s} and χ_{B1s}. On the other hand, a similar wave function expressed as

$$\phi_{NE}(H\text{-}H) = \chi_{A1s}(1)\chi_{B1s}(2) \quad \text{or} \quad \chi_{A1s}(2)\chi_{B1s}(1) \quad (2\text{-}9)$$

can hardly explain the observed bond energy.

The above results about the bond energy indicate that $\phi_{HL}(H\text{-}H)$ is a much better wave function than $\phi_{NE}(H\text{-}H)$. The essential difference between them is that $\phi_{HL}(H\text{-}H)$ consists of the superposition of two equivalent wave functions, $\chi_{A1s}(1)\chi_{B1s}(2)$ and $\chi_{A1s}(2)\chi_{B1s}(1)$, in which electrons 1 and 2 are exchanged between χ_{A1s} and χ_{B1s}, while $\phi_{NE}(H\text{-}H)$ only consists of either of them. The superposition of wave functions in quantum mechanics means that superposed wave functions are fused into unity and simultaneously realized, as mentioned earlier. Therefore, in the wave function $\phi_{HL}(H\text{-}H)$, electron 1 exists *simultaneously* both in χ_{A1s} and χ_{B1s}, i.e. it exists simultaneously in all possible spatial positions. The same holds for electron 2. On the other hand, in $\phi_{NE}(H\text{-}H)$, both electrons 1 and 2 only exist either in χ_{A1s} or in χ_{B1s}, i.e. they do not exist simultaneously in all possible spatial positions. Thus, we can say that the simultaneous existence of electrons in all possible spatial positions plays a critically important role in the formation of chemical bond.

If we say in somewhat more detail, the increase in the bond energy for $\phi_{HL}(H\text{-}H)$ is due to an increase in the attractive interaction between electrons and atomic nuclei. The simultaneous existence of electrons both in χ_{A1s} and χ_{B1s} in $\phi_{HL}(H\text{-}H)$ leads to an increase in the electron density in the region between two atomic nuclei and leads to an increase in electrostatic attractive energy between electrons and atomic nuclei and hence leads to an increase in the bond energy.

In quantum mechanics, not individual things in the external visible world but a wave function describing a whole about a system is the basic governing reality

It was argued in Section 1.3.2 and 1.4.1 that the formation of a continuous internal connection of things leads to a great revolution in the way that they exist. Similarly, the introduction of a wave nature to constituent particles in quantum mechanics leads to a great revolution in the way that they exist. Such a good correspondence also indicates that a wave function in quantum mechanics and a continuous internal connection in Section 1.3.2 and 1.4.1 have the same meaning.

The most important aspect of such a great revolution is that the basic governing reality in a system is transferred from individual things in the external visible world to a continuous internal connection of them. It was argued in Section 1.3.2 that for a person who is at the stage of "understanding by words", constituent elements of "the ability gained by the attainment of *Tai-toku*" exist individually and separately and are only logically connected with one another. Namely, they always have individual unchanging properties and exist as the basic governing reality. For this reason, such a person can behave only one by one. On the other hand, for a person who has attained *Tai-toku*, such constituent elements are internally continuously connected with one another and fused into unity. Then, they incessantly change their structures, properties and functions so that they as a whole realize a free independent harmonious state. Accordingly, for such a person, not individual constituent elements but a continuous internal connection of them, which controls individual constituent elements, becomes the basic governing reality. In this way, a person who has attained *Tai-toku* can display free independent harmonious spirit and wisdom.

Quite the same arguments apply to a system described by Newtonian mechanics and one described by quantum mechanics. A system described by Newtonian mechanics corresponds to a person who is at the stage of "understanding by words" while a system described by quantum mechanics corresponds to a person who has attained *Tai-toku*. In Newtonian mechanics, in which no wave nature is taken into account, individual constituent particles exist as the basic governing reality, as will be explained later. On the other hand, in quantum mechanics, not individual constituent particles but a wave function that describes a system as a whole (or describes a whole about a system) is the basic governing reality. This can be explained as follows.

First of all, the fundamental law of motion in quantum mechanics, i.e. the Schrödinger equation (2-1) or (2-4) contains only the wave function, $\Psi(q,t)$ or $\Phi_m(q)$, which describes a whole about a system. In addition, observed physical quantities of a system such as momentum, angular momentum and energy are also completely determined by the wave function, $\Psi(q,t)$ or $\Phi_m(q)$, which describes a whole about a system, as can be seen from equations (2-4) and (2-7).

The above feature of a system in quantum mechanics becomes much clearer if we consider how we solve the Schrödinger equation. The Schrödinger equation, both equation (2-1) and equation (2-4), cannot be solved analytically for all practical systems, except some idealized simple systems such as a free electron in vacuum or an isolated hydrogen atom in vacuum. Therefore, the Schrödinger equation, for example, equation (2-4) is usually solved under an approximation that weak interactions are neglected. Amazingly, under such an approximation, equation (2-4) gives a variety of solutions which display the existence of various individual particles or systems we can be conscious of, such as atoms, molecules, crystals and their aggregates, depending on what approximation is used.

On the other hand, if equation (2-4) is solved under no approximation, for example, by numerical calculation, it gives only one continuous wave function $\Phi_m(q)$. Here is no individual particle or system we can be conscious of. Thus, equation (2-4) clearly indicates that *any system (including nature itself) exists in a fully continuous state with no discrepancy, distinction and separation. Individual particles or systems such as atoms, molecules, crystals or their aggregates emerge from the continuum only in an approximate form.* This conclusion is in exact agreement with the one described in Section 1.4.1, 1.4.2, etc. The same conclusion is also obtained from equation (2-1).

How does the introduction of a wave nature affect the way that things exist?

Interestingly, quantum mechanics teaches us how the introduction of a wave nature to constituent particles affects the way that things exist. In the first place, the introduction of a wave nature leads to Heisenberg's uncertainty principle, which says that the spatial position and the momentum of a particle can no longer be accurately determined simultaneously. The principle is described as follows

$$\Delta x \, \Delta p_x \geq h/4\pi \qquad (2\text{-}10a)$$

where Δx and Δp_x refer to uncertainties of the spatial position and the momentum of a particle in the x direction, respectively. Equation (2-10a) can be proved by a hypothetical experiment [1,2]. It indicates that a moving particle cannot be correctly described by words, i.e. a moving particle is beyond word-based understanding, as discussed in Section 1.2.3. Equation (2-10a) can be converted to the following form by a mathematical transformation

$$\Delta E \, \Delta t \geq h/4\pi \qquad (2\text{-}10b)$$

where ΔE and Δt stand for an uncertainty of the energy of a microscopic state of an individual particle or system and the lifetime of it, respectively. An energy uncertainty ΔE emerges from the fact that an individual particle or system interacts with other particles or systems and is of an approximate character, as argued in Section 2.1.1 and 2.1.2. Equation (2-10b) indicates that a change in the qualities of a system is again beyond word-based understanding because the energy of a system specifies its quality, as also discussed in Section 1.2.3.

In close relation to the above uncertainty principles, the introduction of a wave nature also leads to another important principle called *the indistinguishableness of particles of the same kind*. This principle says that individual particles of the same kind can no longer be distinguished from one another because the wave nature of a particle makes it impossible to follow a trace of a particle's motion. We cannot distinguish this electron and that electron or this proton and that proton.

The basic qualities of atoms, molecules and crystals are changed by their interaction

In the second place, the introduction of a wave nature leads to incessant changes in the basic qualities of a system by its interaction with others. In fact, interaction among electronic systems such as atoms, molecules and crystals necessarily leads to the superposition of their wave functions and hence necessarily leads to changes in their basic qualities.

As a simple example, let us consider an interaction between two hydrogen atoms. It was mentioned earlier that a wave function $\phi_{NE}(\text{H-H})$ described by equation (2-9), $\chi_{A1s}(1)\chi_{B1s}(2)$ or $\chi_{A1s}(2)\chi_{B1s}(1)$, is not the correct wave function for a hydrogen molecule. However, this function is the correct wave function for two hydrogen atoms lying far apart from each other because no electron exchange occurs in this case. Now, when two hydrogen atoms lying far apart approach each other, electron exchange starts at a certain distance through the overlap between χ_{A1s} and χ_{B1s}. At this moment, $\phi_{NE}(\text{H-H})$ comes to be an incorrect wave function and the Heitler-London wave function $\phi_{HL}(\text{H-H})$ described by equation (2-8) comes to be the correct wave function.

In this way, interaction between two hydrogen atoms induces the superposition of wave functions for them and leads to a change in the basic qualities of two hydrogen atoms. The same things happen for other atoms and for molecules and crystals and furthermore for macroscopic visible substances such as a stone, a living thing and a human being. They interact with one another and incessantly change their structures, properties and functions, as argued in Section 1.1.1, 1.3.2 and 1.4.1. Note here that such superposition of wave functions and following changes in the basic qualities of things happen because electrons (or in general electrons and atomic nuclei) have a wave nature, i.e. they have the basic quality of existing simultaneously in all possible spatiotemporal positions under given conditions. Thus, this is a phenomenon peculiar to quantum mechanics. Interestingly, *quantum mechanics definitely asserts that holism is correct.*

The difference between Newtonian mechanics and quantum mechanics

Contrary to quantum mechanics, in Newtonian mechanics in which no wave nature is taken into account, the spatial position and the momentum of an individual particle can be accurately determined simultaneously. Particles of the same kind can also be clearly distinguished from one another. Besides, individual particles do not change their structures and properties when they interact with one another. Therefore, the behavior of a system in Newtonian mechanics can be accurately described by using the spatial coordinates and the momentums of individual constituent particles. In Newtonian mechanics, individual constituent elements exist as the basic governing reality and their properties determine properties of a system, in the same way as a person who is at the stage of understanding by words (see Section 1.3.2).

It is worth noting here that the above characteristics of Newtonian mechanics apply to a system in which an electromagnetic field is present. This fact clearly indicates that an electromagnetic field has an essentially different character from a wave function, as already mentioned earlier. In harmony with this argument, electromagnetic interaction among constituent particles is included in the Hamiltonian operator of the Schrödinger equation and clearly separated from the wave function.

Properties of individual particles, substances or systems of an approximate character can be observed

We have thus far considered the internal world in quantum mechanics, described by the Schrödinger equation and the wave function. Next, let us consider the external visible world in quantum mechanics, embodied in observed physical quantities. The discussions described below also show that a wave function in quantum mechanics and a continuous internal connection introduced in Section 1.3.2 and 1.4.1 have the same meaning.

It was stated earlier that in quantum mechanics individual particles, substances or systems such as atoms, molecules, crystals and their aggregates are of an approximate character. Namely, the Schrödinger equation, if it is solved under no approximation, gives only one continuous wave function $\Psi(q,t)$ or $\Phi_m(q)$ describing a whole about a system. Individual particles, substances or systems only emerge under an approximation that weak interactions are neglected. However, strikingly, we can experimentally observe properties of such individual particles, substances or systems of an approximate character. In fact, we can observe various properties of atoms, molecules and crystals or their aggregates such as morphological, mechanical, electrical, optical and chemical properties. This fact clearly indicates that *natural things outwardly look as if they existed individually and separately though they are internally continuously connected with one another and fused into unity*, as mentioned in Preface. This conclusion is also in good agreement with discussions in Section 1.3.2, 1.4.1 and 1.4.2.

Another striking point is that indeed physical quantities of individual particles, substances or systems can be observed but observed values are often scattered from experiment to experiment even though they are situated under the same conditions, as mentioned in Section 2.1.1. Namely, observed values of physical quantities show uncertainties. This is a natural result because individual particles, substances or systems are of an approximate character, only appearing under an approximation that weak interactions are neglected, as mentioned above (see also Section 2.1.1 and 2.1.2). In fact, the stronger the neglected interaction becomes, the larger the uncertainty.

In this way, in quantum mechanics, only the probability of observing a particular value or the average observed value for a physical quantity follows a strict inevitable law, as mentioned in Section 2.1.1. Thus, a system in quantum mechanics displays individual unchanging properties only in an averaged form, in agreement with the argument in Section 1.4.2.

An example of actual observation of properties of individual particles, substances or systems

For getting better understanding of the above argument, let us consider the measurement of a light absorption spectrum of water vapor (see Figure 2-1(A)) as an example of actual observation. When light is irradiated to water vapor, it is partly absorbed by water molecules. Thus, the comparison of intensities of incident and transmitted light at various wavelengths or frequencies of light gives a light absorption spectrum of water vapor. The absorption of light is caused by a quantum mechanical transition (jump) from a quantum state m of a water molecule with energy E_m to another quantum state l with energy E_l. For E_m and E_l, the law of energy conservation, $E_l - E_m = h\nu$, holds, where h is the Planck constant, ν the frequency of light and $h\nu$ the photon energy. What kind of transition happens depends on the wavelength or frequency of light and the transition probability for a water molecule. When ultraviolet light with large photon energy is irradiated, a transition between electronic states of a water molecule is induced. On the other hand, when infrared light with relatively small photon energy is irradiated, a transition between vibrational states (states of atomic vibrations) of a water molecule is induced.

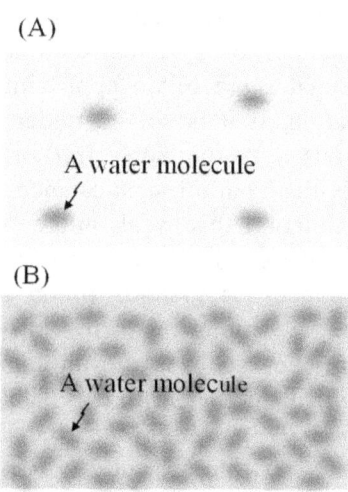

Figure 2-1 Schematic illustrations of (A) water vapor and (B) liquid water. A dark gray area refers to an area where the electron density for a water molecule is high.

Now, when infrared light is irradiated to water vapor, we obtain an infrared absorption spectrum. The spectrum consists of a large number of spectral lines as a function of photon energy. The energy of each line corresponds to an energy difference ($E_l - E_m$) between two vibrational states of an *individual* water molecule accompanied by rotational levels for it. The fact that the spectrum consists of spectral lines indicates that vibrational-rotational states of a water molecule in water vapor have discrete energy values, as expected for a system lying in a bound state (see Section 2.1.1). Each spectral line has a definite width, indicating that vibrational-rotational states of a water molecule have energy uncertainties [*1]. For water molecules in water vapor, the width of a spectral line is very small, indicating that there are only small energy uncertainties or only weak intermolecular interactions, in harmony with a spatially well-separated distribution of water molecules in water vapor (see Figure 2-1(A)).

> [*1]Monochromatic light used for a spectral measurement consists of a huge number of photons and therefore each spectral line represents results of a huge number of quantum mechanical transitions. Therefore, the width of a spectral line represents the degree of energy uncertainties in vibrational states of a water molecule.

When a measured sample is changed from water vapor to liquid water, an infrared absorption spectrum no longer consists of lines but consists of broad bands. Rotational structures completely disappear. In addition, some new bands appear. The broadening of bands indicates that rotational-vibrational states of individual water molecules in liquid water have large energy uncertainties because of stronger inter-molecular interactions in liquid water due to a spatially dense distribution of water molecules in it (Figure 2-1(B)). The appearance of new bands is due to the motion of water molecules as a whole in liquid water.

Thus, the above examples confirm that (1) an energy difference between microscopic states of individual molecules of an approximate character can be really observed, (2) an observed energy difference between microscopic states shows an uncertainty because of an approximate character of individual molecules, and (3) an uncertainty becomes larger as interaction between individual molecules gets stronger. The above examples also indicate that the basic qualities of individual molecules are changed by their mutual interaction or interaction with others.

2.1.3 The Internal Ability in Quantum Mechanics

According to the arguments in Section 1.4.1, a truly real thing in the very truth has two important features: (1) it exists as a continuous internal connection of all other things and (2) it has the internal ability to organize itself so that it realizes a free independent harmonious stable state. The first feature was verified by the discussions in the preceding section. Then, this section examines the second feature.

Electrons and atomic nuclei are controlled by the internal power to make them exist simultaneously in all possible spatiotemporal positions

As argued in the preceding section, microscopic particles such as electrons and atomic nuclei have a wave nature. Namely, they have the basic quality of existing simultaneously in all possible spatiotemporal positions under given conditions. Thus, we can say as follows.

Electrons and atomic nuclei are controlled by the internal power to make them exist simultaneously in all possible spatiotemporal positions with an appropriate weight distribution under given conditions.

The presence of the internal power has quite an important philosophical meaning because it makes every phenomenon in nature and the world happen in one direction in an irreversible manner. This will be discussed in detail in Section 2.2 and 3.2.1.

According to the argument in Section 2.1.2, making electrons and atomic nuclei exist simultaneously in all possible spatiotemporal positions is equivalent to forming as widely-spreading a continuous internal connection of electrons and atomic nuclei as possible. Therefore, we can say as follows.

Electrons and atomic nuclei are controlled by the internal power to form as widely-spreading a continuous internal connection of them as possible under given conditions.

An electronic system such as an atom, a molecule and a crystal has the internal ability to organize itself so that it realizes a free independent harmonious stable state

Now, if electrons and atomic nuclei are controlled by the internal power such as mentioned above, how is an electronic system such as an atom, a molecule and a crystal produced? Electrons and atomic nuclei have electric charges, magnetic moments and the mass as well as a wave nature. Therefore, we can say that an electronic system such as an atom, a molecule and a crystal is produced in the following way.

An electronic system such as an atom, a molecule or a crystal is produced by electrons and atomic nuclei which exist simultaneously in all possible spatial positions in the presence of their electrostatic and other interactions.

An electronic system such as an atom, a molecule or a crystal exists as an infinitely-spreading fully-continuous internal connection of electrons and atomic nuclei, formed in the presence of their electrostatic and other interactions.

A wave function for an electronic system such as an atom, a molecule and a crystal describes such a state of electrons and atomic nuclei.

Next, if an electronic system exists in the above way, what character does it have? It realizes a free independent harmonious stable state both in the interior and in relation to the outer world. This is because an infinitely-spreading fully-continuous internal connection of electrons and atomic nuclei, formed in the presence of their electrostatic and other interactions, has no discrepancy, distinction, roughness, separation and restriction and thus achieves a complete independent form in itself. In addition, nevertheless, such a continuous internal connection has a freely changing dynamic character and moreover achieves full harmony in every respect.

There is another important point to be noted. The above infinitely-spreading fully-continuous internal connection is formed by the wave nature of electrons and atomic nuclei (or the aforementioned internal power to control electrons and atomic nuclei) together with electrostatic and other interactions among them, both of which are always active. Therefore, such a continuous internal connection has the internal ability to organize itself so that it always realizes a free independent harmonious stable state. Accordingly, we can say as follows.

An electronic system such as an atom, a molecule or a crystal realizes a free independent harmonious stable state and besides has the internal ability to organize itself so that it realizes a free independent harmonious stable state.

The above argument is supported by our experiences. For example, a water molecule in water vapor or liquid water actually exists in a free independent harmonious stable state. In addition, water molecules demonstrate various properties such as morphological, electrical, optical and chemical properties. This fact also indicates that a water molecule realizes a free independent harmonious state. As discussed in Section 1.4.2, the appearance of various individual unchanging properties in an averaged or common form comes from the realization of a free independent harmonious state.

Electronic systems in nature and the world are controlled by the grand internal ability

In the above argument, we paid attention to an individual electronic system such as an atom, a molecule and a crystal. If we direct our eyes to a set of various electronic systems such as one in nature, the situation becomes more complex. Firstly, various electronic systems have free motion, i.e. kinetic energy, apart from interactions, i.e. potential energy. In addition, the superposition of wave functions leads to the formation of a repulsive force as well as an attractive force, depending on conditions. We have to take into account all these factors. It was mentioned in

Section 1.4.1 that all things in nature and the world are controlled by the grand internal ability. By bearing this statement in mind, we can say that electronic systems in nature and the world are produced in the following way.

Electronic systems in nature and the world are controlled by the grand internal ability to make electrons and atomic nuclei interact with one another according to their inherent properties and motion, form as widely-spreading fully-continuous internal connections of them as possible under given conditions in diverse quality levels and aspects, and complete the formation of such continuous internal connections. In this way, the grand internal ability produces a new thing existing as an infinitely-spreading fully-continuous internal connection of other things.

In the above statement, the internal ability to make electrons and atomic nuclei interact with one another comes from various kinds of interaction among electrons and atomic nuclei and their motion (kinetic energy). On the other hand, the internal ability to form as widely-spreading fully-continuous internal connections as possible and to complete the formation of them comes from the wave nature of electrons and atomic nuclei (or the aforementioned internal power to control electrons and atomic nuclei). The fact that the superposition of wave functions leads to the formation of a repulsive force as well as an attractive force is regarded as included in the words of "in diverse quality levels and aspects".

It is important to note that the above statement is definitely supported by the Schrödinger equation (2-1). At first, the Hamiltonian operator $\hat{H}(q,t)$ (equation (2-1a)) can be interpreted as representing the act of making electrons and atomic nuclei interact with one another according to their inherent properties (electrostatic and other interactions) and motion (kinetic energy). Then, the Schrödinger equation (2-1) says that the act of the Hamiltonian operator $\hat{H}(q,t)$ on an electronic system leads to the production of a state described by a wave function $\Psi(q,t)$, in which as widely-spreading a fully-continuous internal connection of electrons and atomic nuclei as possible is formed in the presence of their various interactions and kinetic motion. Thus, the Schrödinger equation (2-1) says that electronic systems are really controlled by the grand internal ability mentioned above.

It was argued earlier that an infinitely-spreading fully-continuous internal connection of electrons and atomic nuclei and hence an electronic system such as an atom, a molecule or a crystal existing as such a continuous internal connection realizes a free independent harmonious stable state and besides has the internal ability to organize itself so that it realizes such a state. This can be explained by using the grand internal ability as follows. Such an infinitely-spreading fully-continuous internal connection is formed by the grand internal ability that comes from the wave nature of electrons and atomic nuclei and their various interactions and motion, all of which are always active. Therefore, it realizes a free independent harmonious stable state and has the internal ability to organize itself so that it realizes such a state.

We can also say as follows. The formation of such an infinitely-spreading fully-continuous internal connection represents a "result" of activities of the grand internal ability. Therefore, it catches the grand internal ability within it, as mentioned in Section 1.4.1. Accordingly, it realizes a free independent harmonious stable state and besides has the internal ability to organize itself so that it realizes such a state.

2.2 The Internal Ability in Macroscopic Visible Systems

The preceding section mainly dealt with a microscopic system such as an atom, a molecule or a crystal. This section deals with a macroscopic visible system such as a gas in a box, a piece of iron metal, water in a cup, a living body, the earth environment, nature itself, and so on. If we regard the wave function as a real existent, as mentioned in Section 2.1.2, we can show that in the very truth a macroscopic visible system also exists as a continuous internal connection of all other things, in agreement with the arguments in Section 1.3.2 and 1.4.1.

The basic principles of thermodynamics

In modern science, a macroscopic visible system is treated by thermodynamics and statistical thermodynamics. An outstanding feature of a macroscopic visible system is that it includes a nearly infinite number ($\geq 1\times 10^{23}$) of microscopic particles (atoms and molecules or atomic nuclei and electrons). Thus, a macroscopic visible system has a nearly infinite number of microscopic (quantum) states even within a small interval of energy. In addition, the rate of microscopic processes is in general quite high, or in other words, the lifetime of a microscopic (quantum) state is very short. For example, the lifetime of an electronic excited state of a normal molecule is about 10^{-9} s or less.

Accordingly, it is actually impossible to deal with a macroscopic visible system exactly. Fortunately, it is known empirically that a macroscopic system has the following important unique characteristic.

An isolated macroscopic system, even if it was initially in a complex state, necessarily changes into a simple stable state which can be defined by a small number of macroscopic physical quantities such as the temperature, the pressure, and the chemical potentials. Such a simple stable state, once it has been reached, no longer spontaneously changes further.

Here, the chemical potential refers to the (free) energy change in a system when one constituent particle or one mole of constituent particles is transferred from the outer world into a system. Therefore, the chemical potential is equal to the (free) energy of one constituent particle or one mole of constituent particles in a system with the outer world taken as the energy standard.

A simple stable state, which a macroscopic system finally reaches, is called the equilibrium state. For example, hot water placed in a room is cooled down and lastly comes to have the same temperature as the ambient. A piece of salt crystal put into a glass of water is gradually dissolved and finally reaches a stable state in which a salt crystal and a homogeneous aqueous salt solution coexist. No further macroscopic change occurs after the equilibrium state is attained as far as no external action is added. A change toward the equilibrium state is spontaneous and irreversible.

Thermodynamics is constructed based on the above empirical fact. The first law of thermodynamics simply expresses the conservation of energy including heat. The second law of thermodynamics describes the above-mentioned spontaneous and irreversible change specific to a macroscopic system. For convenience, thermodynamics uses three idealized model systems: (1) an isolated system in which the energy, the volume and the amounts of constituent particles are all kept constant, (2) a closed system in which energy or heat can transfer across the interface between a system and the outer world, the amounts of constituent particles being kept constant, and (3) an open system in which both energy and constituent particles can transfer across the interface between a system and the outer world. The volume of a system in (2) and (3) is kept either able to change or fixed.

A closed and an open system can be regarded as a partial system of a large isolated system, as illustrated in Figure 2-2. For example, a gas in a closed box placed in a room is "a closed system" because heat (energy) can transfer across the wall of a box. On the other hand, water in a glass placed in a room is "an open system" because not only heat (energy) but also water molecules can transfer across the interface between water and the ambient. It is interesting to note that a living body including a human body can be regarded as an open system.

Now, the second law of thermodynamics for an isolated system is described as follows: An isolated system spontaneously and irreversibly changes so that its entropy increases, and the equilibrium state is attained when the entropy takes a possible maximum value. The meaning of entropy will be explained later. For a closed system, the concept of free energy is used instead of entropy. The second law of thermodynamics in a closed system is then described as follows. A closed macroscopic system under a constant temperature and pressure spontaneously and irreversibly changes so that its free energy decreases, and the equilibrium state is attained when the free energy takes a possible minimum value. The meaning of free energy will also be explained later. The decrease in the free energy of a closed system is completely equivalent to the increase in the entropy of an isolated system. The second law of thermodynamics for an open system will be explained in Section 3.1.1.

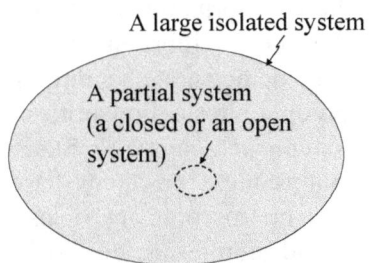

Figure 2-2 Schematic illustration of a large isolated system and a closed or an open system as a partial system.

It is important to note that the second law of thermodynamics says that structures and properties of a closed or open macroscopic system in equilibrium are determined so that the entropy of a whole consisting of "a closed or open macroscopic system and its surroundings" takes a maximum value (see Figure 2.2). This means that *a closed or open macroscopic system in equilibrium exists in infinite dependence on the surroundings and incessantly changes its structures, properties and functions so that it realizes a stable state in harmony with the surroundings.* A closed or open macroscopic system in equilibrium is in a stable state but not in an unchanging state. *It has autonomous dynamic self-organizing ability.* In fact, the above-mentioned way of existing is the same as the way a living thing exists, discussed in Section 1.1.1.

The explanation of the law of increase of entropy by statistical thermodynamics in the present form[4,5)]

Why does a macroscopic system spontaneously and irreversibly approach the equilibrium state? Why does no further macroscopic change occur after the equilibrium state is attained, and why does the equilibrium state demonstrate a simple stable state defined by a small number of macroscopic physical quantities such as the temperature, the pressure and the chemical potentials? Statistical thermodynamics in the present form explains these problems based on the motion and interaction of constituent microscopic particles such as atoms and molecules or electrons and atomic nuclei and by introducing Boltzmann's statistical interpretation.

Let us consider an isolated macroscopic system with given values of the energy E, the volume V, and the numbers N_i of constituent microscopic particles, where the subscript, i, expresses the kinds of constituent microscopic particles. Microscopic (or quantum) states of an isolated macroscopic system can be obtained by solving Schrödinger equation (2-4) for this system. Note here that all microscopic states in an isolated macroscopic system are of an approximate character and have energy uncertainties. This is because the observation of the energy of an isolated macroscopic system needs the connection of an observing system to it and such connection causes a perturbation in an observed isolated macroscopic system, as mentioned in Section 2.1.1. In addition, originally, an isolated system is only an idealized theoretical model and any real system interacts with the surroundings[1]. Such a perturbation by an observing system and interactions with the surroundings lead to energy uncertainties.

[1] An isolated system in thermodynamics is only an approximate model imagined in the external visible world. In quantum mechanics, there is only a continuous wave function, namely, there is no individual thing, as argued in Section 2.1.2. Therefore, no isolated system can in principle exist.

Now, let us assume that the energy of an isolated macroscopic system lies between E and $E + \delta E$, where δE represents an energy uncertainty of a microscopic state having energy E. Indeed, the amount of δE is very small but a macroscopic system in general has a tremendous or nearly infinite number of microscopic states between E and $E + \delta E$. In such a situation, statistical thermodynamics in the present form assumes that *all* allowed microscopic states having energy between E and $E + \delta E$ in an isolated macroscopic system are realized at the *equal* probability. This is called *the assumption of equal a priori probability*. Under this assumption, the entropy S for an isolated macroscopic system in equilibrium is defined[4] as

$$S = k \ln W \qquad (2\text{-}11)$$

where k is the Boltzmann constant, "ln" refers to natural logarithm, and W is the number of *all* allowed microscopic states having energy between E and $E + \delta E$ [*2].

[*2] The size of δE is not fixed. However, thermodynamic quantities and their relations, derived from equation (2-11), are independent of the size of δE.

The definition of entropy in the form of equation (2-11) was first given by L. Boltzmann. The emergence of a small number of macroscopic physical quantities in equilibrium such as the temperature (T), the pressure (p), and the chemical potential (μ_i), which characterize the equilibrium state, can be explained based on this definition [4]. An isolated macroscopic system in equilibrium achieves maximum entropy with respect to energy distribution, volume distribution, and particle-number distribution within a system. Analyses of these conditions by the use of equation (2-11) lead to the conclusion that T, p, and μ_i are constant within a system, where T, p, and μ_i are defined as follows:

$$\partial S/\partial E = 1/T \qquad (2\text{-}12a)$$

$$\partial S/\partial V = p/T \qquad (2\text{-}12b)$$

$$\partial S/\partial N_i = -\mu_i/T \qquad (2\text{-}12c)$$

This definition of T, p, and μ_i agrees with that in thermodynamics.

It is important to note that W in equation (2-11) is the number of *all* allowed microscopic states having energy between E and $E + \delta E$ for an isolated macroscopic system, as mentioned earlier, and thus includes the number of microscopic states allotted to a non-equilibrium state. Nevertheless, S in equation (2-11) gives the entropy of an isolated macroscopic system *in equilibrium*. This is because the overwhelming majority of allowed microscopic states having energy between E and $E + \delta E$ for an isolated macroscopic system are allotted to the equilibrium state, only a negligible number of allowed microscopic states being allotted to non-equilibrium states, if N_i is extremely large. The discovery of this fact is a key achievement of statistical thermodynamics in the present form [4].

Based on the above argument, we can easily explain why an isolated macroscopic system spontaneously and irreversibly approaches the equilibrium state, by using Boltzmann's statistical interpretation. Under the assumption of equal a priori probability for all allowed microscopic states, the probability of realizing a certain macroscopic state is in proportion to the number of allowed microscopic states allotted to this macroscopic state. Thus, the probability of realizing the equilibrium state is extremely high because the overwhelming majority of allowed microscopic states are allotted to this state, as mentioned above. Accordingly, any non-equilibrium state *nearly necessarily* approaches the equilibrium state and once it is realized, it *nearly necessarily* shows no further macroscopic change.

Statistical thermodynamics in the present form also allows us to calculate structures and properties of a macroscopic system in equilibrium. For example, for an isolated macroscopic system in equilibrium, all allowed microscopic states are realized at the equal probability, as mentioned earlier. Therefore, structures and properties of such a system can be calculated as a simple average of structures and properties of all allowed microscopic states. Similarly, structures and properties of a closed or an open macroscopic system in equilibrium can be calculated by constructing a distribution function which gives a probability of finding an allowed microscopic state as a function of energy and other physical quantities[*3]. Great successes in materials science such as solid-state physics, chemistry and biology are attained by this principle.

[*3] In statistical thermodynamics, an observed physical quantity, for example, an observed energy value is in general a doubly averaged value. This is because a macroscopic system in equilibrium consists of a nearly infinite number of allowed microscopic states and besides each microscopic state shows a distribution of an observed energy value due to Heisenberg's uncertainty principle, as discussed in Section 2.1.1.

A new interpretation of the law of increase of entropy

Indeed, statistical thermodynamics in the present form has achieved great successes in materials science, as mentioned above, but it still has serious problems in a theoretical aspect. Firstly, it uses the assumption of equal a priori probability for all allowed microscopic states in an isolated macroscopic system, as mentioned earlier. It also uses the assumption that a time average is equal to a phase average. The validity of these assumptions, in particular, the assumption of equal a priori probability has not been fully proved[4]. Secondly, statistical thermodynamics in the present form explains thermodynamic laws such as the law of increase of entropy by introducing Boltzmann's statistical interpretation. As a result, thermodynamic laws are not directly connected with mechanical laws for constituent particles.

It is well known that Boltzmann tried to deduce the law of increase of entropy directly from Newtonian mechanics. However, the law of increase of entropy has an irreversible character with respect to the passage of time while Newtonian mechanics has a reversible character.

Therefore, Boltzmann encountered severe paradoxes called reversibility paradox and recurrence paradox [5]. In order to avoid such paradoxes, Boltzmann was finally forced to introduce a statistical interpretation. The statistical interpretation has been retained in quantum mechanics-based statistical thermodynamics in the present form. This is most probably because only observed physical quantities, which have individual unchanging qualities, have been regarded as real existents, as argued in Section 2.1.2. If only observed physical quantities are regarded as real existents, Newtonian mechanics-based statistical thermodynamics and quantum mechanics-based statistical thermodynamics can be constructed in the same theoretical framework.

In contrast to such a traditional interpretation, we in this book regard the wave function as a real existent, as emphasized in Section 2.1.2. Namely, we consider that the wave function represents the way that truly real things exist. According to this interpretation, observed physical quantities only represent the external appearance of truly real things and are of an approximate character. Importantly, such a new interpretation leads to the complete resolution of all the problems which statistical thermodynamics in the present form has faced, as will be explained later.

Now, if the wave function is regarded as a real existent, an isolated macroscopic system is treated as follows. We again consider an isolated macroscopic system with energy E. As already mentioned earlier, the energy of a microscopic (quantum) state for an isolated macroscopic system lying at energy E has an energy uncertainty δE. Indeed, the amount of δE is very small but an isolated macroscopic system has a nearly infinite number of microscopic states between E and $E + \delta E$. Such microscopic states can all be obtained by solving the following Schrödinger equation (see equation (2-4))

$$\hat{H}_{\text{ims}}(q)\,\Phi_n(q) = E_n \Phi_n(q) \quad n = 1, 2, \cdots\ W \quad\quad (2\text{-}13)$$

where $\hat{H}_{\text{ims}}(q)$ is the Hamiltonian operator of an isolated macroscopic system. $\Phi_n(q)$ and E_n represent, respectively, a wave function and an observed energy value for a microscopic state designated by n. W is the number of all allowed microscopic states having energy between E and $E + \delta E$ and is the same as W in equation (2-11).

In statistical thermodynamics of the present form, all allowed microscopic states $\Phi_n(q)$ are distinguished by the quantum number n and have been treated individually, as discussed earlier. This is probably because only observed physical quantities have been regarded as real existents, as mentioned above. Namely, an isolated macroscopic system has been dealt with in the external visible world (see a footnote #1 of this section). Now, if we regard the wave function as a real existent, we have to treat $\Phi_n(q)$ in a different way.

As mentioned above, all microscopic states $\Phi_n(q)$ ($n = 1, 2, \cdots W$) have energy between E and $E + \delta E$. In addition, they all each have an energy uncertainty δE. Therefore, they completely overlap one another in energy and we can never distinguish them. They are fused into unity [*4]. Furthermore, in quantum mechanics, any system is expressed by one continuous wave function spreading to an infinite distance and here is no individual thing, as argued in Section 2.1.1 and 2.1.2. Accordingly, we have to say that an isolated macroscopic system is described by one of the following wave functions $\Psi_m(q)$ ($m = 1, 2, \cdots W$).

$$\Psi_m(q) = \Sigma_n C_{mn} \Phi_n(q) \qquad m, n = 1, 2, \cdots W \qquad (2\text{-}14)$$

$$\Sigma_n |C_{mn}|^2 = 1 \qquad (2\text{-}14a)$$

where the square of the absolute value of the coefficient C_{mn}, $|C_{mn}|^2 = C_{mn}{}^* C_{mn}$, represents the probability of observing a microscopic state $\Phi_n(q)$. If we consider the above considerations in the reverse direction, equation (2-14) explains why microscopic states $\Phi_n(q)$ have energy uncertainties δE [*5].

[*4] In quantum mechanics, overlapping microscopic states are not mixed states. They form a new single unified state. As an example of such coalescence, it is reported in NMR (nuclear magnetic resonance) spectra that two sharp spectral peaks due to two different species arising from intra-molecular isomerization (proton transfer) become broad, coalesce and finally gets one sharp peak as the temperature is raised. The appearance of one sharp peak indicates the formation of a new single unified state.

[*5] Note here that we can only observe energy E_n for $\Phi_n(q)$ or energy for a modified state of $\Phi_n(q)$ arising from interaction between an isolated macroscopic system and the surroundings. Namely, we cannot observe the energy of a microscopic state for an isolated macroscopic system, described by $\Psi_m(q)$, because it is a state (or a wave function) emerging from interaction between an observed system and an observing one (see the argument given at the end of Section 2.1.1).

A next problem is which of $\Psi_m(q)$ ($m = 1, 2, \cdots W$) or what set of C_{mn} in equation (2-14) describes the *equilibrium* state of an isolated macroscopic system. Quantum mechanics provides a method of determining the coefficients C_{mn} for equation (2-14) but the above problem is difficult to solve by a mathematical treatment alone. Therefore we follow another principle. As discussed in Section 2.1.3, electrons and atomic nuclei are controlled by the internal power to make them exist simultaneously in all possible spatial positions under given conditions. It is certain that this argument holds for electrons and atomic nuclei in an isolated macroscopic system. The equilibrium state is the *final* stable state which a non-equilibrium state reaches. Therefore, electrons and atomic nuclei in the equilibrium state should exist simultaneously in all possible spatial positions under given conditions.

From this argument, we can say that all $|C_{mn}|^2$ ($n = 1, 2, \cdots W$) are equal to one another for an isolated macroscopic system in equilibrium. This is because all wave functions $\Phi_n(q)$ in equation (2-14) are Eigen (proper) functions of the Hamiltonian operator $\hat{H}_{\text{ims}}(q)$ and thus are mathematically independent of one another. Thus, only $\Psi_m(q)$ with equal $|C_{mn}|^2$ meets the condition that electrons and atomic nuclei exist simultaneously in all possible spatial positions. Accordingly, we can say as follows.

The equilibrium state for an isolated macroscopic system exists as a continuous internal connection of all allowed microscopic states with the equal weight.

The equilibrium state for an isolated macroscopic system realizes all allowed microscopic states simultaneously at the equal weight.

Problems in statistical thermodynamics of the present form can be completely overcome by a new interpretation in this book

The above new interpretation allows us to solve all the problems which statistical thermodynamics of the present form has faced. Firstly, the principle of equal a priori probability is a natural result of the above argument that electrons and atomic nuclei exist simultaneously in all possible spatial positions in the equilibrium state. We can rather say that the new interpretation proves the principle of equal a priori probability and hence proves the validity of the definition of entropy by equation (2-11). Accordingly, mathematical techniques in the present statistical thermodynamics can be used without any essential alteration.

Secondly, the law of increase of entropy is also a natural result of the above argument. $\Psi_m(q)$ with *unequal* $|C_{mn}|^2$ does not meet the condition that electrons and atomic nuclei exist simultaneously in all possible spatial positions. Therefore, it spontaneously and irreversibly changes into $\Psi_m(q)$ with *equal* $|C_{mn}|^2$ because electrons and atomic nuclei are controlled by the internal power to make them exist simultaneously in all possible spatial positions under given conditions. This means that $\Psi_m(q)$ with *unequal* $|C_{mn}|^2$ represents a non-equilibrium state. Thus, a change from a non-equilibrium state into the equilibrium one is expressed as a change from a linear combination of $\Phi_n(q)$ with *unequal* $|C_{mn}|^2$ to one with *equal* $|C_{mn}|^2$.

$$\Psi_m(q) = \Sigma_n C_{mn} \Phi_n(q) \text{ with unequal } |C_{mn}|^2$$
$$\to \Psi_m(q) = \Sigma_n C_{mn} \Phi_n(q) \text{ with equal } |C_{mn}|^2 \qquad (2\text{-}15)$$

Once such a change has been completed, the resultant $\Psi_m(q)$ no longer changes. Thus, *equation (2-15) correctly describes the second law of thermodynamics or the law of increase of entropy for an isolated macroscopic system.* Note that in the above argument the law of increase of entropy is directly deduced from quantum mechanics without using Boltzmann's statistical interpretation. Therefore, this book calls quantum mechanics-based thermodynamics developed here "quantum thermosdynamics", as already mentioned in Section 1.1.2.

Now, from the above consideration, we can obtain the following important conclusions:

The origin of the law of increase of entropy is not probability, as has been explained to date, but the wave nature of electrons and atomic nuclei or the internal power to make them exist simultaneously in all possible spatiotemporal positions under given conditions

The law of increase of entropy is not a probability law, as has been explained to date, but an inevitable natural law coming from the wave nature of electrons and atomic nuclei.

The law of increase of entropy is an expression of the internal power introduced in Section 2.1.3, which comes from the wave nature of electrons and atomic nuclei.

Such successful results clearly indicate the importance of regarding the wave function as a real existent.

Why does only the second law of thermodynamics have an irreversible character with respect to the passage of time?

Though discussions may slightly wander from the main subject, it is interesting to consider here why the second law of thermodynamics has an irreversible character with respect to the passage of time, in contrast to many other natural laws. This problem has attracted much attention of many philosophers and scientists [6,7].

According to the new interpretation in this book, *all* phenomena or processes in nature have an irreversible character with respect to time because electrons and atomic nuclei in all natural things are controlled by the internal power to make them exist simultaneously in all possible spatial positions under given conditions. Now, why do many natural laws other than the second law of thermodynamics have a reversible character with respect to time? This is simply because these laws are constructed by the use of words created by picking out common or repeatedly observed parts of truly real motion or changes, unrepeated or irreversible parts being neglected, as discussed in Section 1.2.3.

On the other hand, thermodynamics treats the whole of a system as it is, without using such a picking-out procedure. Therefore, an irreversible character of natural phenomena is correctly reflected in the second law of thermodynamics. Quantum mechanics also deals with a system as a whole, as discussed in Section 2.1.1. For this reason, the fundamental law of motion in quantum mechanics, i.e. the Schrödinger equation (2-1) also has an irreversible character with respect to time. The Schrödinger equation can be regarded as having a reversible character if the complex conjugate is taken into account. Interestingly, the complex conjugate is taken into consideration when observed physical quantities are considered.

The internal power and the internal ability in a macroscopic system

To return to the original subject, let us consider characteristics of the equilibrium state of an isolated macroscopic system in more detail. We argued earlier that the equilibrium state of such a system is characterized by the simultaneous realization of all allowed microscopic states with the equal weight. The point of this conclusion is that electrons and atomic nuclei in the equilibrium state of an isolated macroscopic system exist simultaneously in all possible spatial positions under given conditions, i.e. they are continuously connected with one another.

This conclusion can be developed further. Namely, we can say that in the equilibrium state of an isolated macroscopic system not only electrons and atomic nuclei but also all constituent elements produced by electrons and atomic nuclei such as atoms and molecules, visible macroscopic systems, etc. are continuously connected with one another. Such a state is achieved by activities of the grand internal ability introduced in Section 2.1.3. Thus, we can rewrite the grand internal ability in the following generalized form.

An isolated macroscopic system is controlled by the grand internal ability to make various constituent elements (electrons and atomic nuclei, atoms and molecules, visible macroscopic systems, etc.) interact with one another according to their inherent properties and motion, form as widely-spreading fully-

continuous internal connections of them as possible under given conditions in diverse quality levels and aspects, and complete the formation of such continuous internal connections.

According to this statement, the law of increase of entropy can be regarded as an expression of activities of the grand internal ability. Thus, the equilibrium state of an isolated macroscopic system represents a "result" of activities of the grand internal ability [*6].

> [*6] This never means that a non-equilibrium macroscopic system has the physical or dynamic power to cause an increase in entropy. The law of increase of entropy only says that when a huge number of microscopic processes happen in an isolted non-equilibrium macroscopic system, entropy in it increases as a result of such microscopic processes. The importance of this law is that it says that entropy *necessarily* increases when a huge number of microscopic processes happen.

An isolated macroscopic system in equilibrium has the internal ability to organize itself so that it realizes a free independent harmonious stable state

Let us consider the equilibrium state of an isolated macroscopic system in further detail. According to the above argument, in this state the formation of as widely-spreading fully-continuous internal connections of constituent elements as possible is completed under given conditions. This means that all possible constituent elements and related actions or all possible states and processes under given conditions are internally continuously connected with one another and simultaneously realized in harmony with one another. Now, an infinitely-spreading continuous internal connection realizes a free independent harmonious stable state and besides has the internal ability to organize itself so that it realizes such a state, as argued in Section 2.1.3. Therefore, we can say as follows.

An isolated macroscopic system in equilibrium realizes a free independent harmonious stable state and besides has the internal ability to organize itself so that it realizes such a state in all possible areas of motion under given conditions.

The words of *"in all possible areas of motion under given conditions"* will be explained later.

In harmony with the above argument, an isolated macroscopic system in equilibrium really realizes a free independent harmonious stable state. In fact, the equilibrium state is a free independent harmonious stable state defined by a small number of macroscopic physical quantities such as the temperature, the pressure and the chemical potentials, as mentioned earlier. In addition, the equilibrium state demonstrates various properties such as morphological, mechanical, electrical, optical and chemical properties in an averaged or common form. This fact also indicates that the equilibrium state realizes a free independent harmonious state because the realization of such a state is responsible for the appearance of various individual unchanging properties in an averaged or common form (see Section 1.4.2).

It is often stated that the equilibrium state is characterized by complete randomness. This statement is not wrong but expresses no whole truth. Indeed, the equilibrium state is characterized by complete randomness if it is viewed from a microscopic point of view. The simultaneous realization of all allowed microscopic states means complete randomness. However, the equilibrium state realizes a free independent harmonious stable state on a macroscopic scale and demonstrates various individual unchanging properties in the external visible world. Thus, the equilibrium state is also characterized by the formation of a free independent harmonious stable state.

It is important to note that complete randomness on a microscopic level, i.e. the completion of the simultaneous realization of all allowed microscopic states and processes under given conditions in the equilibrium state leads to the achievement of a free independent harmonious stable state on a macroscopic level. Namely, the former is a necessary condition for the latter. They are two faces of one thing. The same situation exists in the formation of stable atoms, molecules, and crystals. Electrons and atomic nuclei in atoms, molecules and crystals exist simultaneously in all possible spatial positions under given conditions, i.e. they are in complete randomness. Such completely random motion of electrons and atomic nuclei is a necessary condition for the formation of atoms, molecules or crystals realizing a free independent harmonious stable state.

In general, in traditional understanding, both the wave nature of electrons and atomic nuclei and the law of increase of entropy have been interpreted in a rather negative way. In fact, the wave nature of an electron has been understood to cause uncertainties in observed physical quantities and make natural laws dependent on probability. The law of increase of entropy has been understood to indicate that nature has a tendency to change toward complete randomness. Such interpretations give us a feeling that we cannot help but admit that natural phenomena are impossible to understand in a strict way. On the contrary, in the new interpretation in this book, both the wave nature of an electron and atomic nuclei and the law of increase of entropy are understood to have a positive meaning. Namely, they both are understood to indicate that *nature has a tendency to realize all possible states and processes under given conditions and achieve a free independent harmonious stable state.* Traditional understanding has overlooked such a positive meaning.

The equilibrium state is achieved only in areas of motion in which changes are kinetically allowed

It was mentioned earlier that an isolated macroscopic system in equilibrium realizes a free independent harmonious stable state *in all possible areas of motion under given conditions*. This statement has the following meaning.

The point is that equilibrium and hence a free independent harmonious state are realized only in areas of motion in which changes are kinetically allowed under given conditions. Let us consider a piece of wood placed in the air as an example. At room temperature and under the atmospheric pressure, translational, rotational and vibrational motion of nitrogen molecules, oxygen molecules, etc. in the air and atomic (or lattice) vibration of wood carry out frequent energy exchange with one another and are in equilibrium. However, wood does not burn, i.e. wood is not oxidized by oxygen molecules in the air under the above conditions. Therefore, wood and oxygen in the air are not in equilibrium. This is because the oxidation reaction of wood has large activation energy and its rate is extremely low, as will be discussed in the next chapter. In this way, a piece of wood placed in the air realizes equilibrium only in areas of motion in which changes are kinetically allowed under given conditions.

From these considerations, we can see that properties of an isolated macroscopic system in equilibrium strongly depend on in what areas of motion changes are kinetically allowed under given conditions.

References

(1) L. I. Schiff, *Quantum Mechanics*, 2nd Edition, McGraw-Hill, NY (1955): The Japanese edition translated by T. Inoue, Yoshioka-shoten (1957).

(2) H. Eyring, J. Walter, G. E. Kimball, *Quantum Chemistry*, John Wiley & Sons (1944): The Japanese edition translated by M. Kotani and K. Tomita, Yamaguchi-shoten (1953).

(3) P. A. M. Dirac, *The Principles of Quantum Mechanics*, 3rd ed. Oxford University Press, Oxford, 1947: The Japanese edition

translated by S. Tomonaga, H. Tamaki, J. Kiniwa and M. Otsuka, Iwanami-shoten (1954).
(4) R. Kubo (editor), *Thermodynamics and Statistical Mechanics for University Exercises* (in Japanese), Shokabo, Tokyo, 1961.
(5) D. ter Haar, *Elements of Statistical Mechanics*, Rinehart & Company, Inc., New York, 1956: The Japanese edition translated by T. Tanaka and K. Ikeda, Misuzu-shobo, Tokyo, 1960.
(6) I. Prigogine, *The End of Certainty: Time, Chaos, and the New Laws of Nature*, Free Press, 1997: The Japanese edition translated by S. Abiko and K. Taniguchi, Misuzu-shobo, Tokyo, 1997.
(7) M. Mitchell, *"Complexity: A Guided Tour"*, Oxford University Press, Inc., Oxford, 2009: The Japanese edition translated by H. Takahashi, Kinokuniya Shoten, Tokyo, 2011.

3. Dynamic Self-organization in Non-equilibrium Systems

The elucidation of structures and properties of non-equilibrium macroscopic systems is of key importance in that the emergence of living things on the primitive earth and their evolution in later ages are events characteristic of non-equilibrium macroscopic systems. The purpose of this chapter is to clarify fundamental characteristics of non-equilibrium macroscopic systems and investigate structures and properties of them. A focus is placed on the disclosure of basic principles that explain how highly ordered structures such as those in living things are produced in non-equilibrium macroscopic systems. Discussions in this chapter offer a theoretical basis for considering the origin and evolution of life in the next chapter.

3.1 Preceding Studies of Non-equilibrium Systems

The equilibrium state has played a key role in constructing thermosdynamics and statistical thermodynamics, as discussed in Section 2.2. A non-equilibrium state has long been regarded as being of a temporary character. However, the situation has largely changed since the latter half of the 20th century. Non-equilibrium systems have now become a central subject in scientific studies. This is natural if we consider that the equilibrium state is only a hypothetical idealized state. In fact, all substances on the earth are in a non-equilibrium state because they are irradiated by solar light and radiate thermal rays into the universe. The universe itself also demonstrates a clear characteristic of non-equilibrium, displaying the concentration and the divergence of energy on a grand scale. In addition, the emergence of living things on the primitive earth and their evolution later are events characteristic of a non-equilibrium system, as mentioned above. Therefore, we cannot understand the basic qualities of nature and the world without gaining detailed understanding of non-equilibrium systems.

In this section we briefly survey studies made thus far about non-equilibrium systems. In the next section we consider a new theory of a non-equilibrium system proposed in this book.

3.1.1 Fundamental Characteristics of Non-equilibrium

Systems

The non-equilibrium structure acts as a driving force for causing a macroscopic change

At first, let us consider fundamental characteristics of non-equilibrium systems. A non-equilibrium macroscopic system spontaneously and irreversibly changes toward an equilibrium system, as discussed in Section 2.2. An equilibrium macroscopic system demonstrates no spontaneous macroscopic change once it has been produced. Therefore, it is the non-equilibrium structure that acts as a driving force for causing a macroscopic change in a system. Dynamic self-organization in a non-equilibrium macroscopic system also arises from this characteristic, as will be discussed later.

Now, how can we describe quantitatively such a driving force the non-equilibrium structure has? In general, a non-equilibrium system is extremely complex. It consists of a nearly infinite number ($\geq 1\times 10^{23}$) of microscopic particles (atoms and molecules or electrons and atomic nuclei), in the same way as an equilibrium system discussed in Section 2.2. In addition, in contrast to an equilibrium system, a non-equilibrium system is impossible to describe by using a small number of macroscopic physical quantities such as the temperature, the pressure and the chemical potentials. The behavior of a non-equilibrium system strongly depends on detailed microscopic structures and interactions in it [*1].

[*1] It is worth noting that such an extremely complex characteristic of a non-equilibrium macroscopic system means that it has a possibility of displaying a huge diversity of properties and functions. In fact, living organisms, which are typical of a non-equilibrium macroscopic system, exhibit a great diversity of properties and functions, as will be discussed in the next chapter.

Accordingly, various approximate approaches have been proposed to date. Of these, a new version of thermodynamics, called thermodynamics of non-equilibrium systems or thermodynamics of irreversible processes, is useful for investigating properties and characteristics of a non-equilibrium system. Therefore, let us here explain it briefly.

Thermodynamics of non-equilibrium systems

As mentioned at the beginning of Section 3.1, thermodynamics was first established for equilibrium systems. Then, it was developed by De Donder, L. Onsager, I. Prigogine, et al. so as to include non-equilibrium systems in the former half of the 20th century [1,2].

Thermodynamics of non-equilibrium systems in general deals with a non-equilibrium system lying *not far from equilibrium*. Let us start with the second law of thermodynamics expressed by the Clausius inequality

$$dS \geq \delta Q / T^e \tag{3-1}$$

where dS is an increase in the entropy of a closed system, δQ is heat supplied to a closed system from the outer world, and T^e is the temperature of the outer world, which is assumed to be in equilibrium. If a system is isolated, $\delta Q = 0$, and therefore we obtain

$$dS \geq 0 \tag{3-2}.$$

This is a mathematical expression of the law of increase of entropy for an isolated system. The sigh of equality applies to the equilibrium state.

No change in entropy happens for a reversible process (i.e. a process which occurs under the condition that equilibrium is maintained) in an isolated system. Therefore, equation (3-2) indicates that an increase of entropy in an isolated system is solely due to irreversible processes within it. The detailed meaning of this statement will be explained later. Accordingly, in thermodynamics of non-equilibrium systems, the second law of thermodynamics (the law of increase of entropy) is expressed as follows [1].

Entropy is produced by irreversible processes in a macroscopic system. The produced entropy always takes a positive value.

This expression holds for a closed or an open system as well as an isolated system. Based on this expression, De Donder and later Prigogine described equation (3-1) as follows.

$$dS = d_oS + d_iS \quad (d_oS = \delta Q/T^e, \quad d_iS \geq 0) \tag{3-3}$$

For a closed non-equilibrium system, which is in thermal equilibrium with the outer world with the temperature T, equation (3-3) can be rewritten as follows.

$$dS = \delta Q/T + \delta Q'/T \tag{3-4}$$

$$d_oS = \delta Q/T, \quad d_iS = \delta Q'/T, \quad \delta Q' \geq 0 \tag{3-4a}$$

Equation (3-4) indicates that the entropy increase dS in a closed system is the sum of the entropy increase d_oS due to the inflow of heat (δQ) from the outer world and the entropy increase d_iS due to the production of heat ($\delta Q'$) by irreversible processes within a system. According to Clausius's naming, $\delta Q'$ is called uncompensated heat.

Now, let us consider equation (3-4) in more detail. The first law of thermodynamics (the law of conservation of energy) is expressed for a closed system with a variable volume (V) as follows.

$$dE = \delta Q - p\,dV \tag{3-5}$$

where dE is a change in the energy of a system (where the kinetic energy of the center of gravity of a system is excluded) and dV is a change in the volume of a system. Thus, $-p\,dV$ is pressure-work done to a system by the outer world. The outer world is assumed to be in equilibrium, as mentioned above, and therefore p is defined in the outer world. On the other hand, the Gibbs free energy G is defined as

$$G = E - TS + pV \tag{3-6}$$

where E, S, and V are the energy, the entropy, and the volume of a system, respectively. Accordingly, by using equations (3-4) and (3-5), a Gibbs free energy change for a closed system under constant temperature and pressure is given by

$$dG = -\delta Q' \leq 0 \tag{3-7}$$

This is a mathematical expression of the second law of thermodynamics for a closed system.

For an open system, various substances enter into or come out of it. Therefore, we can assume that a change in the Gibbs free energy for an open system under constant temperature and pressure is expressed as follows.

$$dG = \Sigma\,\mu_i\,dN_i + \Sigma F_j\,df_j - \delta Q' \qquad (\delta Q' \geq 0) \tag{3-8}$$

where μ_i is the chemical potential of a substance i, dN_i is the number or the amount of substance for a substance i injected from the outer world into a system, F_j is the force other than the pressure, and f_j is the variable conjugate with F_j. The subscript i refers to the kinds of substances while the subscript j stands for the kinds of forces. The quantities μ_i and F_j are defined in the outer world, which is assumed to be in equilibrium. In a similar way to T, p and μ_i, F_j is given by the equation of $\partial S/\partial f_j = -F_j/T$ (see equation (2-12)). Therefore, $\Sigma \mu_i dN_i$ represents the free energy injected into a system by the inflow of substances from the outer world and $\Sigma F_j df_j$ does the thermodynamic work done to a system by the outer world other than the pressure work ($-p\,dV$).

Equation (3-8) means that the free energy ($\Sigma\,\mu_i\,dN_i$) injected into a system by the inflow of substances is used partly for doing the thermodynamic work $(-\Sigma F_j \, df_j)$ to the outer world and partly for driving irreversible processes which produce uncompensated heat $(\delta Q')$ in a system, the rest dG being stored in a system.

Local equilibrium, transport phenomena, the law of linear response, and coupling

Now, how can we express $\delta Q'$ quantitatively by thermodynamic quantities? For clarifying this problem, we have to understand the important theoretical equation that L. Onsager first proposed in 1931. It is said that this equation acted as a key basis for constructing thermodynamics of non-equilibrium systems.

As mentioned earlier, thermodynamics of non-equilibrium systems deals with a non-equilibrium system lying not far from equilibrium. In such a system, we can assume *local equilibrium*, which says that equilibrium or quasi-equilibrium is achieved in a sufficiently small macroscopic area. By this assumption, we can define thermodynamic quantities such as the temperature T, the pressure p, the electric potential ϕ and the chemical potentials μ_i as functions of the spatial position. Thus, a non-equilibrium system can be described as a system with an inhomogeneous distribution of T, p, ϕ, μ_i and so on.

The presence of an inhomogeneous distribution of T, p, ϕ, μ_i, etc. brings about various kinds of flow, such as heat flow, volume expansion, the flow of charged species, and the flow of chemical substances, which are usually called *transport phenomena*. Thus we can expect that the rate of flow per unit area, i.e. the flux, J_k, is given as a function of gradients of T, p, ϕ, μ_i and so on. Here we introduce the concept of the thermodynamic force, X_k, which is roughly in proportion to a gradient of T, p, ϕ, μ_i and so on [*2]. Then, the flux, J_k, is expected to be given as a function of the thermodynamic force, X_l, where the subscript k or l refers to the kind of flow, i.e. what quantity flows. X_k is the thermodynamic force conjugate with J_k. For a non-equilibrium system not far from equilibrium, thermodynamic forces X_l are small. Thus, by taking into account that $J_k = 0$ when $X_k = 0$, J_k can be expanded with respect to X_l as follows [2].

$$J_k = \Sigma_l \, (\partial J_k / \partial X_l) \, X_l + \text{(the secondary terms for } X_l) + \cdots \quad (3\text{-}9)$$

Equation (3-9) is called the constitutional equation. If a system lies close to equilibrium and X_l are very small, the secondary and higher terms for X_l can be neglected. Then we obtain [1,2]

$$J_k = \Sigma_l L_{kl} X_l \qquad (3\text{-}10).$$

This is the equation first proposed by L. Onsager in 1931 and called the phenomenological constitutional equation. Equation (3-10) indicates that J_k is in proportion to the thermodynamic force X_l, i.e. *the law of linear response* holds. This equation also indicates that the non-equilibrium structure acts as a driving force for causing a macroscopic change, as mentioned earlier.

[*2] The quantities, T, p, ϕ, μ_i and so on, are defined in the form of $\partial S/\partial a$, where a refers to E, V, q (electric charge), N_i and so on (see equation (2-12)). Based on this definition, the thermodynamic force X is in general given as a gradient of $\partial S/\partial a$ [2]. Accordingly, for example, the thermodynamic force X for heat flow (thermal conduction) and substance diffusion is given as $\nabla(1/T)$ and $\nabla_i(-\mu/T) = -(1/T)\nabla\mu$, respectively.

Equation (3-10) has another important meaning. In this equation, coefficients, $L_{kl} = (\partial J_k/\partial X_l)$, are called phenomenological coefficients or transport coefficients. Onsager deduced the following important relation about off-diagonal coefficients

$$L_{kl} = L_{lk} \qquad (3\text{-}11)$$

based on the theory of fluctuation in statistical dynamics and the principle of microscopic reversibility [1]. Relation (3-11) is called Onsager's reciprocity theorem. The existence of off-diagonal coefficients, L_{kl} with $k \neq l$, in equation (3-10) expresses that *coupling* occurs between processes k and l, where coupling means that two different flow processes simultaneously occur and give a result which can no longer be divided into individual processes. Accordingly, equation (3-10) indicates that coupling happens between various flow processes.

The rate of entropy production

Based on equation (3-10), thermodynamics of non-equilibrium systems has revealed that the rate of entropy production (or the rate of uncompensated-heat production) σ per unit volume for a non-equilibrium system lying not far from equilibrium is given as follows [1,2].

$$\sigma = d_i S/dt = (1/T)\,\delta Q'/\delta t = \Sigma\, J_k X_k \geq 0 \qquad (3\text{-}12)$$

where J_k is the rate of flow per unit area, i.e. the flux and X_k is the thermodynamic force conjugate with J_k (see a footnote #2 of this section). The rate of entropy production in a system σ_{system} is thus given by integrating equation (3-12) over the whole volume of a system.

$$\sigma_{\text{system}} = \int \Sigma J_k X_k \, dxdydz \tag{3-13}$$

If all processes obey the law of linear response (equation (3-10)), equation (3-12) can be rewritten as follows.

$$\sigma = d_i S / dt = (1/T) \, \delta Q'/\delta t = \Sigma\Sigma \, L_{kl} X_k X_l \tag{3-14}$$

Prigogine deduced that the entropy production rate becomes a minimum in a stationary state based on this equation, as will be explained in the next section.

Here, let us consider the meaning of entropy production or uncompensated-heat ($\delta Q'$) production. A non-equilibrium system with an inhomogeneous distribution of T, p, ϕ, μ_i and so on brings about various kinds of flow, as mentioned earlier. This means that such a system has the ability (or free energy) to do work either to a system itself or to the outer world by utilizing a gradient of T, p, ϕ, μ_i and so on. The ability is often called effective energy, available energy or exergy.

A non-equilibrium state can move to another state without losing such ability (or free energy) to do work. Such a process is called a reversible process. A quasi-static process, i.e. a process which proceeds at an infinitesimal rate under the condition that equilibrium is maintained is an example of a reversible process, as mentioned earlier. However, a reversible process is only a hypothetical process by theoretical idealization. In an actual process, a non-equilibrium state moves to another state in the presence of a finite gradient of T, p, ϕ, μ_i and so on and at a finite rate. In this case, part of the ability (or free energy) to do work is converted to heat. This heat is uncompensated heat. Thus, an actual process is an irreversible process. In this way, irreversibility arises from the conversion of the ability (or free energy) to do work into heat. The heat of friction, Joule's heat in electric conduction and the heat of reaction are examples of uncompensated heat. Therefore, the total energy of a system is kept unchanged by an irreversible process even though uncompensated heat is produced. The extent of irreversibility is determined by how much ability (or free energy) to do work is converted to heat, or in other words, under how large gradients of T, p, ϕ, μ_i and so on an irreversible process happens.

There is another important point to be noted. *Uncompensated heat can be interpreted as the energy required for driving an irreversible process at a finite rate in the presence of a gradient of T, p, ϕ or μ_i.* In general, stationary flow following the law of linear response (equation (10)) is observed when a thermodynamic force acting as a driving force for such flow balances a resistance to the flow. Therefore, equation

(3-14) indicates that the rate of uncompensated-heat production becomes larger as the resistance to flow gets stronger. This means that uncompensated heat represents the energy required for driving an irreversible process. This argument is expected to roughly apply to flow which does not follow the law of linear response.

A non-equilibrium system has the ability to produce an ordered structure

We can see from the foregoing arguments that the non-equilibrium structure acts as a driving force for causing a macroscopic change, as mentioned earlier. The non-equilibrium structure also acts as a driving force for producing an ordered structure under an appropriate condition. In fact, the emergence of living things is an event characteristic of a non-equilibrium system, as mentioned at the beginning of this chapter. Now, how can the non-equilibrium structure produce an ordered structure?

This problem has not been sufficiently solved to date, as briefly discussed in Section 1.1.2. A fundamental problem we meet in dealing with this problem is the existence of the second law of thermodynamics, which says that nature has a tendency to spontaneously and irreversibly change into complete disorder. Therefore, it has been regarded as a key subject to clarify how highly ordered structures in living organisms are produced *against* the second law of thermodynamics, as mentioned in Section 1.1.2. Let us look back on past studies on this issue.

Firstly, it is famous that Schrödinger claimed, "A living thing keeps its life by eating negative entropy" [1,2]. If a living body is regarded as lying in a stationary state, an entropy change in it should be zero, i.e. $dS = d_oS + d_iS = 0$ (see equation (3-3)). On the other hand, irreversible processes in a living body necessarily produce entropy and hence $d_iS > 0$. Accordingly we obtain $d_oS < 0$ for a living body, namely, an entropy change by entropy transport in a living thing should be negative, in agreement with Schrödinger's claim.

In relation to Schrödinger's claim, attention was also paid to the concept of information. For example, Shannon took negative entropy as giving a measure of information [1-3] and in 1949 defined information I in the form of $I = -S/k$ where S is Boltzmann's entropy in a general form and k is the Boltzmann constant. This equation means that a decrease in entropy is equivalent to an increase in information. As entropy can be regarded as giving a measure of uncertainty, a decrease in entropy means an increase in certainty, i.e. an increase in information.

With the progress of study, it has become clear that it is important to divide non-equilibrium systems into two regions: the linear-response region and the nonlinear-response region, as shown in Figure 3-1. In the linear-response region, the law of linear response given by equation (3-10) holds while in the nonlinear-response region it does not hold. Then, much attention has been paid to the nonlinear-response region. This is most probably because experimental studies have revealed that non-equilibrium systems in this region demonstrate a variety of dynamic self-organizing phenomena such as oscillations and spatiotemporal pattern formations *against* the second law of thermodynamics [2-7], as will be explained in the next section. Non-equilibrium systems in the linear-response region demonstrate no such phenomena.

Furthermore, based on such studies, complexity science has recently proposed that life exists near the edge of chaos [2-4], where chaos

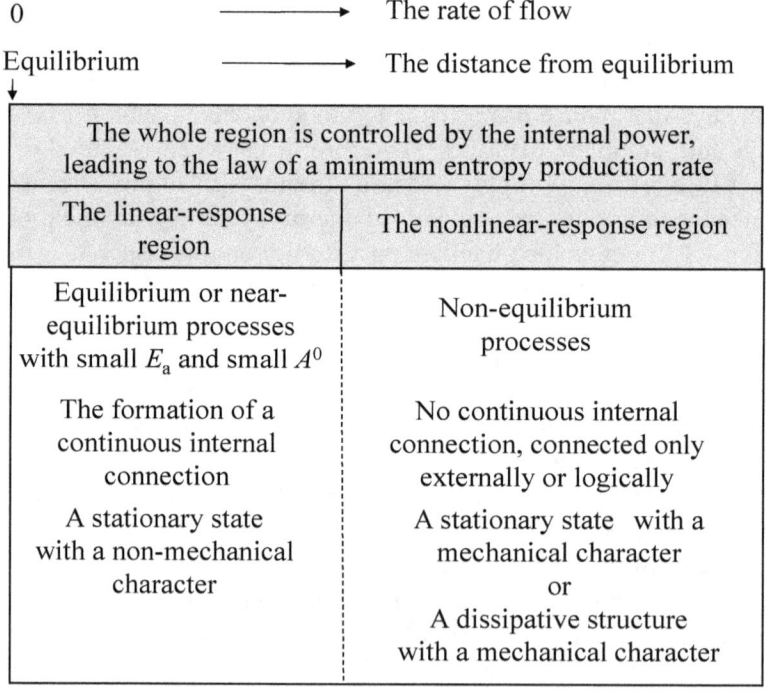

Figure 3-1 The linear-response region and the nonlinear-response region of a non-equilibrium system, depicted as a function of the distance from equilibrium (Details of the figure are explained in the text).

refers to entirely random behavior demonstrated by a system that obeys a deterministic law (see Section 3.1.2). This proposal can be explained as follows. Through the study of chaos, complexity science has revealed that an increase in complexity leads to the conversion of the physical deterministic world to the chaotic one under an appropriate condition. Naturally, life cannot exist in the physical deterministic world of a mechanical character. Life also cannot exist in the chaotic world in which there is no information. However, just before reaching the chaotic world, there is a world with a non-chaotic and non-deterministic character, in which information can play a leading role. This is the edge of chaos. Complexity science asserts that life exists here.

In contrast to such preceding studies, this book proposes a novel idea that highly ordered structures in living organisms are produced in the linear-response region according to the second law of thermodynamics, as already mentioned in Section 1.1.2. Such a drastic change has been brought about by taking into account the internal invisible world. Details will be explained in Section 3.2.

3.1.2 A Stationary State and a Dissipative Structure

A stationary state in a non-equilibrium system placed under constant thermodynamic forces

When a non-equilibrium system is not controlled externally, i.e. when there is no constant thermodynamic force in it, it spontaneously changes into an equilibrium system, as discussed in Section 2.2. On the other hand, when a non-equilibrium system is placed under constant thermodynamic forces, it spontaneously changes to *a stationary state* if no dissipative structure appears (see Section 3.2.1 for the validity of this argument). Here, a stationary state refers to a macroscopically unchanging stable state lying in harmony with the surroundings, in which all kinds of flow are kept constant everywhere in a system. A dissipative structure is explained later. When a non-equilibrium system includes various kinds of flow processes, they are usually coupled with one another in a stationary state.

As a simple example, let us consider a stationary state in which heat conduction and substance diffusion are coupled. Some empirical laws are known for transport phenomena with no coupling. As to thermal conduction, Fourier's law is known, which says that heat flux, J_h, is in proportion to the gradient of temperature, T.

$$J_h = -\kappa \, (\partial T / \partial x) \qquad (3\text{-}15)$$

where κ is the heat conductivity and x the spatial coordinate. This law holds for simple heat transport in a system in which no substance diffusion occurs. On the other hand, as to substance diffusion, Fick's law is known. It says that the flux of substance diffusion, J_d, is in proportion to the gradient of the substance density, N.

$$J_d = -D \, (\partial N/\partial x) \qquad (3\text{-}16)$$

where D is the diffusion coefficient. This law holds for diffusion in a constant temperature in which no heat transport occurs.

The coupling of heat conduction and substance diffusion occurs, for example, when a nitrogen gas, sealed in a glass tube, is placed between two heat baths with different temperatures, as schematically illustrated in Figure 3-2. In this system, both heat conduction and substance diffusion can occur. Just after a sealed nitrogen gas has been placed between heat baths, the rate of heat flow in it changes in a complex manner. However, it soon becomes constant. Namely, a stationary state is attained in heat flow. Interestingly, a gradient in the nitrogen-gas density is induced in such a stationary state (see Figure 3-2), indicating that coupling between heat conduction and substance diffusion really occurs. This result can be explained by using equation (3-10), i.e. by using the following equations.

$$J_h = L_{hh} X_h + L_{hd} X_d \qquad (3\text{-}17a)$$

$$J_d = L_{dh} X_h + L_{dd} X_d \qquad (3\text{-}17b)$$

where X_h and X_d are thermodynamic forces for heat conduction and gas diffusion, respectively. In a stationary state, X_h and X_d and hence J_h and J_d are constant everywhere, as mentioned earlier. Besides, in this case

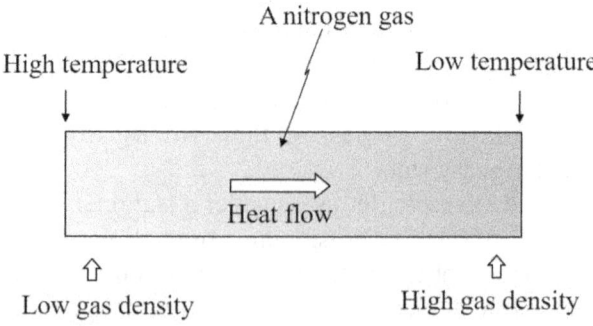

Figure 3-2 A stationary state achieved when a sealed nitrogen gas is placed between two heat baths with different temperatures.

they do not depend on the spatial position because of the symmetry of a system. The off-diagonal terms in equations (3-17a) and (3-17b) represent the coupling of substance diffusion and heat conduction. For example, the second term of the right-hand side of equation (3-17a) indicates that heat flow is caused by the diffusion of heated or cooled nitrogen gas. In a stationary state, no net substance diffusion occurs, i.e. $J_d = 0$ and thus $X_d = -(L_{dh}/L_{dd})X_h \neq 0$ is obtained from equation (3-17b). The emergence of non-zero X_d indicates that a gradient in the nitrogen-gas density is produced by coupling.

The law of a minimum entropy production rate in a stationary state

One of the most important achievements in thermodynamics of non-equilibrium systems will be the discovery of the law of a minimum entropy production rate in a stationary state [1,2,8]. Prigogine showed in 1947 that a stationary state of a non-equilibrium system placed under constant thermodynamic forces is characterized by a minimum entropy production rate. Namely, based on the law of linear response (equation (3-10)) and Onsager's reciprocity theorem (equation (3-11)), Prigogine showed that the following equations hold for variations in all thermosdynamic forces X_k (except X_l which are kept constant) in a stationary state.

$$\partial \sigma_{system} / \partial X_k = 0 \qquad (3\text{-}18)$$

Furthermore, Prigogine revealed that under the same assumptions as the above the rate of entropy production σ_{system} by irreversible processes *within* a closed or open system spontaneously decreases with time and finally becomes zero at a stationary state.

$$d\sigma_{system} / dt \leq 0 \qquad (3\text{-}19)$$

Equations (3-18) and (3-19) indicate that σ_{system} takes a minimum in a stationary state and also that a stationary state is a stable state. Then, we can say as follows.

> *In a macroscopic non-equilibrium closed or open system placed under constant thermodynamic forces, the rate of entropy production by irreversible processes within it spontaneously decreases and attains a minimum when it has reached a stationary state.*

This is called Prigogine's theorem or the law of a minimum entropy production rate. According to Prigogine's treatment, the theorem should only hold for a stationary state lying close to equilibrium because it is deduced based on Onsager's reciprocity theorem that is obtained by the use of the theory of fluctuation. However, actually, this theorem holds for every stationary state including one lying far from equilibrium, as will be discussed in Section 3.2.1 (see also Figure 3-1). The law of a minimum entropy production rate is not specific to the linear-response region.

Interestingly, equation (3-18) can be easily proved for a sealed nitrogen-gas system placed between two heat baths. In this case, the temperatures of two heat baths are kept constant and hence X_h is constant. In addition, X_h and X_d and hence J_h and J_d do not depend on the spatial position, as mentioned earlier. Therefore, by using the relation of $L_{hd} = L_{dh}$ (Onsager's reciprocity theorem), we obtain from equation (3-13)

$$\sigma_{system} = \{(L_{hh} X_h + L_{hd} X_d) X_h + (L_{dh} X_h + L_{dd} X_d) X_d\} V$$
$$= \{L_{hh} X_h^2 + 2 L_{hd} X_h X_d + L_{dd} X_d^2\} V \qquad (3\text{-}20a)$$

$$\partial \sigma_{system} / \partial X_d = 2(L_{hd} X_h + L_{dd} X_d) V = 2 J_d V \qquad (3\text{-}20b)$$

where V is the volume of a system. In a stationary state, $J_d = 0$ and hence $\partial \sigma_{system} / \partial X_d = 0$.

As another example of a stationary state, Figure 3-3 schematically illustrates a stationary electric current formed when a constant voltage is applied between two local spots in a metal (e.g. iron) plate. In this case, the thermodynamic force X (i.e. the gradient of the electric potential or the electric field, divided by the temperature T, see a footnote #2 of Section 3.1.1) and the flux J (the electrical current density) depend on the spatial position. Thus, a stationary electrical current is formed so that the rate of entropy production given by

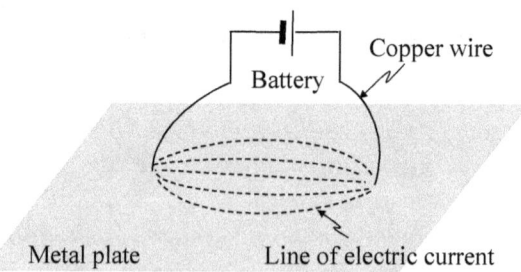

Figure 3-3 A stationary electric current formed in a metal plate (a main part).

$$\sigma_{\text{system}} = \int JX\,dx\,dy\,dz \qquad (3\text{-}21)$$

takes a minimum. Interestingly, we can prove that σ_{system} in equation (3-21) really takes a minimum by making use of Thomson's theorem for the electrostatic field [9]. W. Thomson showed that the energy U of the electrostatic field defined by

$$U = (1/2)\int DE\,dx\,dy\,dz \quad (D = \varepsilon_0 E) \qquad (3\text{-}22)$$

takes a minimum when the electric field E is given by the Coulomb field, where D is the electric flux density or the electric displacement and ε_0 the permittivity of vacuum. As is well known, the electric field X and the electrical current density J in equation (3-21) are expressed mathematically in the same way as E and D in the Coulomb field for a pair of positive and negative point charges placed apart.

The rate of a chemical reaction and the law of linear response

Chemical reactions are main processes in a living organism. Therefore, it is crucially important to clarify properties of a reaction system in detail from various points of views.

The rate of a chemical reaction in general does not follow the law of linear response (equation (3-10)), in contrast to the rates of transport phenomena. However, under a particular condition, even the rate of a chemical reaction follows the law of linear response. In general, a non-equilibrium reaction system placed under constant thermodynamic forces displays largely different behavior, depending on whether it is in the linear-response region or not, as will be explained later (see also Figure 3-1). Therefore, it is quite important to clarify under what condition the rate of a chemical reaction follows the law of linear response.

For clarifying the condition, let us consider a chemical reaction expressed as follows.

$$\nu_p R_p + \nu_q R_q + \cdots\cdots + \nu_r R_r \rightarrow \nu_s R_s + \cdots\cdots + \nu_t R_t \qquad (3\text{-}23)$$

This reaction equation shows that chemical substances, R_p, R_q, $\cdots\cdots R_r$, react with one another and produce chemical substances R_s, $\cdots\cdots R_t$. In the discussion given below, the stoichiometric coefficient ν_m is taken as a positive integer for reaction products, R_s, $\cdots\cdots R_t$, and a negative integer for reactants, R_p, R_q, $\cdots\cdots R_r$. The subscripts m ($= p, q, r, s, t$, etc., a positive integer) express the kinds of chemical substances. In general, a chemical reaction such as equation (3-23) consists of a number of partial chemical reactions and proceeds via many intermediate compounds. A chemical reaction that can no further be divided into partial reactions is called *an elementary reaction*. An important characteristic of an elementary reaction is that a reaction equation expresses a reaction mechanism. For example, if reaction (3-23) is an elementary reaction, it indicates that molecules R_p, R_q, ······ R_r *collide* with one another and produce molecules R_s, ······ R_t. Therefore, in this case, we can say that the rate of reaction (3-23) is in proportion to the concentrations of reactants, R_p, R_q, ······ R_r.

For a chemical reaction, the rate of flow (or the flux), J, in equation (3-9) is given by the rate of a chemical reaction, $\mathcal{R}$, and the thermodynamic force, X, is given by the reaction affinity, A, divided by T, as will be explained below. The rate of a chemical reaction, $\mathcal{R}$, is defined as follows[1,2].

$$\mathcal{R} \equiv (1/V)(d\xi/dt) \qquad (3\text{-}24)$$

where V is the volume of a reaction system and t is time. The quantity ξ expresses the degree of extent of reaction and is defined for reaction (3-23) as follows

$$\Delta\xi = \Delta n_m / \nu_m \qquad (3\text{-}25)$$

where Δn_m is a change of the amount of substance for a chemical substance R_m by reaction (3-23). Thus, if reaction (3-23) consists of the forward and the backward processes, both of which are an elementary reaction, the rate of this reaction, $\mathcal{R}$, is given as

$$\mathcal{R} = k_f \prod_{m=\text{reactants}} (n_m/V)^{|\nu_m|} - k_b \prod_{m=\text{products}} (n_m/V)^{\nu_m} \qquad (3\text{-}26)$$

where k_f and k_b are the rate constant for the forward and the backward processes, respectively. They are independent of the concentrations of reactants and products, only depending on reaction conditions such as the temperature, the pressure, the used solvent, and the presence or the absence of a catalyst. By using these ξ and $\mathcal{R}$, the reaction affinity A is defined as follows[1,2].

$$\delta Q' = T\, d_i S = (1/V)\, A\, d\xi \geq 0 \qquad (3\text{-}27a)$$

$$\sigma = d_i S / dt = (1/T)\, \delta Q'/\delta t = (1/T)\, A\mathcal{R} \geq 0 \qquad (3\text{-}27b)$$

Equation (3-27a) is called De Donder's inequality. The comparison of equation (3-27a) and equation (3-7) or (3-8) indicates that the reaction affinity A can be expressed as follows.

$$A = -(\partial G/\partial \xi)_{T,p} = -\sum \nu_m (\partial G/\partial n_m)_{T,p} = -\sum_m \nu_m \mu_m \qquad (3\text{-}28).$$

The second equality of this equation is obtained by using equation (3-25). Also, the comparison of equations (3-27b) and (3-12) indicates that (A/T) represents the thermodynamic force X for reaction (3-23).

Equation (3-27b) indicates that if A is positive ($A > 0$), $\mathcal{R}$ is also positive ($\mathcal{R} > 0$), i.e. reaction (3-23) occurs in the forward direction. If A is negative ($A < 0$), $\mathcal{R}$ is negative ($\mathcal{R} < 0$), and reaction (3-23) occurs in the backward direction. If $A = 0$, no reaction happens, i.e. equilibrium is achieved. On the other hand, when reaction (3-23) occurs simultaneously with other reactions, i.e. it is coupled with other reactions, σ is given as follows[1].

$$\sigma = d_i S / dt = (1/T)\, \delta Q'/\delta t = (1/T) \sum_k A_k\, \mathcal{R}_k \geq 0 \qquad (3\text{-}29)$$

In such a case, reaction (3-23) can occur in the forward direction even if A for it is negative if the condition of $\sum_k A_k\, \mathcal{R}_k \geq 0$ is met for all coupled reactions. This implies that an up-hill reaction can occur by coupling with down-hill reactions. This mechanism plays an important role in reactions in living bodies (see Section 4.2).

Now, let us consider the condition that the law of linear response holds. In equilibrium, $\mathcal{R} = 0$, and thus from equation (3-26) we obtain

$$K_{\text{kin}} = [\prod_{m=\text{products}} (n_m/V)^{vm} / \prod_{m=\text{reactants}} (n_m/V)^{|vm|}]_{\text{eq}} \qquad (3\text{-}30\text{a})$$

$$= k_f / k_b \qquad (3\text{-}30\text{b}).$$

K_{kin} is the equilibrium constant for reaction (3-23), where the subscript, kin, means that it is the equilibrium constant obtained from chemical kinetics. The subscript, eq, indicates that the concentrations (n_m/V) in brackets are values at equilibrium.

The chemical potential, μ_A, for molecule A is expressed, under an approximation of using the concentration instead of the activity, as follows.

$$\mu_A = \mu_A^0 + RT \ln\{(n_A/V)/C^0\} \qquad (3\text{-}31)$$

where μ_A^0 is the chemical potential of a chemical substance A at the standard state, C^0 the unit concentration, and R the gas constant. Thus, the reaction affinity A for reaction (3-23) is expressed, from equation (3-28), as follows.

$$A = A^0 - RT \ln [\prod_{m=\text{products}} (n_m/V)^{vm} / \prod_{m=\text{reactants}} (n_m/V)^{|vm|}] \qquad (3\text{-}32)$$

$$A^0 = -\sum_m v_m \mu_m^0 \qquad (3\text{-}32\text{a})$$

In equation (3-32), the terms connected with C^0 is omitted. At equilibrium, $A = 0$, and therefore we obtain from equation (3-32)

$$A^0 = RT \ln [\prod_{m=\text{products}} (n_m/V)^{\nu m} / \prod_{m=\text{reactants}} (n_m/V)^{|\nu m|}]_{\text{eq}} \quad (3\text{-}33\text{a})$$

$$= RT \ln K_{\text{kin}} = RT \ln (k_f/k_b) \quad (3\text{-}33\text{b})$$

From equations (3-32), (3-33a) and (3-33b), we obtain a relation

$$\prod_{m=\text{products}} (n_m/V)^{\nu m} / \prod_{m=\text{reactants}} (n_m/V)^{|\nu m|} = (k_f/k_b) \exp(-A/RT)$$
$$(3\text{-}34).$$

Therefore, the rate of reaction (3-23), $\mathcal{R}$, given by equation (3-26), is expressed as follows.

$$\mathcal{R} = k_f \prod_{m=\text{reactants}} (n_m/V)^{|\nu m|} \{1 - \exp(-A/RT)\} \quad (3\text{-}35)$$

This gives the relation between the rate of flow (the rate of a chemical reaction), $\mathcal{R}$, and the thermodynamic force, A/T, for reaction (3-23). Evidently, $\mathcal{R}$ is not in proportion to A/T, i.e. the law of linear response does not hold. However, if $|A/RT| \ll 1$ is met, equation (3-35) is converted to the following equation.

$$\mathcal{R} = \{k_f \prod_{m=\text{reactants}} (n_m/V)^{|\nu m|}/R\} (A/T) \quad (3\text{-}36)$$

This equation indicates that $\mathcal{R}$ is in proportion to A/T, i.e. the law of linear response holds. The condition of $|A/RT| \ll 1$ is met if a reaction system is close to equilibrium, as can be seen from equations (3-32) and (3-33a). Thus, we can conclude that *the rate of a chemical reaction, $\mathcal{R}$, obeys the law of linear response when a reaction system is close to equilibrium.*

Now, what conditions does a chemical reaction have to meet when it keeps equilibrium or near-equilibrium in a non-equilibrium system? The following two conditions need to be met.

1. The rate constants for the forward and the backward processes, k_f and k_b, are enough large to keep the concentration ratio of reactants to products, $\prod_{m=\text{reactants}} (n_m/V)^{|\nu m|}/\prod_{m=\text{products}} (n_m/V)^{\nu m}$, nearly constant.
2. The reaction affinity in the standard state, A^0, is not large in the absolute value, i.e. $|A^0|$ is not large, for example, $|A^0/RT| \leq 1$.

Condition 1 is explained as follows. In general, the concentrations of reactants and products of a chemical reaction are affected by various processes other than the chemical reaction in question, such as preceding or following reactions, competitive reactions, diffusion, and so on. Here, if the rate constants for the forward and the backward processes, k_f and k_b, are very large and if the rates of the forward and the backward processes are very high, reactants and products keep a good balance and the concentration ratio of reactants to products, $\prod_{m=\text{reactants}} (n_m/V)^{|vm|} / \prod_{m=\text{products}} (n_m/V)^{vm}$, is always kept nearly constant. This means that a chemical reaction is in or near equilibrium.

Condition 2 provides an additional condition. From equations (3-33a) and (3-33b), we obtain

$$k_f/k_b = \exp(A^0/RT) \qquad (3\text{-}37).$$

This equation indicates that if A^0 is positive and large, the ratio (k_f/k_b) becomes extremely large. This means that k_b becomes extremely small because the rate constant of any elementary chemical reaction and hence k_f cannot exceed a critical value, A_f, as explained just later (see equation (3-38)). The same argument holds when A^0 is negative and $|A^0|$ is large. Accordingly, if $|A^0|$ is large, it is impossible that both k_f and k_b are simultaneously very large, thus leading to the breakdown of condition 1. This is why condition 2 is necessary.

What chemical reaction has a large rate constant k? The rate constant k for an elementary reaction is given by the Arrhenius equation

$$k = A_f \exp(-E_a/RT) \qquad (3\text{-}38)$$

where A_f is the frequency factor and E_a is the activation energy, both being quantities specific to an individual reaction. In general, molecular structures of reactants are largely distorted in the course of reaction. The activation energy E_a refers to energy required for such molecular distortion. Namely, it represents an energy barrier in chemical reaction. When E_a is sufficiently small compared with RT, the rate constant k is very large though it has the maximum value, A_f, as mentioned above Thus, *small E_a and small A^0 are key factors for obtaining a chemical reaction whose rate stably obeys the law of linear response.*

The most effective way to decrease E_a is to use a suitable catalyst. Various enzymes in living bodies are a typical example of effective catalysts. In fact, it is known that almost all chemical reactions in living bodies have quite large rate constants.

The difference in properties of a non-equilibrium system between the linear-response region and the nonlinear-response one

A non-equilibrium reaction system placed under constant thermodynamic forces displays largely different behavior, depending on whether it is in the linear-response region or not, as mentioned earlier. There are two aspects in the difference. Firstly, a system in the linear-response region demonstrates only a stationary state while a system in the nonlinear-response region demonstrates a stationary state or a dissipative structure, depending on conditions (see Figure 3-1). Secondly, properties of a stationary state are essentially different between the linear-response and the nonlinear-response regions (see Figure 3-1).

At first, let us briefly consider the second aspect, i.e. the difference between a stationary state in the linear-response region and one in the nonlinear-response region. Details of this issue will be discussed in Section 3.2.2. The foregoing discussion of the rate of a chemical reaction teaches us what difference stationary states in these two regions have. As mentioned just earlier, the rate of a chemical reaction obeys the law of linear response when it is close to equilibrium. Now, a chemical reaction is close to equilibrium when the rates of the forward and the backward processes are high enough to keep the concentration ratio of reactants to products nearly constant.

Accordingly, we can say that for a chemical reaction which obeys the law of linear response or which lies in the linear-response region, reactants and products are connected with each other by high-rate forward and backward reactions and exist simultaneously in a nearly constant concentration ratio in any situation. This means that they are fused into unity or are continuously connected with each other. Thus, the linear-response region is well characterized by the formation of a continuous internal connection of reactants and products, as illustrated in Figure 3-1.

In relation to the above argument, it is important to note that transport processes such as heat conduction, substance diffusion, and so on in a gas or a liquid in general follow the law of linear response, as discussed earlier. This means that microscopic processes such as intra- and inter-molecular energy transfer in translational, rotational and vibrational motion of constituent molecules in a gas or a liquid, spatial movement of them, etc. have enough high rates and thus constituent molecules are in or near equilibrium with respect to such modes of motion and are continuously connected with one another. This argument is reasonable because it is expected that such modes of mode have very small activation energy E_a and very small reaction affinity A^0 at the standard state.

The formation of a dissipative structure in the nonlinear-response region

Next, let us consider the first aspect, i.e. the formation of a dissipative structure in the nonlinear-response region. A large number of studies have been made on non-equilibrium, nonlinear and open reaction systems [*1] in the latter half of the 20th century [5-7]. Interestingly, such reaction systems sometimes demonstrate spontaneous formation of a diversity of self-sustaining dynamic order such as oscillations and spatiotemporal pattern formation. Such spontaneously formed dynamic order is called a "dissipative structure" or "self-organization" according to Prigogine's proposal [5]].

> [*1] The term of nonlinearity is in general used in two different meanings: (1) nonlinear response and (2) nonlinear kinetics. Nonlinear response means that the rate of a chemical reaction or in general the rate of flow does not follow the law of linear response (equation (3-10)). On the other hand, nonlinear kinetics means that the rate equation for a chemical reaction, such as equation (3-26), is described by a nonlinear differential equation with respect to the concentrations of reactants. Autocatalysis described just later is an example of nonlinear kinetics.

A famous example of chemical reactions showing dissipative structures in a homogeneous solution is Belousov-Zhabotinsky (BZ) reaction. It is an oxidation reaction of an organic acid by bromate ions in the presence of a catalyst such as ceric ions in an aqueous acidic solution. For example, the oxidation of malonic acid is roughly described as follows.

$$5CH_2(COOH)_2 + 6BrO_3^- + 2H^+ \rightarrow 3BrCH(COOH)_2 + 2HCOOH + 4CO_2 + 5H_2O \quad (3\text{-}39)$$

This reaction is very complex, including many intermediate compounds and proceeding via a large number of steps, and its mechanism has not yet been fully clarified.

The oscillating behavior of such a reaction was for the first time discovered by B. P. Belousov in the 1950's. He found that an oxidation reaction of citric acid by bromate ions with ceric ions as a catalyst showed a periodic change in solution color for a long time. However, unfortunately, his finding looked too strange in those years. His paper was not accepted for publication in any academic journal dur-

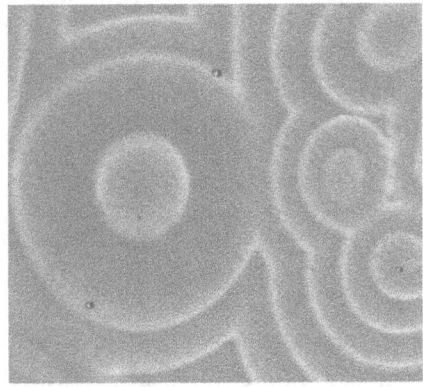

Figure 3-4 A propagating pattern of solution color in a BZ reaction.

ing his life. The reason was that it was improbable for a chemical reaction to oscillate spontaneously. It is said that Belousov died in low spirits. Fortunately, his result was reinvestigated and reproduced by the same Russian scientist, A. M. Zhabotinsky, in the 1960's. In addition, Zhabotinsky discovered that Belousov's reaction showed a propagating spatiotemporal pattern of solution color when a reaction solution was placed in a glass vessel in the form of a thin layer. Figure 3-4 shows an example of such a propagating pattern.

Later detailed studies have revealed that spontaneous formation of dynamic order called a dissipative structure or self-organization in a reaction system is attained if it meets the following conditions: [5-7]
1. The coupling of chemical reactions and diffusion occurs.
2. Reaction kinetics includes at least one autocatalytic process.
3. A reaction system is in the nonlinear-response region.

The term of coupling was already explained. For a chemical reaction in a homogeneous solution, the coupling of a chemical reaction and substance diffusion in general always occurs. Autocatalysis in condition (2), which is also called a positive feedback mechanism, means that a product of a chemical reaction has an effect of enhancing the rate of the original chemical reaction. For example, chemical reaction (3-39) is reported to include the following autocatalytic process.

$$HBrO_2 + BrO_3^- + 2\ Ce^{3+} + 3\ H^+$$
$$\rightarrow \cdots \rightarrow 2\ HBrO_2 + 2\ Ce^{4+} + H_2O \qquad (3\text{-}40)$$

In this reaction, one $HBrO_2$ molecule and one BrO_3^- molecule react and produce two $HBrO_2$ molecules and therefore the concentration of $HBrO_2$ and hence the reaction rate shows an accelerative increase with the progress of the reaction if sufficient amounts of BrO_3^- and Ce^{3+} are present. Namely, $HBrO_2$, which is a reaction product, has an effect of enhancing the rate of the original chemical reaction. Reaction (3-40) can be rewritten as follows.

$$BrO_3^- + 2\ Ce^{3+} + 3\ H^+ \rightarrow \cdots \rightarrow HBrO_2 + 2\ Ce^{4+} + H_2O \qquad (3\text{-}41)$$

In this equation, $HBrO_2$ acts as a catalyst for increasing the rate constant of the reaction, though this effect is not expressed in the reaction equation.

Condition (3) indicates that a dissipative structure or self-organization appears only when a reaction system lies in the nonlinear-response region. Now, let us assume that a reaction system is gradually shifted from the equilibrium state to a non-equilibrium state far from equilibrium (i.e. in the right direction in Figure 3-1). A dissipative structure suddenly starts to appear at a critical point at which a reaction system moves from the linear-response region to the nonlinear-response one. Before reaching the critical point, a reaction system is in a stationary state even if it includes autocatalytic processes.

Apart from oscillations and propagating spatiotemporal patterns, BZ reactions show a non-propagating non-oscillating spatial pattern, called the Turing pattern, under an appropriate condition. An English mathematician, A. Turing, who proposed the "Turing machine", tried to mathematically simulate various patterns appearing on bodies of animals such as zebras, leopards and tropical fishes. Then, he predicted in 1952 that the coupling of chemical reactions and diffusion can produce similar periodic spatial patterns. The prediction was verified experimentally in a BZ reaction in 1991.

Furthermore, BZ reactions are reported to show chaos. Here, chaos is defined as entirely random behavior demonstrated by a system that obeys a deterministic law, as already mentioned in Section 3.1.1. Such a strange situation occurs because a small difference in the initial condition increases with time in a chaotic system, resulting in an unexpectedly large difference in a later time. For this reason, it is impossible to predict results on a long-time scale because actually we cannot know the initial condition with the absolute accuracy.

The appearances of oscillations, propagating spatiotemporal patterns, steady Turing patterns and chaos, together with the synchronization of oscillations called the entrainment, are characteristic of nonlinear chemical reaction systems. These phenomena have attracted growing attention as models for explaining high functions in living bodies or others.

3.1.3 A Reaction Network

Apart from dynamic order such as self-organization or a dissipative structure, a chemical reaction system demonstrates another interesting feature in that it often forms a large network of chemical reactions. In this section we consider two representative examples of a reaction network reported thus far.

The reaction network with collective catalytic actions

S. Kauffman and his group in Santafe institute, U.S.A., investigated properties of a reaction network from the standpoint of complexity science, by using computer simulation, and disclosed [4] that a reaction network comes to have "collective catalytic actions" when it becomes sufficiently complex, as already briefly explained in Section 1.1.2. An important characteristic of a reaction network with collective catalytic actions is that it has the ability to replicate itself and sustain itself. Thus, Kauffman asserts that such a reaction network can act as a precursor of the first living organism on the primitive earth.

According to Kauffman's book (1995)[4], the reaction network with collective catalytic actions is explained as follows. He starts with an explanation of a game called the "random graph". Let us assume that there are large numbers of buttons and threads. Two buttons are randomly picked out from a set of buttons, they are connected with each other with a thread, and then they are put back to the original set of buttons. Next, in the same way, two buttons are randomly picked out from the set of buttons, connected with each other with a thread, and put back to the original set of buttons. Such a procedure is repeated many times. What happens by the procedure?

In case the number of threads (N_t) is much smaller than the number of buttons (N_b), most of buttons are in no connection or only in a pair. As the ratio N_t/N_b is in- creased, the number of paired buttons increases and in addition small clusters of buttons in which three or more buttons are connected start to be formed. Surprisingly, when the ratio N_t/N_b is further increased and approaches 1/2, a so-called "phase transition" happens, as schematically illustrated in Figure 3-5. Namely, the number of buttons in the *largest* cluster steeply increases in the region of $N_t/N_b \cong 1/2$ and comes to be nearly equal to the number of all buttons N_b in the region of $N_t/N_b \geq 1/2$. This means that almost all buttons are connected in the form of one cluster in the region of $N_t/N_b > 1/2$, in sharp contrast to the region of $N_t/N_b < 1/2$. Another remarkable fact is that the slope of increase of the number of buttons in the largest cluster near $N_t/N_b = 1/2$ becomes steeper as N_b increases. Namely, the characteristic of phase transition becomes more prominent as N_b increases. This result indicates that N_t/N_b and N_b play a critically important role in the phase transition.

Based on the result of the "random graph", Kauffman next con-

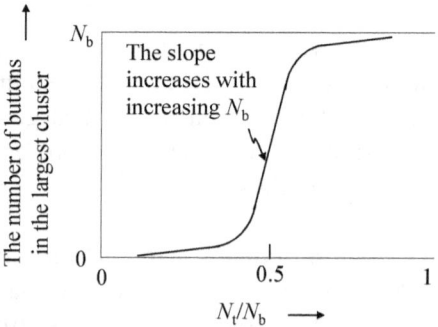

Figure 3-5 The number of buttons in the largest cluster vs. the ratio N_t/N_b.

siders the properties of a reaction network called the "reaction graph" in which chemical substances are connected with one another by chemical reactions. Chemical substances and chemical reactions in the reaction graph play the same role as buttons and threads in the random graph, respectively. Therefore, the same conclusion as in the random graph is expected to be obtained in the reaction graph. However, there is a little difference. A thread in the random graph only connects two buttons, while chemical reactions connect chemical substances in various ways. This is because there are various types of chemical reactions such as one-molecular reactions (A → B, A → B + C, and so on) and bimolecular reactions (A + B → C, A + B → C + D, and so on), where A, B, C and D refer to different chemical substances. Therefore, chemical substances in the reaction graph are connected with one another in a complex manner.

Now, a set of chemical substances connected by chemical reactions forms a reaction network. Then, Kauffman investigated a possibility of forming a special type of reaction network, i.e. "the reaction network with collective catalytic actions" by using computer simulation. A reaction network of this type meets the following conditions: (1) all chemical substances in the reaction network are connected with others by chemical reactions affected by catalysts and (2) all catalysts are supplied by chemical substances in the reaction network or by source substances, where source substances are chemical substances entering into the reaction network from the outer world.

Figure 3-6 schematically illustrates a simple model of the reaction network with collective catalytic actions. Chemical substances are

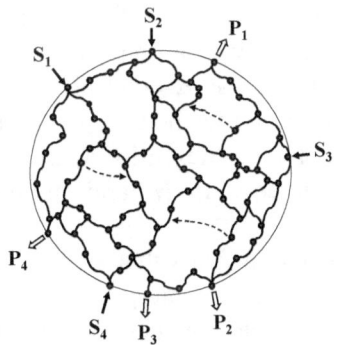

Figure 3-6 A schematic model of a reaction network with collective catalytic actions.

depicted by solid dots and curves connecting solid dots express chemical reactions. Arrays with dashed curves indicate some examples of catalytic acts of chemical substances on other reactions. S_1, S_2, S_3, $\cdots$ are source substances entering into the reaction network from the outer world, while P_1, P_2, P_3, $\cdots$ are reaction products which are discharged from the reaction network.

For investigating a possibility of forming the reaction network with collective catalytic actions, Kauffman used the following model system: (1) monomers M and N and dimers MM, NN, MN, and NM are injected from the outer world as source substances, where M and N refer to, for example, two kinds of nucleotides, (2) source substances and their reaction products all react with one another and form adducts such as MNN, MMNN and MMNNMNN, and (3) some of chemical substances in the reaction network act as catalysts for chemical reactions in the reaction network. By the second condition, various chemical substances are produced and a large reaction network is formed. By this model, Kauffman reached the conclusion that the reaction network with collective catalytic actions necessarily emerges if the number of the kinds of chemical substances in a reaction network becomes sufficiently large.

Kauffman explains this conclusion as follows. Let us designate the number of the kinds of chemical substances, that of the kinds of chemical reactions, and that of the kinds of catalyzed chemical reactions N_s, N_r, and N_r^c, respectively. As N_s in a reaction network increases, the ratio N_r/N_s always increases. As the ratio N_r/N_s increases, N_r^c also increases if it is assumed that the probability (P) at which a chemical substance in a reaction network acts as a catalyst for a chemical reaction in the same reaction network is constant. Now, when the ratio N_r^c/N_s approaches a certain critical value, say, 1/2, a phase transition occurs and a large network of chemical reactions with collective catalytic actions suddenly emerges, just in the same way as a phase transition in the random graph of Figure 3-5. Kauffman emphasizes that the ratio N_r/N_s rapidly increases with an increase in N_s. For this reason, even if the probability (P) is extremely small, say, $1/10^6$, the ratio N_r^c/N_s can approach a certain critical value by an increase in N_s and therefore a phase transition necessarily occurs.

In this way, Kauffman affirms that a reaction network with collective catalytic actions necessarily emerges if the number of the kinds of chemical substances in a reaction network becomes sufficiently large, namely, if a reaction network becomes sufficiently "complex". This is the only necessary condition for the emergence of collective catalytic actions. Then, Kauffman's final conclusion is that a precursor of the first living organism on the primitive earth emerged when a great variety of compounds spontaneously gathered together.

The autopoiesis system

In the 1970's, Maturana and Varela proposed a reaction network called the autopoiesis system, where autopoiesis means "self-production". All living things consist of cells, i.e. sphere-like compartments, each of which has the boundary made of a semi-permeable membrane. In a cell, a large number of biopolymers such as proteins, RNA and DNA carry out metabolic reactions. The autopoiesis system represents a simplified conceptual model of such a cell. To be accurate, the autopoiesis system is defined by Varela as follows [10]: (1) it has the semi-permeable boundary, (2) substances forming the boundary are produced in the inside of the system, and (3) the body of the system surrounded by the boundary includes reactions for reproducing components of the system. According to Varela, all reaction networks which reproduce themselves can be called the autopoiesis system.

The importance of the autopoiesis system is that a simple form of it can be synthesized in laboratories and the properties can be investigated experimentally. P. Luisi emphasizes this point in his book (2006) [10]. Figure 3-7 illustrates an example of an experimentally produced autopoiesis system of the simplest form, reported in Luisi's book. It is made of a reverse micelle of oleic acid, RCOOH, where R refers to $CH_3(CH_2)_7CH=CH(CH_2)_7-$. Oleic anhydride $(RCO)_2O$ in organic solvent enters into a water pool inside a reverse micelle and is hydrolyzed into

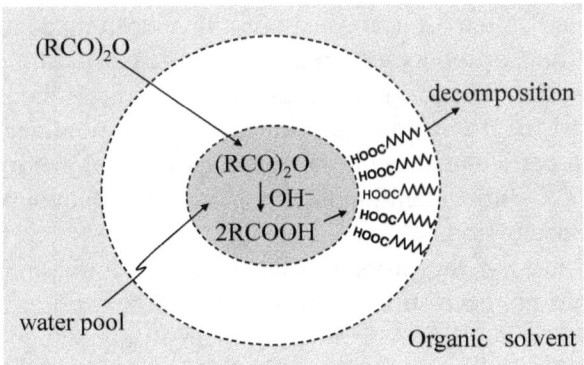

Figure 3-7 An autopoiesis system made of a reverse micelle of oleic acid.

oleic acid with OH⁻ as a catalyst. Resultant oleic acid is used as an element of a membrane forming the boundary. Some of oleic acid molecules in a membrane are decomposed by oxidation with osmium oxide (OsO_4). Let us designate the rate of the production of oleic acid as v_p and the rate of its decomposition as v_d. If v_p is equal to v_d, a reverse micelle maintain itself and realizes homeostasis. If v_p is larger than v_d, a reverse micelle grows, leading to self-reproduction, while if v_p is smaller than v_d, a reverse micelle diminishes and finally disappears. All cases can be realized experimentally by controlling the conditions of a system.

Luisi emphasizes that the autopoiesis system (i.e. a self-reproductive reaction network) has a diversity of important properties. It has the boundary made of a membrane and thus has the self. In addition, it reproduces its components by reactions in it. Thus, the autopoiesis system forms a reaction network with a recurrent inner structure and sustains itself. Such a system is an open system with respect to substance transfer but is a closed system from the point of view of system organization. Autonomous or self-structured properties of the autopoiesis system are determined by inner rules present in a reaction network. Thus, "emergence" can arise from here.

3.2 A New Approach to Non-equilibrium Systems

In the preceding sections we briefly surveyed previous studies about fundamental characteristic and structures and properties of non-equilibrium systems, with the focus placed on the clarification of basic principles about how living things produce ordered structures. As mentioned in Section 3.1.1, most studies have been made under the belief that living things produce ordered structures *against* the second law of thermodynamics. Therefore, main attention has been paid to dissipative structures arising from nonlinear kinetics in the nonlinear-response region [2-7] or a particular structure appearing from complexity in a network connection [2-4], little attention being paid to a stationary state in the linear-response region.

Unfortunately, the problem of how living things produce ordered structures has not been sufficiently solved to date, as briefly discussed in Section 1.1.2 and 3.1.1. A serious problem in preceding studies is that dynamic self-organization in the nonlinear-response region is of a mechanical character (see Section 3.2.3) and cannot explain the emergence of free independent harmonious spirit and wisdom in living things.

In this section we again consider fundamental characteristics and structures and properties of non-equilibrium systems, with the focus placed on the clarification of basic principles about how living things produce ordered structures. In sharp contrast to previous studies, this section shows that dynamic self-organization happens in a stationary state in the linear-response region according to the second law of thermodynamics, not *against* it. Interestingly, Prigogine, who discovered the law of a minimum entropy production rate in a stationary state in the linear-response region, had told nothing about the importance of a stationary state in this region. As is well known, he rather emphasized the importance of dissipative structures in the nonlinear-response region. It seems that the conclusions obtained in this section were difficult to reach by using mathematical equations alone. It was important to pay attention to the internal world we cannot be conscious of, which works as the origin of mathematical equations.

3.2.1 The Internal Ability in a Stationary State

The generalization of the law of a minimum entropy production rate to the nonlinear-response region

As mentioned in Section 3.1.2, Prigogine's theorem should only hold for a stationary state lying close to equilibrium according to his theoretical treatments. This is because he deduced the theorem based on Onsager's reciprocity theorem, which was obtained by the use of the theory of fluctuation. Amazingly, later experimental studies have disclosed that the law of a minimum entropy production rate holds even for a stationary state lying far from equilibrium, i.e. lying in the nonlinear-response region [1,2].

This finding is important, indicating wide applicability of the law of a minimum entropy production rate. However, it has simultaneously brought us a serious problem of how we can explain it theoretically. Naturally, the finding is impossible to explain theoretically by developing Prigogine's treatments. First of all, we cannot know how J_k depends on X_k in the nonlinear-response region (see equation (3-9)). To our delight, we can explain the above finding if we take into account the internal invisible world discussed in Chapter 2. Furthermore, this explanation leads to a striking conclusion that dynamic self-organization happens in a stationary state in the linear-response region lying close to equilibrium, as mentioned above.

Now, the above finding can be explained as follows. As argued in Section 2.1.3 and 2.2, electrons and atomic nuclei in a macroscopic system are controlled by the internal power to make them exist simultaneously in all possible spatiotemporal positions under given conditions. This conclusion does not depend on whether a macroscopic system is in equilibrium or not and whether it is isolated, closed or open, because the internal power comes from the wave nature of electrons and atomic nuclei. Accordingly, electrons and atomic nuclei in a closed or open non-equilibrium system placed under constant thermodynamic forces spontaneously spread in every direction and finally come to exist simultaneously in all possible spatiotemporal positions under given conditions if no dissipative structure appears. This means that such a non-equilibrium system spontaneously changes and finally reaches a stationary state, i.e. a macroscopically unchanging stable state lying in harmony with the surroundings.

Thus we can say that electrons and atomic nuclei exist simultaneously in *all* possible spatial positions under given conditions in a stationary state. This means that a closed or open non-equilibrium system placed under constant thermodynamic forces attains maximum entropy under given conditions in a stationary state.

> The above conclusion can be explained as follows. For a non-equilibrium macroscopic system lying not far from equilibrium, we can assume local equilibrium. Thus we can define the entropy of such a system by dividing it into small areas. The entropy of such a system can also be calculated in the following way. The entropy S of an equilibrium isolated system is defined by equation (2-11) as $S = k \ln W$, where W is the number of all allowed microscopic states for an isolated system. On the other hand, equation (2-15) indicates that the entropy of a *non-equilibrium* isolated system is given by the number of *really realized* allowed microscopic states instead of W. Therefore, the entropy S_{non} of a closed or open non-equilibrium system placed under constant thermodynamic forces can be calculated by the following equation.
>
> $$S_{non} = k \ln W_{non} \tag{3-42}$$
>
> where W_{non} is the number of *really realized* allowed microscopic states in a non-equilibrium isolated system, which has the same structure and properties as a closed or open non-equilibrium system placed under constant thermodynamic forces. W_{non} is in proportion to the size of the spatial area in which electrons and atomic nuclei simultaneously exist in a non-equilibrium system. Therefore, we can say that W_{non} and hence S_{non} take a maximum in a stationary state because electrons and atomic nuclei exist simultaneously in all possible spatial positions under given conditions in a stationary state, as mentioned above.

In a non-equilibrium system, entropy production necessarily occurs. An important point is that the attainment of the maximum entropy in a non-equilibrium system means the achievement of the minimum entropy production rate under the same condition. Thus, we can conclude from the above argument that the entropy production rate in a closed or open non-equilibrium system placed under constant thermodynamic forces takes a minimum in a stationary state.

A closed or open non-equilibrium macroscopic system placed under constant thermodynamic forces (and lying near a stationary state) spontaneously changes so that the entropy production rate within it decreases and reaches a stationary state when the entropy production rate takes a minimum under given conditions, irrespective of whether it is in the linear-response region or not, if no dissipative structure emerges.

The condition of "lying near a stationary state" is added because we can only say that the entropy production rate takes a minimum in a stationary state, as argued above. The above statement is the generalized form of the law of a minimum entropy production rate, deduced independently of the law of linear response (equation (3-10)) and Onsager's reciprocity theorem (equation (3-11)). Prigogine's theorem given in Section 3.1.2 can be regarded as a special case of the above generalized form.

A stationary state has the internal ability to organize itself so that it realizes a free independent harmonious stable state

What state is realized when a minimum entropy production rate is achieved? As can be seen from the above argument, the law of increase of entropy (or the second law of thermodynamics) is responsible for the attainment of a minimum entropy production rate in a stationary state. In other words, the law of a minimum entropy production rate is an expression of the law of increase of entropy when it is applied to a closed or open non-equilibrium system placed under constant thermodynamic forces. Thus, a stationary state of such a non-equilibrium system has essentially the same properties as the equilibrium state discussed in Section 2.2. Accordingly, all the statements about the equilibrium state given in Section 2.2 apply to a stationary state.

Naturally, a stationary state is characterized by the realization of a minimum entropy (uncompensated-heat) production rate. Now, uncompensated heat can be interpreted as the energy required for driving an irreversible process at a finite rate in the presence of a gradient of T, p, ϕ or μ_i, as argued in Section 3.1.1. Therefore, the achievement of a minimum entropy production rate means that all processes in a system occur smoothly and efficiently with the lowest resistance. Thus, based on this conclusion together with the discussions in Section 2.2, we can say as follows.

(1) *A closed or open non-equilibrium macroscopic system placed*

under constant thermodynamic forces is controlled by the grand internal ability. Thus, a stationary state of it represents a "result" of activities of the grand internal ability.

(2) *Accordingly, in a stationary state, the formation of as widely-spreading fully-continuous internal connections of constituent elements (electrons and atomic nuclei, atoms and molecules, visible macroscopic systems, etc.) as possible is completed, i.e. an infinitely-spreading fully-continuous internal connection is formed under given conditions. Then, all possible constituent elements and related actions or all possible states and processes under given conditions are fully continuously connected with one another and simultaneously realized in harmony with one another.*

(3) *Because of the formation of an infinitely-spreading fully-continuous internal connection, a stationary state attains possible maximum entropy and hence achieves a minimum entropy production rate under given conditions. It also has the internal ability to organize itself so that it attains a minimum entropy production rate.*

(4) *For the same reason as the above, a stationary state realizes a well-organized free independent harmonious stable state with the lowest resistance under given conditions and besides has the internal ability to organize itself so that it realizes such a state.*

The important meaning of the formation of an infinitely-spreading fully-continuous internal connection was argued in Section 1.3.2, 1.4.1, 2.1.3 and 2.2. The internal ability of a stationary state mentioned above comes from the formation of such a continuous internal connection.

There are some important points to be noted about the above statements. Firstly, little attention has been paid to a stationary state in previous studies, as mentioned in Section 1.1.2 and 3.1.1. This is most probably because previous studies have overlooked the existence of the internal ability in a stationary state. The importance of the internal ability is evident if we consider that the emergence of the first living organism on the primitive earth and its evolution later can be reasonably explained based on it, as will be discussed in Section 4.2 to 4.4.

Secondly, the law of a minimum entropy production rate in a stationary state never means that a non-equilibrium system placed under constant thermodynamic forces has the physical or dynamic power to bring about a decrease in the entropy production rate. The law only says that when a huge number of microscopic processes happen in such a non-equilibrium system, the entropy production rate decreases as a result of such microscopic processes (see a footnote #6 of Section 2.2). The importance of this law is that it says that the entropy production rate *necessarily* decreases and reaches a minimum when a huge number of microscopic processes happen.

Thirdly, in a stationary state of a non-equilibrium macroscopic system placed under constant thermodynamic forces, *only microscopic processes which have high rates and can achieve equilibrium or near-equilibrium under given conditions have the internal ability to organize a system so that it as a whole attains a minimum entropy production rate.* This is because only such microscopic processes can form a continuous internal connection of substances participating in them (for example, of reactants and products in a chemical reaction) (see Section 3.1.2, page 142). In this case, thermodynamic-force-driven non-equilibrium slow processes are controlled by the above-mentioned high-rate microscopic processes so that the former each realize a minimum entropy production rate individually. This is because high-rate microscopic processes can adapt quickly to any new situation.

The above argument implies that the internal ability of a stationary state comes from high-rate processes in it, which can achieve equilibrium or near-equilibrium under given conditions, and thus properties of a stationary state strongly depend on what constituent elements and related actions or what states and processes can achieve equilibrium or near-equilibrium under given conditions. In particular, we should note that the achievement of equilibrium or near-equilibrium, i.e. the formation of a continuous internal connection leads to the realization of a well-organized free independent harmonious stable state in a new quality-level and aspect. Therefore, we obtain the following important conclusion.

Properties of a stationary state are raised to a higher level as the number of continuous internal connections in it increases.

It is by this principle that the emergence of the first living organism on the primitive earth and its later evolution happened, as will be discussed in Section 4.2 to 4.4.

Fourthly, characteristics and properties of a stationary state mentioned above are in harmony with our experiences. For example, a sealed nitrogen gas placed between two heat baths (Figure 3-2) and a stationary electrical current in a metal plate (Figure 3-3) indicate that a stationary state really realizes a well-organized free independent harmonious stable state with the lowest resistance. A stationary state also demonstrates various properties such as morphological, mechanical, electrical, optical and chemical properties in an averaged or common form. This fact also indicates that a stationary state realizes a free independent harmonious state, as discussed in Section 1.4.2.

3.2.2 A Reaction Network with the Internal Ability to Organize Itself

A stationary state of a reaction network placed under a constant thermosdynamic force

For getting better understanding of the arguments in the preceding section, let us consider properties of a stationary state of a reaction network placed under a constant thermodynamic force as an example. A stationary state of such a reaction network is also of much interest because it can be regarded as a precursor of the first living organism on the primitive earth, as will be discussed in Section 4.2.

A reaction network we consider here is schematically illustrated in Figure 3-8. Reactions forming a reaction network occur within a drop of organic compounds, placed in an aqueous solution containing source substances, S_1, S_2, S_3, $\cdots$, and reaction products, P_1, P_2, P_3. Such a reaction system is placed at room temperature and under the atmospheric pressure. Let us assume that the concentrations of source substances in an aqueous solution are high while those of products are low. Thus, source substances in an aqueous solution enter into a drop of organic compounds, react in it, and resulting products are discharged from a drop into an aqueous solution.

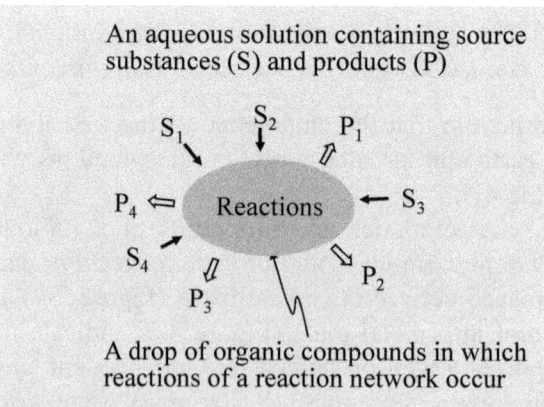

Figure 3-8 A model of a reaction network placed in a solution of source substances and products.

$S_1, S_2, S_3, \cdots \rightarrow$ reactions forming a reaction network in a drop
$\rightarrow P_1, P_2, P_3, \cdots$

The total reaction in such a reaction network can be expressed as follows

$$\Sigma_S \nu_S S_S \rightarrow \Sigma_P \nu_P P_P \qquad (3\text{-}43)$$

where subscripts $S = 1, 2, 3, \cdots$ and $P = 1, 2, 3, \cdots$ express the kinds of source substances and products, respectively, and ν_S and ν_P are the stoichiometric coefficients.

If a reaction network such as illustrated in Figure 3-8 is in local equilibrium, the reaction affinity, A, for reaction (3-43) is given by $A = -\Sigma_m \nu_m \mu_m$ ($m = S$ or P, $\nu_S < 0$, $\nu_P > 0$, see equation (3-28)), where μ_S and μ_P are the chemical potential for S_S and P_P existing just outside the boundary between an organic drop and an aqueous solution, respectively. The reaction affinity divided by the temperature, (A/T), acts as the thermodynamic force for a reaction network (see equation (3-27b)). Therefore, reaction (3-43) can be regarded as a thermodynamic force-driven reaction. If the concentrations of source substances and products just outside the boundary between an organic drop and an aqueous solution are kept nearly constant, the reaction affinity A is also kept nearly constant. In such a case, a reaction network expressed by reaction (3-43) is placed under a nearly constant thermodynamic force. Accordingly, it reaches a stationary state if no dissipative structure emerges. In discussions given below, we assume that a reaction network is in a stationary state.

A reaction network placed under a constant thermosdynamic force reaches a stationary state when the entropy production rate takes a minimum, irrespective of whether it is in the linear-response region or not, as argued in Section 3.2.1. For reference in later discussions, let us briefly explain how we can calculate the entropy production rate for a reaction network lying in the linear-response region. Readers can understand later discussions even if they skip this part of discussions.

For simplicity, let us assume that a reaction system such as illustrated in Figure 3-8 is in equilibrium except for chemical reactions in a reaction network and transport processes for source substances and products at the boundary between an organic drop and an aqueous solution. An organic drop includes a set of chemical substances expressed as S_S, R_c, P_P, where R refers to chemical substances other than S and P in a reaction network. The subscripts, S, c and P, are positive integer and express the kinds of chemical substances. Thus, the k-th chemical reaction in a reaction network can be described as follows

$$\nu_{kp} R_p + \nu_{kq} R_q + \cdots + \nu_{kr} R_r \rightarrow \nu_{ks} R_s + \cdots + \nu_{kt} R_t \qquad (3\text{-}44)$$

where the subscript, p, q, r, s or t is positive integer and refers to one of the above-mentioned subscript c. Chemical reactions including S_S and/or P_P can be described similarly. In the same way as the case of reaction (3-23), the degree of extent of reaction, ξ_k, for the k-th chemical reaction can be defined as

$$\Delta \xi_k = \Delta n_{kc} / \nu_{kc} \tag{3-45}$$

where Δn_{kc} is a change of the amount of substance for a chemical substance R_c by the k-th chemical reaction in an organic drop. Equation (3-45) indicates that one chemical reaction is described by one ξ. By using this ξ_k, the rate of the k-th chemical reaction R_k is defined as

$$R_k = (1/V)\, d\xi_k/dt \tag{3-46}$$

where V is the volume of an organic drop. Thus, the rate of entropy production in a reaction network in an organic drop σ_{drop} can be expressed as follows [1].

$$\sigma_{\text{drop}} = \Sigma\, X_S(dn_S/dt) + V \Sigma_k R_k (A_k/T) + \Sigma\, X_P(dn_P/dt) \geq 0 \tag{3-47}$$

where A_k is the reaction affinity for the k-th chemical reaction and given by $A_k = -\Sigma_c \nu_{kc} \mu_c$ (the sign of ν_{kc} is negative for reactants and positive for products). If the law of linear response holds for chemical reactions (see equation (3-10) and (3-36)), R_k is expressed as follows.

$$R_k = \Sigma_l L_{kl} (A_l/T) \tag{3-48}$$

X_S or X_P in equation (3-47) refers to the thermodynamic force for transport of a source substance S_S or a product P_P at the boundary between an organic drop and an aqueous solution and is given as [1]

$$X_S = -(1/T)(\mu^{\text{in}}_S - \mu^{\text{out}}_S) \tag{3-49a}$$

$$X_P = -(1/T)(\mu^{\text{in}}_P - \mu^{\text{out}}_P) \tag{3-49b}$$

where μ^{in}_S and μ^{out}_S are the chemical potentials of S_S inside and outside an organic drop, respectively, and μ^{in}_P and μ^{out}_P are those of P_P. The quantities, n_S and n_P, in equation (3-47) stand for the amount of substance for S_S and P_P in an organic drop, respectively.

What stationary state is achieved strongly depends on what coupling happens among chemical reactions and transport processes in a reaction system. Such a coupling can be described, for example, by the following equations.

$$dn_S/dt = L_S X_S + \Sigma_k L_{Sk} V (A_k/T) \quad (S = 1, 2, 3, \ldots\ldots) \tag{3-50a}$$

$$dn_c/dt = \Sigma_k L_{ck} V (A_k/T) \quad (c = 1, 2, 3, \ldots\ldots) \tag{3-50b}$$

$$dn_P/dt = L_P X_P + \Sigma_k L_{Pk} V (A_k/T) \quad (P = 1, 2, 3, \ldots\ldots) \tag{3-50c}$$

A reaction network in which chemical reactions are far from equilibrium

In discussions given below, we assume that a reaction network is in a stationary state, as already mentioned earlier. Now, we can divide the above reaction network into two types: a reaction network lying in the nonlinear-response region and one lying in the linear-response region. In the former reaction network, reactions are far from equilibrium and are not continuously connected with one another, while in the latter reaction network, reactions are in or near equilibrium and are continuously connected with one another (see Section 3.1.2, see also Figure 3-1). Thus, the latter reaction network has one more continuous internal connection than the former one. Accordingly, if properties of a stationary state are raised to a higher level as the number of continuous internal connections in it increases, as mentioned in the preceding section, the latter reaction network should demonstrate properties of a higher level. Certainly, this is really the case.

At first, let us consider properties of the former reaction network, i.e. a reaction network in which chemical reactions are far from equilibrium and are in the nonlinear-response region. Even in such a reaction network, transport processes such as heat conduction, substance diffusion, etc. follow the law of linear response and lie in the linear-response region, as mentioned in Section 3.1.2. Namely, microscopic processes such as intra- and inter-molecular energy transfer in translational, rotational and vibrational motion of solute and solvent molecules, spatial movement of them, etc., which occupy the great majority of microscopic processes in a reaction system, have high rates and are in or near equilibrium and are continuously connected with one another. Therefore, such microscopic processes work to control a reaction network so that it as a whole attains a minimum entropy production rate [1], as argued in Section 3.2.1. In this case, non-equilibrium slow chemical reactions are controlled by such high-rate microscopic processes so that they each realize a minimum entropy production rate individually, as also argued in Section 3.2.1.

[1] High-rate microscopic processes such as mentioned above control a reaction network so that it as a whole attains a minimum entropy production rate. This is because even chemical substances taking part in non-equilibrium slow chemical reactions are in or near equilibrium with other chemical substances with respect to microscopic processes such as intra- and inter-molecular energy transfer in translational, rotational and vibrational motion, spatial movement, and so on.

The above conclusion is supported by the quantum mechanical theory of the rate of a chemical reaction. In quantum mechanics, the rate of a chemical reaction is calculated using the potential energy surface (i.e. a profile of the electronic energy of a reaction system including solvent expressed as a function of the spatial coordinates of atoms of reactants). The electronic energy of a reaction system is obtained so that it takes the lowest value according to the variation method of quantum mechanics under the assumption that all microscopic processes except the atomic motion of reactants are in equilibrium. A chemical reaction occurs along a path with the lowest value of the electronic energy of a reaction system between reactants and products. These arguments mean that a reaction system is controlled so that a chemical reaction occurs with the lowest resistance, i.e. at a minimum entropy production rate.

Indeed, chemical reactions in a reaction network are connected with one another in the form of reaction network. However, when reactions are far from equilibrium, they are only connected externally. Namely, they are only controlled so that they each realize a minimum entropy production rate individually, as mentioned above, and thus actually occur individually, independent of others. Accordingly, for example, when a certain chemical substance other than source substances enters into a reaction network in this case, it only responds to such a foreign chemical substance locally. Namely, only a part of a reaction network, attacked by a foreign chemical substance, is changed.

A reaction network in which chemical reactions are far from equilibrium is only a set of individual reactions which are connected externally. Therefore, even if it may behave in an organized manner, it only behaves mechanically like a machine. Thus, a reaction network of this type is of a mechanical character (see Figure 3-1).

A reaction network in which reactions are in or near equilibrium

Quite different behavior emerges in a reaction network in which reactions are in or near equilibrium, i.e. they are in the linear-response region [*2]. In this case, chemical reactions themselves have high rates and are continuously connected with one another and thus have the internal ability to organize a reaction network so that it as a whole attains a minimum entropy production rate. The above-mentioned reaction network in which chemical reactions are far from equilibrium has no such internal ability.

[*2] A reaction network of this type is expected to have collective catalytic actions such as proposed by Kauffman and his group (see Section 3.1.3). This is because equilibrium or near-equilibrium reactions in a reaction network placed under a constant thermodynamic force are expected to have sufficiently large rate constants and hence be catalyzed by chemical substances included in a reaction network. On the other hand, a reaction network with collective catalytic actions is not necessarily a reaction network in which reactions are in or near equilibrium.

The emergence of the internal ability mentioned above can be explained in the following way as well. In a reaction network in which reactions are in or near equilibrium, the concentrations of chemical substances are entirely controlled by a network of equilibrium or near-equilibrium reactions and are always kept in harmony with one another. Accordingly, for example, when a chemical substance other than source substances has suddenly entered into such a reaction network and attacked a chemical substance in it, the effect is immediately transmitted to the whole of a reaction network. In addition, for the same reason, such a reaction network responds to an attack of a foreign chemical substance so that it as a whole realizes a minimum entropy production rate. This means that reactions in a reaction network in this case happen so that they as a whole realize a minimum entropy production rate, i.e. a reaction network has the internal ability to organize itself so that it as a whole realizes a minimum entropy production rate.

The internal ability of a reaction network gives a convincing explanation to the origin and the evolution of life

As mentioned earlier, the internal ability of a reaction network placed under a constant thermodynamic force is of key importance in that it gives a convincing explanation to the origin and the evolution of life (see Section 4.2 to 4.4). In particular, the internal ability of a reaction network in which reactions are in or near equilibrium is of crucial importance in that it gives a clear explanation to the origin of autonomous dynamic self-organizing ability or free independent harmonious spirit and wisdom (creativity) of living things.

Let us remember Mitchell's words in her book [3], cited in Section 1.1.2. She said, "How does nature determine the standard of information treatment in the adaptive and evolutional behavior of a complex system such as a living thing? How does nature recognize the meaning of an actual situation of a complex system and how does nature choose a next suitable action? These problems are left quite uncertain at present." To our delight, we can now find solutions to these problems. A reaction network in which chemical reactions are in or near equilibrium has the internal ability to organize itself so that it as a whole realizes a minimum entropy production rate. This means that such a reaction network is aware of in what situation it is placed and in what direction it should change. Strikingly, such a reaction network has *self-referentiability* (the ability of a system to observe itself and act on itself) and *adaptability* to the environment.

It will be worth noting here that there is a good similarity between properties of a reaction network in which chemical reactions are in or near equilibrium and those of a person who has attained *Tai-toku* discussed in Section 1.3.1. When reactions in a reaction network are in or near equilibrium, the concentrations of chemical substances in it are entirely controlled by a network of equilibrium or near-equilibrium reactions and are always kept in harmony with one another, as mentioned earlier. This means that reactions in a reaction network are controlled so that they occur simultaneously in harmony with one another. This way that chemical substances exist is the same as the way that constituent elements of "the ability gained by the attainment of *Tai-toku*" exist. In addition, when a reaction network is in contact with the outer world, the concentrations of chemical substances in a reaction network are kept in harmony with the outer world. This means that a reaction network exists in infinite dependence on the outer world and incessantly changes its structures, properties and functions so that it realizes a harmonious state with the outer world. This way of existing is also the same as the way that a person who has attained *Tai-toku* exists or the way that a living thing exists (see Section 1.1.1).

We cannot be conscious of the origin of the internal ability of a reaction network in which chemical reactions are in or near equilibrium

There is another important point to be noted. The foregoing argument indicates that a big revolution happens in a reaction network when reactions in it come close to equilibrium. Such a revolution is clearly reflected in the external visible world as the emergence of the internal ability to organize a reaction network so that it as a whole realizes a well-organized free independent harmonious stable state. An important point is that we cannot be conscious of the origin of the internal ability. This means that we cannot be conscious of the origin of autonomous dynamic self-organizing ability or free independent harmonious spirit and wisdom of living things.

Let us consider again the difference between a reaction network in which chemical reactions are far from equilibrium and one in which chemical reactions are in or near equilibrium. In the former reaction network, chemical reactions are connected only externally and actually happen individually, independent of others. Accordingly, a reaction network of this type behaves one by one like a machine, as mentioned earlier. Now, if chemical reactions happen individually, there is no problem even if we pick out individual chemical reactions and describe them by reaction kinetics. Thus, the behavior of the former reaction network can be correctly described by reaction kinetics. This also means that the former reaction behaves mechanically, as can be seen from the discussions in Section 1.1.1.

On the other hand, in the latter reaction network, chemical substances and reactions are internally continuously connected with one another and fused into unity. Accordingly, reactions in this case are controlled so that they as a whole realize a well-organized free independent harmonious stable state with the lowest resistance. Therefore, we can no longer correctly describe chemical reactions by reaction kinetics. If some chemical reactions are described by reaction kinetics which consists of words with individual unchanging meanings, they are separated from internally continuously connected chemical reactions working as a whole. Thus, they lose the true way of existing and hence lose the true qualities (see Section 1.4.1). In this way, the latter reaction network is impossible to correctly describe by reaction kinetics. This means that we cannot be conscious of the origin of the internal ability of the latter reaction network.

Interestingly, the relation between the former reaction network and the latter one resembles the relation between a person who remains at the stage of "understanding by words" and one who has attained *Tai-toku*, discussed in Section 1.3.1 and 1.3.2. In fact, a person who remains at the stage of "understanding by words" behaves one by one like a machine while a person who has attained *Tai-toku* behaves freely, skillfully and unconsciously in harmony with the outer world. In addition, we cannot be conscious of the origin of the internal ability of a person who has attained *Tai-toku*, as discussed in Section 1.3.3.

3.2.3 Can We Construct a Humanoid Robot with Free Spirit and Creativity?

In order to get better understanding of the foregoing discussions in this chapter, let us consider here the problem of whether or not we can construct an artificial system with free independent spirit and wisdom (creativity). Complexity science has revealed a possibility that a variety of systems with high functions that can never be explained simply by the sum of parts exists, as discussed in Section 1.1.2. Recent advanced technology has also brought us a diversity of high-performance humanoid robots or artificial intelligence (AI). For example, AI called AlphaGo won at an extremely complex *Go* game against one of the strongest professional players in the world. This fact has proved that AI can indeed display entirely high functions though at present it lacks the ability to carry out free independent creative thinking. Now, does progress in science or technology in the future bring us a humanoid robot or AI with free spirit and wisdom (creativity)?

A possible principle of constructing a humanoid robot or AI with free independent spirit and creativity

The above problem was already discussed in Section 1.2.2. The conclusion was that we can construct a humanoid robot or AI with free spirit and creativity like a human being. Truly real things (or natural things themselves) have qualities and functions more than those expressed by scientific knowledge. Therefore, we can construct such a humanoid robot or AI by combining a variety of truly real things (acting as constituent elements) with one another properly so that their "extra qualities and functions" work effectively. It is now clear that the "extra qualities and functions" of truly real things refer to continuous internal connections and the internal ability. Thus, we can say as follows.

We can construct a humanoid robot or AI with free spirit and creativity by properly combining a variety of things (constituent elements) with one another so that underlying continuous internal connections work effectively.

The above conclusion was given support by later arguments. For example, it was argued in Section 1.3.2 that the attainment of *Tai-toku* by forming a continuous internal connection of a variety of things (constituent elements) leads to the emergence of free independent harmonious spirit and wisdom (creativity). It was also argued in Section 3.2.2 that the formation of a reaction network in which chemical reactions are internally continuously connected with one another leads to the emergence of the internal ability to organize itself so that it realizes a well-organized free independent harmonious stable state.

A serious problem in the above approach is that it is quite uncertain how to combine a variety of things with one another so that they form a continuous internal connection that works effectively. For pseudo-artificial life and intelligence mentioned in Section 1.2.2, it is relatively easy to find how to combine things with one another because scientific study of already existing living things and intelligence provides information about it. However, for truly artificial life and intelligence such as humanoid robots and AI, it is entirely unknown how to combine things. Some clues may be found in the discussions of the attainment of *Tai-toku* given in Section 1.3.1 and 1.3.2, those of the emergence of the internal ability in a reaction network given in Section 3.2.1 and 3.2.2, and those of the origin and the evolution of life given in Section 4.2 to 4.4.

Can we construct a system with free independent harmonious spirit and wisdom based on a dissipative structure?

A dissipative structure is often observed in a reaction system involving nonlinear kinetics such as an autocatalytic process when it is in the nonlinear-response region (see Section 3.1.2). Much attention has been paid to a dissipative structure most probably because it demonstrates a variety of self-organizing dynamic order against the second law of thermodynamics. Now, can we construct a system with free independent harmonious spirit and wisdom based on a dissipative structure?

A dissipative structure is observed in the nonlinear response region where chemical reactions are far from equilibrium (see Section 3.1.2). Therefore, chemical reactions in a dissipative structure have the same properties as those in a stationary state in the nonlinear-response region (see Figure 3-1). Namely, they are connected only externally and actually occur individually, independent of others, according to fixed rules given by nonlinear kinetics. Therefore, a dissipative structure has no internal ability to organize individual reactions so that they as a whole realize a free independent harmonious stable state. Accordingly, it can display no free independent spirit and wisdom.

Can we construct a humanoid robot or AI with free independent spirit and creativity without taking into account the formation of a continuous internal connection?

A high-performance humanoid robot and AI in the present days is constructed by a logical connection of constituent elements with individual unchanging properties, without taking into account the formation of a continuous internal connection. Now, can we construct a humanoid robot or AI with free independent spirit and wisdom (creativity) by direct development of the present technology? In other words, can we construct an artificial system with free independent spirit and wisdom (creativity) using constituent elements with individual unchanging properties, based only on improvement in device technology and algorithm?

In a system constructed by constituent elements with individual unchanging properties (see Figure 1-3(A)), individual constituent elements exist as the basic governing reality. They by no means change their basic qualities during interaction with others. Thus, they only work mechanically one by one according to given algorithm. Therefore, a system constructed by such constituent elements has no internal ability to organize constituent elements so that they as a whole realize a free independent harmonious stable state. Thus, it can display no free independent spirit and wisdom.

Indeed, a humanoid robot in the present days, for example, a humanoid robot which freely rides a bicycle demonstrates free and skillful behavior. Such excellent abilities are achieved by high technology which allows quite fast signal transmittance and device operation occurring within a nanosecond or less. This means that high technology is sufficient for producing free skillful *mechanical* behavior. However, signals are still simply transmitted one by one according to given rules even in this case. Here is no internal ability to organize processes in a system so that they as a whole realize a free independent harmonious stable state. Therefore, a humanoid robot such as mentioned above can display no free independent spirit and wisdom.

The situation in AI is similar to that in a humanoid robot. Certainly, AI can display entirely high functions, as mentioned at the beginning of this section. However, we cannot find any mechanism by which a system consisting of constituent elements with individual unchanging qualities generates free independent harmonious spirit and wisdom. For example, let us assume that we have succeeded in constructing "a wonderful system" which can freely improve its ability by self-reproduction so that it adapt to changes in the surroundings. Does such "a wonderful system" demonstrate free independent spirit and wisdom?

Note first that even such "a wonderful system" has no internal ability to organize constituent elements so that they as a whole realize a free independent harmonious stable state because constituent elements have individual unchanging properties. In addition, even such "a wonderful system" can only work according to algorithm and data which human beings in advance give based on known knowledge. Therefore, after all, what such "a wonderful system" can do is to pick out all possible situations under given conditions, investigate properties of them and choose the most effective one. It is probably by this way that AI called AlphaGo won at a complex *Go* game against one of the strongest professional players in the world [*1]. However, it is obvious that the behavior of "a wonderful system" is restricted within known facts and cannot go forward beyond them. This means that "a wonderful system" can demonstrate no free independent harmonious spirit and wisdom (creativity).

[*1] Why was AI able to win at a *Go* game against one of the strongest professional players in the world? In my opinion, this is because a *Go* game is played under definite rules and within a limited area, namely, an action at the next step is always found within cross points on a wooden plate. A machine is strong in such a case, as discussed in Section 1.2.2. However, a machine is weak when strategy needs to extend into an unknown area.

The above conclusion becomes much clearer if we compare the above-mentioned "wonderful system" and a living thing or a human being, i.e. a system existing as a continuous internal connection of constituent elements with a freely changing dynamic quality (see Figure 1-3(B)). As can be seen from the discussions of a stationary state of a reaction network in which reactions are in or near equilibrium, a system existing as a continuous internal connection of constituent elements has the internal ability to organize itself so that it realizes a well-organized free independent harmonious stable state with the lowest resistance. In addition, such a system has no separation between the known and the unknown or between the finite and the infinite, as argued in Section 1.3.2. Free independent harmonious spirit and wisdom (creativity) in a living thing or a human being come from these factors. The above-mentioned "wonderful system" lacks both of them.

If we use a metaphor, a humanoid robot or AI with improved algorithm is like a *countable* infinite set. In this case, we can clearly discriminate individual constituent elements even though a whole is very complex. Therefore, we can understand such a set on a logical basis. On the other hand, a system existing as a continuous internal connection of constituent elements such as a living thing and a human being is like an *uncountable* infinite set or a continuum. In this case, we can no longer discriminate individual constituent elements. Only a whole is visible. Such a set is impossible to correctly understand based on individual constituent elements on a logical basis. In other words, logical understanding of such a set is only of an approximate character. We can thus say that the essential difference between a machine such as a humanoid robot and AI and a living thing or a human being lies in the difference between the finite and the infinite or between discreteness and continuity.

References

(1) M. Seno, *Introduction to Thermodynamics of Irreversible Processes* (in Japanese), Tokyo-kagaku-dojin, Tokyo, 1964.
(2) H. Tanaka, *Life and Complex Systems* (in Japanese), Baihukan, Tokyo, 2002.
(3) M. Mitchell, *"Complexity: A Guided Tour"*, Oxford University Press, Inc., Oxford, 2009: The Japanese edition translated by H. Takahashi, Kinokuniya Shoten, Tokyo, 2011.
(4) S. Kauffman, *At Home in the Universe: The Search for Laws of Self-organization and Complexity*, Oxford University Press, Inc., Oxford, 1995: The Japanese edition translated by F. Yonezawa, Nihon-keizai-shinbun-sha, Tokyo, 1999.

(5) I. Prigogine, G. Nicolis, *Self-Organization in Non-Equilibrium Systems*. Wiley, 1977.
(6) K. Yoshikawa, *Nonlinear Science – Rhythms and Shapes of Molecular Ensembles* (in Japanese), Gakkai-shuppan-center, Tokyo, 1992.
(7) H. Mi-ike, Y. Mori, and T. Yamaguchi, *Science of Non-equilibrium Systems – Dynamics of Reaction Diffusion Systems* (in Japanese), Kodansha, Tokyo, 1997.
(8) P. Gransdorf and I. Prigogine, Thermodynamic Theory of Structure, Stability and Fluctuations, Wiley-International, 1972: The Japanese edition translated by H. Matsumoto and K. Takeyama, Misuzu-shobo 1997.
(9) H. Takahashi, *Electromagnetism 9th Ed.* (in Japanese), Shokabo, Tokyo, 1966.
(10) P. L. Luisi, *The Emergence of Life From Chemical Origins to Synthetic Biology*, Cambridge University Press, Cambridge, 2006: The Japanese edition translated by T. Shirakawa, P. Y. Gunji, NTT Shuppan, Tokyo, 2009.

4. A New Theory of the Origin and the Evolution of Life

The origin and the evolution of life have attracted intense attention of many people. A great deal of information has been accumulated and a large number of books have been published on these issues. However, no convincing theory has yet been proposed, as briefly discussed in Section 1.1.2. Thus, the elucidation of mechanisms for the origin and the evolution of life is one of the biggest subjects of research in the 21st century. This chapter shows that we can successfully deal with these issues based on a novel theory of a stationary state of a reaction network placed under a constant thermodynamic force, proposed in the preceding chapter.

4.1 A Brief Survey of Previous Studies on Life

4.1.1 Previous Studies on the Evolution of Life

Controversies started by the advancement of science
To begin with, let us briefly survey previous studies about the origin and the evolution of life for clarifying main problems involved in these issues. A living thing is characterized by a variety of unique concepts such as self-maintenance, autonomy, adaptability, self-replication, multiplication, evolution, and so on. In particular, a living thing has an outstanding feature in that it demonstrates autonomous dynamic self-organizing ability or free independent harmonious spirit and wisdom (creativity). Therefore, a diversity of opinions has been proposed about what life is and thus the study of living things has been full of controversies. This situation has continued up to now.

In particular, two opposing opinions have stood face to face to each other. A group of people, who paid attention to free independent harmonious spirit and wisdom of living things, had opinions connected with vitalism, teleology and holism. On the other hand, another group of people, who attached high importance to the fact that even living things are natural things and should obey inevitable natural laws, had opinions based on the standpoint of atomism, mechanism and reductionism. It appears that in ancient times the former opinion prevailed probably because it was in agreement with a human intuitive view. However, as science advanced, the latter opinion gained strength and had a nearly complete win in the middle of the 20th century. Recently, however, the situation is again largely changing because of many new discoveries about the evolution of life. It appears that teleology and holism are now gaining a revival.

Let us look at some examples of opposing opinions in the past. Vitalism originates from the Aristotelian opinion that characteristic qualities of living things result from the vital power peculiar to living organisms and cannot be explained by physical laws [1]. In the Renaissance age, the development of machines became prominent and a mechanistic view of nature got dominant. Therefore, even living things came to be understood to be machines. In fact, Descartes asserted that animals were machines. He also asserted, "Every natural phenomenon should be explained by existing materials and physical power alone, without adding any mental reason and evaluation" (see Section A3.1.1). Reductionism started from here (see Section 1.1.2 for reductionism and holism).

Later advancement of science has supported the idea of Descartes. For example, a German embryologist, H. Driesch, discovered in 1891 that when an embryo of a sea-urchin at the two-cell stage was divided into two pieces, each piece grew to a young sea-urchin of a complete form. He regarded this phenomenon as being beyond understanding by physicochemical laws and proposed the concept of "entelechy" as a natural factor exerting a particular action on living bodies [1]. Driesch's theory was called neo-vitalism. The concept of entelechy was, however, not supported by later studies and is now completely denied.

Teleology, which says that living things evolves so that they achieve their aims, has suffered a rather similar fate to vitalism. J. B. Lamarck in France was the first person to propose the concept of the evolution of life. He proposed this concept about fifty years earlier than Darwin and explained it by assuming that traits individual living things acquired during their life could be transmitted to next generations [1,2]. For example, the emergence of a long neck of a giraffe was explained as follows: the ancestors of this animal made efforts to eat leaves of tall trees. This habit continued from generation to generation and the accumulation of such efforts finally led to the production of a long neck. Lamarck's theory of evolution is thus regarded as teleology. A severe confrontation about teleology arose from the fact that it included the concept of "the inheritance of acquired traits". This concept is of a fascinating character and much attention has been paid to it. However, no scientific support has long been given to it.

Three big scientific discoveries in the middle of the 19th century

Scientific study of life based on experimental investigations started from three big discoveries in the middle of the 19th century [2]. In the first place, L. Pasteur in France discovered that living things were born only from living things. Before this discovery, it was believed that living things were spontaneously born from nature. Pasteur proved that this idea was wrong, showing that no living bodies were born in a cleaned glass vessel with a lid. However, this discovery caused the new problem of how the first living thing on the primitive earth emerged.

Secondly, C. R. Darwin in England collected many data about properties and functions of animals and discovered that all living things (animals) on the earth came into existence by evolution. He explained this discovery by the theory of mutation and natural selection[*1]. Here, mutation refers to a spontaneously occurring or externally induced accidental alteration either in the phenotype (the external appearance of a living thing) or in the genetic material, both of which are transmitted to a child or an offspring. Natural selection stands for the idea that living things with mutations favorable for their survival are multiplied from generation to generation, compared with those with no or other mutations.

[*1] It is said that Darwin's original idea did not necessarily exclude the concept of teleology, in contrast to the Darwinism established in the 20th century. For example, Mitchell says in her book [3] that Darwin accepted the thought that a mutation occurred so that it improved adaptation, similar to Lamarck.

Thirdly, G. J. Mendel in Austria discovered Mendel's law of heredity. Unfortunately, this law was forgotten for a while but the correctness and importance of the law were confirmed by three scientists in 1900.

The win of Darwinism in controversy about the evolution of life in the middle of the 20th century

The confirmation of Mendel' law led to active study about the problem of whether this law was in harmony with Darwin's theory of evolution in the early 20th century [2,3]. This problem was successfully treated by population genetics, started by G. H. Hardy and W. Weinberg at the beginning of the 20th century and established by R. A. Fisher, J. B. S. Haldane and S. Wright in the 1930's. Population genetics dealt with the evolution of genes in a population of animals (i.e. a group of a large number of animals belonging to the same species), based on Darwin's theory of evolution and Mendel's law of heredity. It used the statistical method to analyze the variation and distribution of genes in a population though it was yet unknown what chemical substances really acted as genes. Then, population genetics clearly showed that the emergence of a diversity of animals and their adaptation to the environment were reasonably explained by Darwin's theory of mutation and natural selection combined with Mendel's law of heredity. This conclusion looked splendid and was called "Modern Synthesis"[3].

The conclusion of population genetics was later confirmed by a series of discoveries in molecular genetics such as the assignment of DNA (deoxyribonucleic acid) as the actual entity of genes (1944), the explanation of self-duplication of genes based on double helix structure of DNA (1953), and the clarification of molecular mechanism for gene expression (1958). The last discovery refers to the clarification of molecular mechanism for the transmission of hereditary traits coded on DNA via m-RNA (messenger ribonucleic acid) and t-RNA (transfer ribonucleic acid) to structures of proteins (i.e. arrangements of amino acids in proteins), which is called "Central Dogma"[4].

On the other hand, according to Kimura's book[2], the concept of "the inheritance of acquired traits" was completely denied by the following two experimental facts. (1) Inherited information on DNA is transmitted in a living body only in one direction from DNA via RNA to proteins. Therefore, there is no chance for acquired traits to cause a change in genes. (2) Mutation occurs completely in random directions. There is no evidence supporting the opinion that a mutation occurs so that it improves adaptation to the environment.

The concept of "the inheritance of acquired traits" was also not supported by its practical application to agriculture. It is famous that I. V. Michurin and his follower, T. D. Lysenko, in U. S. S. R. tried to produce new types of fruit trees based on the concept of "the inheritance of acquired traits" in the former half of the 20th century[1]. Unfortunately, it seems that the trial ended in failure. In this way, Darwinism gained a complete win in controversy about the evolution of life in the middle of the 20th century.

Recent great advancement of molecular biology has brought about a new revolution in the theory of evolution

However, a new revolutionary change has happened in the theory of evolution since the 1970's. The incompleteness of Darwinism was first pointed out by S. Gould and N. Eldredge[3]. They found that patterns of evolution seen in records of fossils showed jumps, contrary to systematic gradualism in Darwinism. Then, they proposed the theory of punctuated equilibrium in 1972, which says that not only natural selection but also historical accidents and biological restrictions play roles in the evolution of life.

A real revolutionary change started from discoveries of various new facts, which were in disagreement with Central Dogma established in the 1950's. According to Mitchell's book [3], it was first revealed that genes and the phenotype (the external appearance of living things) do not necessaryily show one-to-one correspondences. For example, one gene has two or more codes for synthesizing different proteins. In fact, a human body is estimated to have more than one hundred thousand proteins though its genome only consists of twenty five thousand genes.

Detailed studies have disclosed that the situation is much more complex [3]. Unexpected new facts have been discovered one after another. For example, a change in gene function can happen simply by the modification of DNA such as methylation with no change in the arrangement of nucleotide bases in a DNA chain. Besides, such a change in gene function by modification is inherited. Moreover, much more surprisingly, the number of base-pairs in DNA or the number of genes has no direct relation to the complexity of living bodies. For example, the number of base-pairs of a unicellular living thing, amoeba, is 225 times as large as that of a human being. Also, the number of genes of a plant (a kind of shepherd purse) is the same as that of a human being. In addition, a diversity of animals including a human being has large similarities of DNA at the ratio of more than 90%.

These results of recent studies in molecular biology clearly indicate that the traditional theory of evolution or the traditional concept of genes should be greatly changed [3]. Now, how can we understand these results? A clue to a solution is given by recent studies of evolutionary developmental biology [3]. In the traditional study of the evolution of life, main attention has been paid to heredity codes on DNA. Therefore, it has long been believed that only a small part of nucleotide-base pairs of DNA is used as genes and the remaining large part, which involves no heredity code, has no significant meaning about heredity. However, evolutionary developmental biology has disclosed that the latter part of DNA, which has been called non-coded areas, involves gene switches and adjustment (master) genes and plays a key role in gene expression. Namely, DNA chains exist as "gene adjustment networks" consisting of functional genes, gene switches and adjustment (master) genes together with suitable proteins. Thus, for example, a human being can become more complex than expected from the number of genes because gene adjustment networks can produce a great number of possibilities in gene expression.

Accordingly, evolutionary developmental biology has now concluded[3] that the diversity of the phenotype of living things in most cases arises not from differences in genes but from differences in gene adjustment networks or gene switching. This means that the evolution of living things has mostly occurred by changes in gene switching rather than changes in genes themselves.

4.1.2 Previous Studies on the Origin of Life

Initial studies about the origin of life

A. I. Oparin in U.S.S.R. and J. S. B. Haldane in England were the first scientists who tried to explain the emergence of the first living organism on the primitive earth as a result of chemical evolution[2,4-8]. In particular, Oparin published a small pamphlet entitled "The Origin of Life" in 1924 and a full book with the same title in 1936 and proposed the idea that the first living organism was born from organic compounds produced under natural conditions on the primitive earth. In those years, Oparin's idea was unique in that it claimed that the first living organism was heterotrophic, where heterotrophic means "living by depending on high-energy chemicals in the outer world with no self-production". This idea was later supported by Haldane and is now recognized to be correct. Oparin also emphasized that the first living organism was characterized as a highly organized molecular system.

Actual experimental studies began at the 1950's [4-8]. In 1952, H. Urey reported, based on geophysical studies, that the atmosphere of the primitive earth was "reductive" and composed of methane (CH_4), ammonia (NH_3), hydrogen (H_2) and water (H_2O). Based on this report, S. L. Miller showed in 1953 that a mixture of CH_4, NH_3, H_2 and H_2O was converted to various amino acids such as glycine, alanine and glutamic acid (which are indispensable chemical substances for living things) when it was kept under electric discharge for several weeks.

After this experiment, a large number of similar experiments were done under various conditions. Namely, a mixture of simple molecules such as CH_4, carbon dioxide (CO_2), NH_3, H_2O and H_2 was placed under an influence of various high-energy sources such as electric discharge, electron-beam impact, heating, and irradiation with far ultraviolet light of 180 – 130 nm in wavelength, each being used as a simulation of lightning, radioactive rays, hot magma in volcano, and solar light[*1], respectively. In harmony with Miller's experiment, all these experiments showed the production of a variety of compounds important for the emergence of the first living organism, such as organic acids, formaldehyde (H_2CO), hydrogen cyanide (HCN) and amino acids. It was also shown later that formaldehyde is converted to sugar such as ribose, amino acids are converted to polypeptides, and HCN is converted to amino acids, polypeptides and adenine. Details of these reactions are reviewed in books by A. Oparin (1966)[6], M. Calvin (1969)[7], and P. Luisi (2006)[8].

> [*1]Simple molecules such as CO_2, NH_3 and H_2O can only absorb light of wavelengths shorter than about 180 nm. H_2 and CH_4 only absorb light of much shorter wavelengths. Such light was not included in solar light even in the primitive age when the atmosphere of the earth contained no oxygen. Therefore, it is improbable that solar light acted as an effective energy source in the primitive earth. Solar light became important when colored compounds such as porphyrins were produced in the primitive sea, as will be discussed later.

Based on these studies, Oparin proposed a scenario that reactions of simple molecules (CH_4, CO_2, NH_3, and H_2O) under influences of solar light or other high-energy sources led to the production of a "primordial soup" containing a variety of organic compounds and polymers in the primitive sea, followed by the formation of a colloidal drop made of proteins, called a "coacervate", acting as a possible precursor of the first living organism[6]. In 1972, S. Fox proposed a similar model, called "protenoid", which is a microsphere made of condensation-polymerized amino acids[4].

Ideas of the origin of life based on the evolution of individual molecules

The success of Darwinism and the establishment of Central Dogma in the middle of the 20th century exerted a strong influence on the study of the origin of life. In particular, the advancement of molecular genetics disclosed the importance of including genetic molecules in a precursor of the first living organism. Thus, several models about a precursor of the primitive life were proposed along this line. The RNA-world hypothesis, proposed by Gilbert (1986), which says that an aggregate consisting of RNA and supporting elements is the origin of the primitive life, is one of such models, though it now loses support for the reason that it is difficult to assume that RNA was synthesized under prebiotic conditions[4,8].

With the advancement of the study of the origin of life, it also became an important problem how highly ordered organization in a living body was formed against the second law of thermodynamics, which says that nature has a tendency to change toward complete disorder, as already mentioned in Section 1.1.2 and 3.1.1. It seems that this problem together with the necessity of incorporating genetic materials was not clearly recognized in the age of Oparin[8]. Probably, it was tacitly assumed that biopolymers with highly ordered structures such as proteins, RNA and DNA were produced by chance. In fact, this idea was dominant even after the above problem was recognized.

For example, Monod proposed in his book[9] cited in Section 1.1.2 that the emergence of the primitive life occurred in the following way. (1) The production of organic compounds such as nucleotides and amino acids, which were indispensable chemical components for living things. (2) The production of biopolymers having the ability to replicate themselves, such as proteins, RNA and DNA, from nucleotides and amino acids. (3) The construction of a molecular device having structures and functions to suit the purpose of living things by arranging organic molecules around such biopolymers, followed by its evolution to a primitive living cell. Monod's basic idea is that small organic molecules reacted in the primitive sea and produced biopolymers with the ability to replicate them themselves by chance. Once such biopolymers were produced, they should have necessarily replicated themselves according to the "invariability of self-reproduction" discussed in Section 1.1.2 against the second law of thermodynamics. Natural selection started to work at this stage.

Monod's model may explain the origin of life but it has a serious problem in that it needs repeated occurrence of lucky accidents and thus cannot give a meaningful probability of the emergence of the primitive life. In fact, it is quite difficult to imagine that biopolymers with the ability to replicate themselves were produced simply by chemical reactions of small organic molecules in the primitive sea against the second law of thermodynamics. In particular, recent fossil studies indicate that the first living organism called progenote or common descent emerged as soon as the primitive earth was cooled to a temperature suitable for the emergence of life, as will be discussed later (see also Figure 4-1). Such quick emergence of the first living organism is nearly impossible to explain by Monod's model.

Ideas of the origin of life based on the evolution of a molecular aggregate

The difficulty in Monod's model or similar ones arises from the fact that attention is paid to the production and the evolution of particular individual molecules such as proteins, RNA and DNA. Therefore, a new way of thinking appeared and has now become the spirit of the age[4,8]. In the new way of thinking, attention is directed to an organized molecular system, not individual molecules. Thus, the main aim is to clarify how various organic molecules can be organized and how such an organized system can evolve.

To date, various ideas about the emergence of the primitive life have been proposed, as reviewed in Luisi's book [8]. A typical example of studies along the above-mentioned new way of thinking will be the formation of "the reaction network with collective catalytic actions" (see Section 3.1.3) proposed by S. Kauffman and his group [10]. They paid attention to an information network in a living organism and disclosed by computer simulation that a reaction network spontaneously comes to have collective catalytic actions if it becomes sufficiently complex. A unique feature of Kauffman's model is that the formation of a molecular aggregate is the basic event and its evolution controls the evolution of individual molecules. Therefore, according to his idea, the production of genetic molecules such as RNA and DNA is not necessarily indispensable for the emergence of the primitive life. Kauffman's model is also important in that it showed a possibility that the primitive life emerged by inevitable natural laws, without relying on lucky accidents. Kauffman himself emphasizes this point.

Another interesting model about the emergence of the primitive life is the autopoiesis system proposed by Maturana and Varela (see Section 3.1.3). Gánti also proposed a similar reaction network, called chemoton, as a model of protocell (1978) [4,8]. In addition, a large number of studies have been made on dissipative structures or self-organization by nonlinear dynamics as models of organized structures in living things (see Section 3.1.2). However, according to Luisi [8], no sufficiently convincing model for the emergence of the primitive life has still been proposed yet.

Luisi says that determinism (the opinion that the primitive life emerged by inevitable natural laws) becomes meaningful only when it can show an evidence for the argument that nature has a tendency to go toward the construction of living organisms, as already mentioned in Section 1.1.2. Thus, he asserts that contingency (the opinion that the primitive life emerged by favorable accidental conditions) is the only reasonable theory on the scientific standard at present. However, he himself acknowledges that contingency has serious difficulty in explaining how biopolymers with highly ordered structures such as proteins and DNA were produced under prebiotic conditions. In my opinion, contingency has another serious problem in that it can never explain the emergence of free independent harmonious spirit and wisdom (creativity) characteristic of living organisms. A reaction system in contingency is in an entirely passive state against an attack of the outer world and has no internal ability to organize itself so that it realizes a free independent harmonious state.

Kauffman's model has a similar problem for a different reason. His model is based on reaction kinetics and hence reactions occur only mechanically according to reaction kinetics, as discussed in Section 3.2.2 and 3.2.3. Thus, a reaction network with collective catalytic actions can demonstrate no free independent harmonious spirit and wisdom (creativity). The same argument holds for other models based on reaction kinetics, including an autopoiesis system and a dissipative structure. No essential change seems to happen by introducing a novel concept of the edge of chaos.

4.2 A New Theory of the Origin of Life

According to my understanding, the difficulty in previous studies on the origin of life, mentioned above, solely arises from the lack of taking into account the internal invisible world. Chemical substances and reactions in the external visible world have individual unchanging qualities and are of a mechanical character. Therefore, the origin of life is impossible to explain as far as we only look at such things. The external visible world is the world of results while the internal invisible world is the world of causes, as mentioned in Section 1.4.1. Therefore, we should pay attention to the internal invisible world. In other words, we should look at natural phenomena from the standpoint of quantum mechanics. Indeed, we cannot directly apply Schrödinger equation to a living thing but we can take into account the grand internal ability that Schrödinger equation demonstrates (see Section 2.1.3).

This section shows that we can successfully explain the emergence of the first living organism on the primitive earth if we use a novel theory of a stationary state of a reaction network developed in Section 3.2.2. Living and non-living things have essentially the same way of existing, as mentioned in Section 1.4.1. The emergence of the first living organism is thus due to improvement in *qualities* of the way of existing by the formation of a new continuous internal connection (see Section 3.2.1).

Modern science has revealed that the first living organism emerged near hydrothermal vents in the deep sea

Now, let us consider systematically how the first living organism emerged on the primitive earth. Most of scientists in the present days will acknowledge that the first living organism emerged spontaneously from nature. Thus, the remaining important problem is by what mechanism it

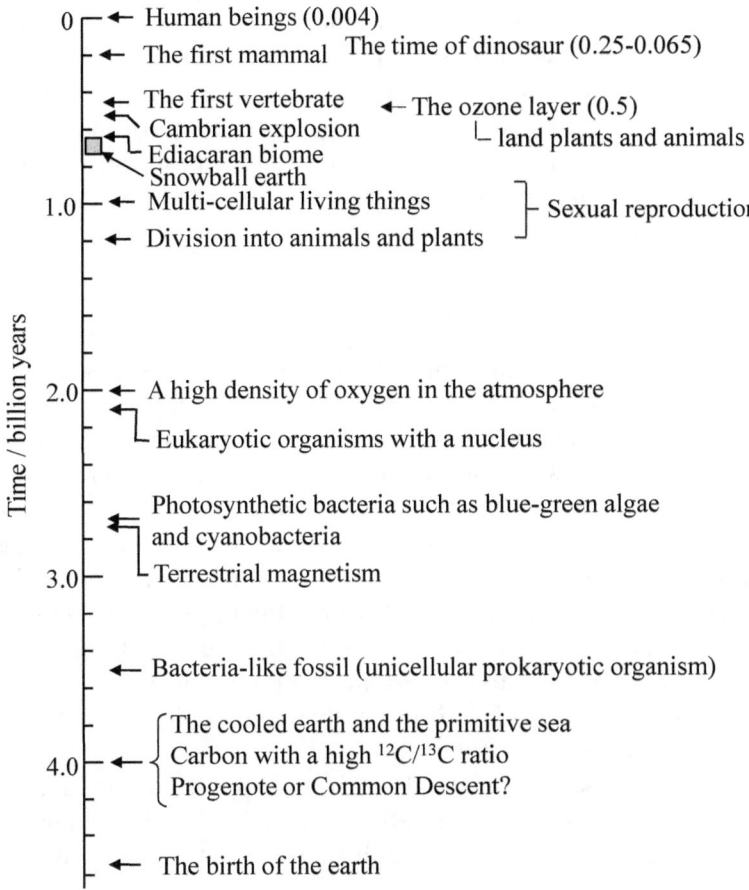

Figure 4-1 Some fundamental events concerning the emergence and the evolution of life on the earth.

emerged.

Figure 4-1 lists some fundamental events concerning the emergence and the evolution of life and their ages according to the literature [2,4,8,10]. The first issue to be tackled is in what situation the primitive earth was and what chemical reactions occurred there. Some studies about this issue in the 1950's to the 1970's were reviewed in Section 4.1.2. However, a drastic change has happened in a basic idea about this issue since the 1990's because of a large advancement of geophysical studies [4,8]. Many ideas proposed till then were greatly altered though the fundamental concept has been retained.

Firstly, the advancement of geophysical studies has revealed that the primitive earth was in a state of the magma sea at a high temperature because of frequent collisions of meteorites. Therefore, the primitive atmosphere was not "reductive" but "neutral" or a little "oxidative" and filled with CO_2, carbon monoxide (CO), nitrogen (N_2) and H_2O. This result suggests that it is difficult to assume that the primitive life emerged near the surface of the primitive sea because it looks impossible to expect that amino acids and other organic compounds were produced from such compounds.

The advancement of geophysical studies has also disclosed a new possibility [4,8,11] that the first living organism emerged in the vicinity of hydrothermal vents in central mountain chains in the deep sea. The new possibility is now given definite support [4] by a number of observed facts. Firstly, it was revealed that a stratum in Western Australia formed at 3.5 billion years ago, in which fossils of the oldest bacteria were found (see Figure 4-1), was a region of central mountain chains in the deep sea in the primitive age. Secondly, it is known that living organisms take in a light carbon isotope (^{12}C) in a higher ratio than a heavy carbon isotope (^{13}C) and thus we can use this fact for investigating the descent of living organisms. According to actual investigations, the oldest carbon with a high $^{12}C/^{13}C$ ratio was discovered in a stratum of black smokers of central mountain chains of 4 billion years ago (see Figure 4-1), which is now located in Isua Island of Green Land.

Studies of bacteria have also given support to the above new possibility [4]. At first, let us briefly explain how living things in the present days are classified, as preliminary information. According to the

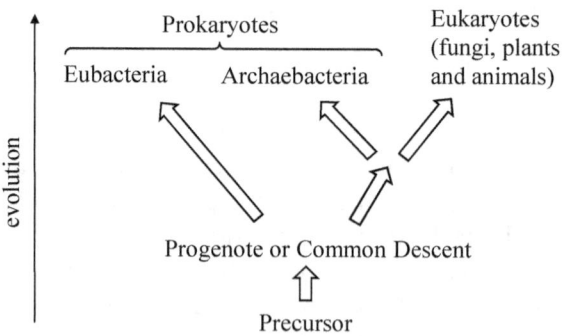

Figure 4-2 The classification of living things in the preset days and their origins.

Woese model in modern biology, living things in the present days are divided into three domains: eubacteria, archaebacteria (or archaea), and eukaryotes (see Figure 4-2). Eubacteria and archaebacteria belong to prokaryotes, where prokaryotes refer to unicellular living organisms with membrane-free genes. On the other hand, eukaryotes are living organisms each having membrane bounded genes, i.e. a nucleus. The majority of eukaryotes exist as multi-cellular organisms such as fungi, plants and animals, though some of them exist as unicellular organisms. Of various bacteria, archaebacteria live in a particular condition such as in a high temperature (80 – 110°C) or in a high salt concentration. On the other hand, eubacteria live in a normal condition. Most of bacteria including blue-green algae and cyanobacteria belong to eubacteria. Archaebacteria and eubacteria have different lipid bilayer membranes, namely, archaebacteria use lipids made of isoprene esters, which are stable in a high temperature, while eubacteria use lipids made of fatty acid esters. Archaebacteria and eubacteria also have different DNA replication systems.

Now, to return to the original subject, let us consider what support studies of bacteria have given to the above-mentioned new possibility. The research of bacterial evolution by the investigation of arrangements of nucleotide-bases in a gene is called ribosomal RNA analysis. It allows us to reveal whether or not some bacteria have a close relation to each other. Actual RNA analyses have shown that it is highly probable that eukaryotes evolved from archaebacteria living in a high temperature or in a high salt concentration, not from eubacteria, as illustrated in Figure 4-2. Thus, it is now believed [4,11] that the first living organism called progenote or common descent divided into eubacteria and archaebacteria and then archaebacteria evolved to eukaryotes by branching. RNA analyses have also revealed [4] that the older the origin of bacteria in the evolutionary system of Figure 4-2, the hotter the sea in which they lived, irrespective of whether they are archaebacteria or eubacteria. This fact strongly suggests that progenote or common descent lived in the hot environment, i.e. in the vicinity of hydrothermal vents.

Furthermore, the aforementioned new possibility is given support by studies of physicochemical processes in the primitive earth [4,8,11,12]. From properties of submarine hydrothermal vents in the present days, it is expected that hydrothermal vents in the primitive earth discharged hot, reductive and weakly alkaline fluid including plenty of H_2, CO, CH_4, NH_3, CN^-, H_2CO, HS^-, COS, CH_3SH and so on, which were important for producing organic compounds indispensable for living organisms. Such small compounds will have reacted with one another under the condition of an elevated temperature (50 – 100°C) and a high pressure in the deep sea near hydrothermal vents and produced various larger-sized organic compounds including oligomers. In addition, it is also expected that solid iron sulfide (FeS) with three-dimensional compartments (pores) of 1 to 100 μm in size was deposited in the vicinity of hydrothermal vents in the primitive earth, in the same way as in the present hydrothermal vents. Such small inorganic compartments will have provided effective sites for the emergence of the primitive life because they were able to keep aggregates of organic compounds stable [11]. Interestingly, it was reported recently [12] that such small inorganic compartments in the vicinity of hydrothermal vents have a function of effectively accumulating organic compounds, in particular, large-sized organic compounds. This is an important finding because the concentrations of organic compounds in the primitive sea are estimated to have been fairly low [11,12]. Moreover, walls of inorganic (FeS) compartments are expected to have worked as catalysts for a variety of organic reactions in the form of Fe-S and Fe-Ni-S clusters. For example, it is reported [11] that FeS-catalyzed reactions of H_2 and CO_2 led to the production of reduced carbon compounds acting as a source of chemical energy.

A plausible place in which a precursor of the primitive life was born

In this way, recent studies suggest [11] that a precursor of the primitive life was born at the bottom of a small inorganic (FeS) compartment

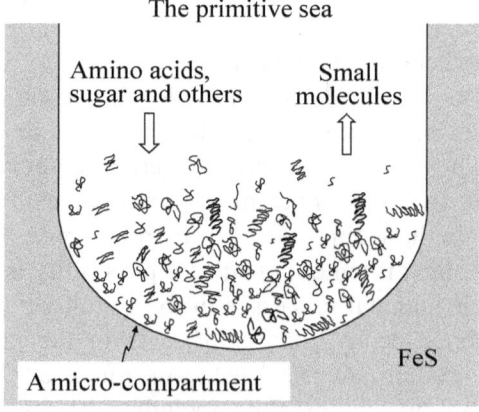

Figure 4-3 A colloidal drop consisting mainly of organic compounds, formed at the bottom of a small inorganic (FeS) compartment.

near a hydrothermal vent and evolved to the progenote or common descent (see Figure 4-2). Based on such a model, Figure 4-3 schematically illustrates a colloidal drop as a precursor of the primitive life, formed at the bottom of a small inorganic (FeS) compartment.

As mentioned earlier, various small compounds such as H_2, CO, NH_3, CH_4, CN^-, H_2CO, and so on, discharged from submarine hydrothermal vents, will have reacted with one another in the deep sea under the condition of an elevated temperature and a high pressure and have produced various larger-sized organic compounds such as sugar (glucose, ribose, etc.), urea, amino acids, alcohols, alkyl and carboxylic acids, pyrimidine, purine, adenine, small polypeptides and porphyrins [8]. Thus, it is probable that a colloidal drop at the bottom of an inorganic compartment mainly consisted of such organic compounds.

It should be emphasized here that the formation of a colloidal drop such as shown in Figure 4-3 was quite an important step toward the emergence of the first living organism. It is very likely that various organic and inorganic compounds entered into such a colloidal drop at a nearly constant rate from the outer world (i.e. the primitive sea), followed by the discharge of small compounds produced by chemical reactions in a colloidal drop, as illustrated in Figure 4-3. Thus, a reaction system in a colloidal drop at the bottom of an inorganic compartment was in nearly the same situation as a reaction network in an organic drop discussed in Section 3.2.2. Namely, *a reaction system in a colloidal drop of Figure 4-3 existed as a non-equilibrium open reaction system placed under a nearly constant thermodynamic force and lay in a stationary or near-stationary state.*

Accordingly, in discussions given below, we assume that a reaction system in a colloidal drop was in a stationary or near-stationary state. The evolution of such a reaction system led to the emergence of the first living organism. Thus, the clarification of the evolution of such a reaction system offers a new theory of the origin of life.

What chemical reactions occurred in a colloidal drop?

Now, the first question is what chemical reactions occurred in a colloidal drop. At the initial stage, only two types of reactions were able to occur spontaneously because of the restriction by the second law of thermodynamics.

In the first place, energy-releasing decomposition reactions of high-energy organic compounds such as sugar and polypeptides were able to occur spontaneously. Organic compounds such as carbohydrates (woods, papers, sugar, etc.) and polypeptides (meat, fish, etc.) spontaneously burn in air, releasing heat, and are converted to small stable compounds such as CO_2 and H_2O. This indicates that organic compounds such as carbohydrates and polypeptides include much energy within them, i.e. they are energy-rich or high-energy compounds. Such compounds can spontaneously decompose into small compounds, releasing energy as heat.

An important point is that no reverse reactions can occur spontaneously. Small compounds such as CO_2 and H_2O can never change spontaneously into high-energy compounds with complex structures. Such reverse reactions can occur only when energy such as heat, electric energy or light energy is injected into a reaction system under an appropriate condition. Miller's experiment mentioned earlier proved that the production of high-energy organic compounds such as amino acids happened only in the presence of a high-energy source such as electric discharge.

In the second place, the accumulation of various organic compounds in high concentrations in a colloidal drop allowed the production of a diversity of organic compounds including fairly large-sized ones. In this case, simple compounds were converted into complex ones, contrary to the above statement. Such a conversion was able to occur because the accumulation of various compounds in high concentrations had the same effect as the presence of a high-energy source. Note that chemical reactions of such a type occurred already in the primitive sea near hydrothermal vents. It was stated just earlier that various small compounds discharged from submarine hydrothermal vents, such as H_2, CO, NH_3, CH_4, CN^-, H_2CO, and so on, reacted with one another in the deep sea and produced various larger-sized organic compounds.

The evolution of a reaction system in a colloidal drop

Next, how did chemical reactions in a colloidal drop evolve later? A reaction system in a colloidal drop, which existed as a non-equilibrium open system placed under a nearly constant thermodynamic force and achieved a stationary or near-stationary state, had the internal ability to organize itself so that it attained a minimum entropy production rate or realized a well-organized free independent harmonious state with the lowest resistance, as argued in Section 3.2.1.

At the initial stage, chemical reactions in a colloidal drop will have been simple, slow and far from equilibrium because of the lack of sufficient organic compounds and effective catalysts. Namely, a reaction system in a colloidal drop at the initial stage was in the nonlinear-response region. Therefore, it only had the internal ability to control individual slow chemical reactions so that they each realize a minimum entropy production rate individually (see Section 3.2.2), i.e. so that they each occur as efficiently as possible individually.

Actual evolution was brought about by the continual inflow of a variety of organic and inorganic compounds including high-energy compounds into a colloidal drop. Namely, such an inflow led to the accumulation of various organic and inorganic compounds at high concentrations in a colloidal drop and hence led to the production of a diversity of organic and inorganic compounds including fairly large-sized ones, as mentioned just earlier. In this way, the kinds of organic and inorganic compounds including catalysts in a colloidal drop became more and more plentiful and the rates of chemical reactions got higher and higher.

Note here that the above argument indicates that a reaction system in a colloidal drop had two kinds of driving forces for evolution: (1) the internal ability to organize it itself so that it attained a minimum entropy production rate and (2) the continual inflow of various chemical substances including high-energy compounds.

The accumulation of chemical energy in a colloidal drop

Importantly, the evolution of a reaction system in a colloidal drop such as mentioned above led to its jump to a new stage. Namely, such evolution led to the initiation of the effective utilization of the energy released from energy-releasing decomposition reactions of high-energy

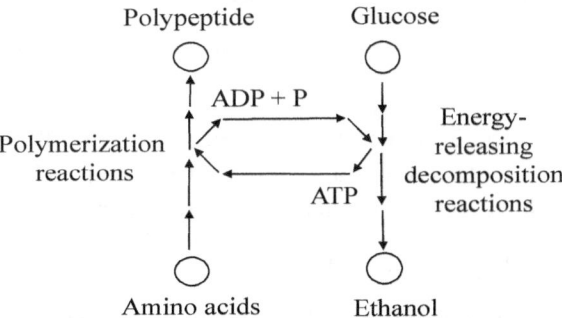

Figure 4-4 Right-hand side: Production of an energy carrier (a high-energy compound), ATP, by coupling with an energy-releasing decomposition reaction of glucose. Left-hand side: Polymerization of amino acids by the use of energy included in ATP.

organic compounds, which had simply been converted to heat and lost till then.

The principle of such effective utilization is schematically illustrated on the right-hand side of Figure 4-4. In this figure, an energy-releasing decomposition reaction of glucose is coupled with an energy-consuming reaction producing a high-energy compound such as ATP (adenosine triphosphate), which acts as an energy carrier in a reaction system. Note that such reaction coupling was made possible by the accumulation of a variety of effective catalysts in a colloidal drop and such accumulation was brought about by the evolution of a reaction system mentioned above. Note also that the effective utilization of the energy released from energy-releasing decomposition reactions, which had been converted simply to heat and lost till then, led to a decrease in the entropy production rate in a reaction system. Therefore, such effective utilization necessarily happened in the course of the evolution of a reaction system because it had the internal ability to organize itself so that it attained a minimum entropy production rate, as mentioned earlier[1].

[1] As emphasized in Section 3.2.1, this does not mean that a reaction system had the physical or dynamic power to bring about such effective utilization. This only means that such effective utilization, i.e. a decrease in the entropy production rate necessarily happened when a huge number of microscopic processes occurred in a reaction system. The same argument holds for all processes throughout the evolution of a reaction system.

According to biochemistry [5], bacteria in the primitive age, which lived in the oxygen-free environment, gained free energy by the fermentation of glucose ($C_6H_{12}O_6$) to ethyl alcohol (C_2H_5OH),

$$C_6H_{12}O_6 \to 2\ C_2H_5OH + 2\ CO_2 \quad -\Delta G_m^0 = 209\ \text{kJ/mol}$$
(4-1)

where $-\Delta G_m^0$ is the free energy per mole, released by this reaction at the standard state. Reaction (4-1) produces two ATP molecules per one glucose molecule. ATP is produced by the following reaction.

$$ADP + P + H^+ \to ATP + H_2O \qquad (4\text{-}2)$$

where ADP refers to adenosine diphosphate and P does phosphoric acid. Therefore, reaction (4-1) can be rewritten as follows.

$$C_6H_{12}O_6 + 2ADP + 2P + 2H^+ \to 2\ C_2H_5OH + 2\ CO_2 + 2ATP$$
(4-1a)

Actually, this reaction proceeds via a complex series of reactions, called the Embden-Meyerhof-Parnas (EMP) route or the Entner-Doudoroff (ED) route.

As already mentioned earlier, the importance of the emergence of such reactions as reaction (4-1a) is that it led to the effective utilization of the energy released by decomposition reactions of high-energy organic compounds. Namely, the emergence of such a reaction led to the accumulation of chemical energy in a reaction system in a colloidal drop in the form of an increase in the concentrations of high-energy compounds such as ATP. ATP molecules carry energy of about 50.2 kJ/mol. Therefore, in the case of reaction (4-1a), the efficiency of energy accumulation is calculated to be $(2 \times 50.2) / 209 \cong 0.5$. The remaining free energy is converted to heat and lost. The EMP or the ED route is known as the most primitive form of energy-acquiring reaction systems investigated to date.

Theoretically, if a reaction system such as illustrated in Figure 4-3 is in local equilibrium, the net free energy, dG_{in}, entering from the outer world (the primitive sea) into a colloidal drop is given by

$$dG_{in} = \Sigma \mu_S \, dN_S - \Sigma \mu_P \, dN_P \qquad (4\text{-}3)$$

where μ is the chemical potential for S_S or P_P existing just outside the boundary between a colloidal drop and the outer world (the primitive sea), N is the number of molecules for S_S or P_P existing in a colloidal drop, and the subscripts, S and P ($S = 1, 2, 3, \cdots\cdots, P = 1, 2, 3, \cdots\cdots$), refer to the kinds of source substances and reaction products, respectively. Thus, the free energy change, dG, in a colloidal drop is given, by using equation (3-8), as follows.

$$dG = dG_{in} - (-\Sigma F_j \, df_j) - \delta Q' \qquad (4\text{-}4)$$

This equation indicates that dG_{in} is partly used to do physicochemical work $(-\Sigma F_j \, df_j)$ and partly used to drive irreversible processes accompanied with uncompensated-heat ($\delta Q'$) production, the rest (dG) being stored in a colloidal drop.

The accumulation of a variety of large-sized high-function organic compounds

The accumulation of chemical energy in a colloidal drop drove a reaction system in it to a further new stage. As mentioned earlier, the reverse reactions of energy-releasing decomposition reactions of high-energy organic compounds, i.e. energy-consuming reactions producing high-energy organic compounds were unable to occur in a colloidal drop at the initial stage. However, the accumulation of chemical energy in a colloidal drop in the form of increases in the concentrations of high-energy compounds such as ATP led to a new situation in which such energy-consuming reactions were able to occur. The principle of producing high-energy organic compounds by the use of ATP is illustrated on the left-hand side of Figure 4-4, by using polypeptide production as an example. Polypeptides are produced as follows.

$$A_n + A \rightarrow A_{n+1} \qquad (4\text{-}5)$$

where A refers to an amino-acid unit and A_n stands for a polypeptide consisting of n amino-acid units. The positive integer n expresses the degree of polymerization. Reaction (4-5) proceeds step by step, leading to an increase in n. The energy necessary for the occurrence of reaction (4-5) is supplied by the reverse process of reaction (4-2). Thus, to be accurate, reaction (4-5) is written as follows.

$$A_n + A + m\ ATP + m\ H_2O \rightarrow A_{n+1} + m\ ADP + m\ P + m\ H^+ \qquad (4\text{-}5a)$$

where m is a positive integer. The free energy change of reaction (4-5a) in the absence of ATP is positive and thus this reaction cannot occur spontaneously, as mentioned earlier. When the concentration of ATP becomes high enough, a free energy change of reaction (4-5a) becomes negative and thus this reaction can occur spontaneously if suitable catalysts are present. If effective catalysts are sufficiently present, reaction (4-5a) exists as the following equilibrium reaction.

$$A_n + A + m\ ATP + m\ H_2O \rightleftharpoons A_{n+1} + m\ ADP + m\ P + m\ H^+ \qquad (4\text{-}5b)$$

Autocatalytic evolution of a reaction system in a colloidal drop

The emergence of reactions such as reactions (4-1a) and (4-5a) had a very important meaning. It led to the production of a variety of high-energy and/or high-function organic compounds with complex structures in a colloidal drop, including a diversity of effective catalysts. Interestingly, the production of a variety of high-function organic compounds led to the emergence of new reactions of the same types as reactions (4-1a) and (4-5a) because they offered effective catalysts for the emergence of such new reactions. On the other hand, the emergence of such new reactions led to the production of a further variety of high-function organic compounds. Thus, here was an autocatalytic mechanism for the development or evolution of a reaction system. Accordingly, it is highly probable that the emergence of reactions such as reactions (4-1a) and

(4-5a) finally led to the accumulation of a huge diversity of large-sized, high-energy and high-function organic compounds with complex structures in a colloidal drop.

It is worth noting here that the accumulation of a diversity of large-sized organic compounds with complex structures led to remarkable increases in the activity of catalysts. There is an interesting example. According to Calvin's book [7], the decomposition reaction of hydrogen peroxide (H_2O_2) into water (H_2O) and oxygen (O_2) is catalyzed by ferric (Fe^{3+}) ions or their complexes. The catalytic activity of naked ferric ions (hydrated ferric ions), $Fe(OH_2)_6^{3+}$, in an aqueous solution, determined by measurements of the reaction rate, is only small, 10^{-5}, while that of iron porphyrin (the complex of Fe^{3+} and porphyrin) is 10^{-2}, one thousand times higher. If iron-porphyrin is incorporated into catalase (protein), the catalytic activity is increased to 10^3. In this way, the catalytic activity of Fe^{3+} ions is increased by a factor of 10^8 by improvement in peripheral and surrounding molecular structures. This example clearly indicates how important the production of a huge diversity of large-sized organic compounds is for the emergence of effective catalysts.

The emergence of a reaction network with the internal ability to organize itself so that it as a whole realized a minimum entropy production rate

The accumulation of a huge diversity of large-sized, high-energy and/or high-function organic compounds in a colloidal drop at last drove a reaction system in it to an epoch-making stage. Such accumulation led to the formation of a reaction network in which reactions were in or near equilibrium and were internally continuously connected with one another. Namely, such accumulation led to the formation of a reaction network which had the internal ability to organize itself so that it as a whole attained a minimum entropy production rate. A reaction system before the emergence of such a reaction network had only the internal ability to organize itself so that individual chemical reactions each realized a minimum entropy production rate individually.

For the formation of a reaction network in which reactions are in or near equilibrium, reactions have to meet two conditions: very large rate constants k_f and k_b and small reaction affinity A^0 at the standard state, as argued in Section 3.1.2. The accumulation of a huge diversity of large-sized, high-energy and/or high-function organic compounds provided a sufficient situation for them. Firstly, such accumulation must have supplied effective catalysts for reactions in a colloidal drop, thus leading to great increases in the rate constants k_f and k_b for them. Also, such accumulation must have allowed the conversion of reactions with large reaction affinity A^0 at the standard state to serially occurring lengthy chains of reactions with small reaction affinity A^0 at the standard state. Thus, all reactions in a colloidal drop were able to have very large rate constants k_f and k_b and small reaction affinity A^0 at the standard state. In addition, the accumulation of a huge diversity of organic compounds will have made reactions in a colloidal drop highly entangled and finally led to the formation of a reaction network. Furthermore, it should be noted that the formation of a reaction network in which reactions were in or near equilibrium led to a decrease in the entropy production rate in a reaction system and hence necessarily happened in the course of the evolution of a reaction system.

In this way, the accumulation of a huge diversity of organic compounds led to the formation of a reaction network in which reactions were in or near equilibrium and were internally continuously connected with one another, as mentioned above.

The evolution of a reaction network with the internal ability to organize itself so that it as a whole realized a minimum entropy production rate

Now, how had such a reaction network as mentioned above evolved later? As mentioned earlier, for a reaction system in a colloidal drop at the initial stage, various organic and inorganic compounds including high-energy organic compounds entered at a nearly constant rate from the outer world (the primitive sea) into it, thus followed by the discharge of small compounds produced by chemical reactions in it. A reaction network mentioned above was in nearly the same situation. Namely, a reaction network mentioned above existed as a non-equilibrium open reaction system placed under a nearly constant thermodynamic force and attained a stationary or near-stationary state. In addition, a reaction network mentioned above, in which reactions were in or near equilibrium and were internally continuously connected with one another, had the internal ability to organize itself so that it as a whole attained a minimum entropy production rate.

Accordingly, a reaction network mentioned above is expected to have evolved in the following way. In general, a sudden inflow of a foreign chemical substance other than source substances into a reaction network lying in a stationary or near-stationary state caused a disturbance in it. Thus, a reaction network was more or less destabilized, being accompanied by an increase in the entropy production rate. If such destabilization was not too large and if a destabilized reaction network was not far from a stationary state, it still had the internal ability to organize itself so that it as a whole attained a minimum entropy production rate (see Section 3.2.1). Accordingly, a destabilized reaction network again returned to a stable stationary or near-stationary state with a reorganized reaction network. The repetition of such destabilization and stabilization led to the evolution of a reaction network.

There will have been various ways that a destabilized reaction network was stabilized, depending on in what state the original reaction network was or what attack of the outer world happened. For example, if a destabilized reaction network achieved stabilization by capturing foreign entering chemical substances and adding them to the original reaction network, a new stationary state with a slightly modified reaction network resulted. In this case, the original reaction network retained the fundamental structure and grew in size and quality. On the other hand, if a destabilized reaction network achieved stabilization by bringing about big reorganization, an essentially new stationary state with a largely reorganized reaction network emerged. It is probable that such large reorganization occurred when a reaction network in advance had grown to a level on which such large reorganization was unavoidable. This argument suggests that a reaction network evolved in most cases by growth with slight modification, sometimes followed by large reorganization.

There is another important aspect about the evolution of a reaction network mentioned above. In what direction had it evolved? As mentioned above, a reaction network evolved individually by reorganizing itself in response to an attack of the outer world. Therefore, a diversity of new reaction networks will have been produced in small inorganic compartments near hydrothermal vents (see Figure 4-3). Thus, it is reasonable to assume that a reaction network which achieved high stability or sustainability survived for a long time and became dominant. This means that a reaction network evolved in the direction of increasing stability or sustainability.

We can also say that a reaction network evolved in the direction of increasing continuous internal connections and decreasing the entropy production rate. This is because a destabilized reaction network was reorganized and stabilized so that it as a whole realized a minimum entropy production rate, as discussed above. In general, a reaction network with a lower entropy production rate has higher stability or sustainability because a reaction network with a low entropy production rate realizes a well-organized free independent harmonious stable state with low resistance. Therefore, the above two conclusions are in harmony with each other.

The emergence of mechanisms for self-stabilization and self-reproduction

Here, let us consider how a reaction network mentioned earlier (i.e. a precursor of the primitive life) constructed mechanisms for self-stabilization and self-reproduction. Unicellular prokaryotic bacteria in the present days use DNA as a genetic substance. Recent studies, however, indicate [4,8] that unicellular bacteria in the primitive age used RNA instead of DNA probably because of easier synthesis. In my opinion, a precursor of the primitive life started to use RNA as a template for controlling arrangements of amino acids in polypeptides, not as a genetic substance.

When a reaction network mentioned earlier (i.e. a precursor of the primitive life) came to have a huge diversity of large-sized high-function organic compounds, the control of their chemical structures must have become crucially important for maintaining a low entropy production rate. For example, reaction (4-5a) produced a large variety of polypeptides but only some of them with particular arrangements of amino acids or particular three-dimensional morphological structures were effective as catalysts. Therefore, it must have become of key importance in a reaction network to construct a way to control reaction (4-5a) and prepare effective polypeptides selectively.

It is well known that coupling between RNA and polypeptide formation is effective for controlling polypeptide structures. Therefore, we can assume that a certain reaction network in a small inorganic compartment happened to succeed in controlling polypeptide structures by the use of such coupling. Then, such a reaction network succeeded in realizing a very low entropy production rate, i.e. quite a well-organized smooth stable state with the lowest resistance. Accordingly, it survived for a long time and became dominant.

In this way, it is probable that a reaction network mentioned earlier (or a precursor of the primitive life) evolved in the direction of controlling polypeptide structures by the use of coupling between RNA and polypeptide formation. The construction of such a way to control polypeptide structures is expected to have necessarily happened because it led to a lower entropy production rate in a reaction network. As mentioned in Section 3.2.1 and in a footnote #1 of this section, a decrease in the entropy production rate (or an increase in entropy) in a reaction network placed under a constant thermodynamic force spontaneously and necessarily happened when a huge number of microscopic processes occurred.

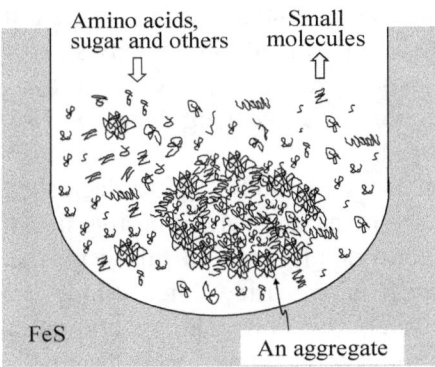

Figure 4-5 The formation of an aggregate of organic compounds taking part in reactions in a reaction network.

The formation of a cell membrane can be explained similarly. Electron micrographs of fossils of primitive bacteria in the age of more than three billion years ago clearly indicate that they had cell membranes for self-stabilization [10]. When a reaction network mentioned earlier (or a precursor of the primitive life) came to have a huge diversity of large-sized high-function organic compounds, as mentioned above, the control of microscopic processes in it must also have become crucially important. Here, let us assume that a certain reaction network happened to succeed in forming a membrane at the boundary between a reaction network and the outer world and controlling the inflow and the outflow of substances across it. Actually, a reaction network will have first formed an aggregate of compounds taking part in reactions in a reaction network, such as schematically illustrated in Figure 4-5, and then formed a membrane around it. Such a reaction network attained a large decrease in the entropy production rate and gained high sustainability and became dominant. In this way, the evolution happened in the direction of utilizing a cell membrane for self-stabilization. The formation of an aggregate such as illustrated in Figure 4-5 and a membrane surrounding it will have necessarily happened because it led to a lower entropy production rate, in the same way as the case of coupling between RNA and polypeptide formation.

The construction of mechanisms for self-stabilization must have led to a large increase in the surviving power of a reaction network (or a precursor of the primitive life). As a result, it will have come to produce a large body. According to electron micrographs of fossils of primitive bacteria [10], they had a body comprised of rods of a dendrite-like shape, the thickness of rods being limited to relatively small values, about 2 to 8 μm. The limited thickness was most probably for avoiding an increase in the entropy production rate by long-distance diffusion of chemical substances within a rod.

The production of a large body led to a decrease in stability. Thus, a reaction network (or a precursor of the primitive life) having a large body must have divided into small ones. In this way, the reproduction of a reaction network by division started. Interestingly, a dendrite-like shape was favorable for such division. The incorporation of RNA as the substance for controlling reactions in a reaction network was also favorable for the reproduction of a reaction network by division. Thus, it is probable that the self-reproduction of a reaction network (or a precursor of the primitive life) by division started when appropriate catalysts (proteins) which allowed the reproduction of RNA emerged. At this stage, RNA came to have a *new* function of transmitting inherited information.

We can say that the first living organism emerged when a reaction network mentioned earlier completed the mechanisms for self-stabilization and self-reproduction.

The verification of a new theory of the origin of life proposed in this book

The foregoing arguments indicate that the emergence of the first living organism on the primitive earth happened by inevitable natural laws. Indeed, the emergence of the first living organism was affected by a variety of factors, including particular geophysical conditions, but it is clear that the law of a minimum entropy production rate (or the law of increase of entropy applied to a reaction system placed under a constant thermodynamic force) played the key role in this event. Namely, the emergence of the first living organism happened by the formation of a reaction system placed under a nearly constant thermodynamic force in a small inorganic compartment, followed by its evolution to a reaction network in which reactions were in or near equilibrium and internally continuously connected with one another. The internal ability of such a reaction network to organize itself, together with the inflow of a variety of chemical substances into it, acted as the driving forces for the emergence of the first living organism.

A new theory of the origin of life proposed in this book is given definite support by the way that living things in the present days exist. According to this theory, the first living organism emerged via the formation of a reaction network in which reactions were in or near equilibrium, as mentioned above. This means that the first living organism emerged via the achievement of very high rates in chemical reactions together with sufficiently small reaction affinity, A^0, in the standard state. Amazingly, such characteristics of chemical reactions in a precursor of the primitive live are clearly embodied in reactions in living things in the present days, indicating that they have really evolved from a precursor of the primitive live mentioned above.

In fact, reactions in living things in the present days are catalyzed by enzymes and have quite large rate constants both in the forward and the backward processes and thus exist as equilibrium or near-equilibrium reactions. In addition, chemical reactions in the present living things have a prominent characteristic in that they in general consist of very lengthy chains of reactions. To date, no explanation has been given to this characteristic. A new theory proposed in this book explains this fact as due to the necessity of decreasing the reaction affinity, A^0, in the standard state for achieving equilibrium or near-equilibrium reactions. Living things have achieved decreases in A^0 by converting a chemical reaction with large A^0 to a large number of serially occurring chemical reactions with small A^0, as discussed earlier. The successful explanation of the presence of lengthy chain reactions in living things also indicates the validity of a new theory proposed in this book.

4.3 How Have Living Things Evolved?

Based on the arguments in the preceding section, let us next consider how the first living organism evolved later. The evolution of life is much more complex than the emergence of life on the primitive earth and thus discussions in this section are much more speculative than in the preceding section. The main purpose of this section is to demonstrate that the evolution of life can also be reasonably explained as due to inevitable natural laws if we take into account the internal invisible world.

The emergence of photosynthetic and aerobic bacteria

The first big event after the emergence of the primitive life (unicellular prokaryotic organisms) will be the emergence of bacteria with photosynthetic ability such as blue-green algae or cyanobacteria. They are estimated to have emerged about 0.8 to 1.3 billion years after the emergence of the first living organism (see Figure 4-1). A representative reaction of photosynthesis can be expressed as follows.

$$6\ CO_2 + 6\ H_2O \rightarrow C_6H_{12}O_6 + 6\ O_2 \quad \Delta G_m^0 = +\ 2{,}870\ \text{kJ/mol}$$
(4-6)

The free energy change for photosynthetic reaction is very large, as indicated above, but photosynthetic bacteria will only have gained energy of 209 kJ/mol from glucose ($C_6H_{12}O_6$) because they gained energy by the fermentation of it (reaction (4-2)).

The importance of the emergence of photosynthetic bacteria was that they acquired the ability to utilize solar light as an energy source and therefore were able to grow quickly and thickly everywhere in the primitive sea. In fact, a large increase in the oxygen content in the atmosphere at around 2.0 billion years ago (Figure 4-1) is regarded as a clear evidence for thick growth of photosynthetic bacteria in those ages. In addition, a large amount of layered deposits of iron oxides was formed by reaction of oxygen and iron ions in the primitive sea. Such deposits, called banded iron formation, are now found in rocks of the primitive age in many places.

Furthermore, an increase in the oxygen content in the atmosphere in turn induced the emergence of oxygen-consuming living organisms such as aerobic bacteria. Reactions of organic compounds with oxygen release a large amount of free energy. For example, the free energy change for the complete oxidation of glucose (which is just the reverse reaction of photosynthetic reaction) is 2,870 kJ/mol and about fourteen times as large as that for the fermentation of glucose.

$$C_6H_{12}O_6 + 6\ O_2 \rightarrow 6\ CO_2 + 6\ H_2O \quad \Delta G^0 = -\ 2{,}870\ \text{kJ/mol}$$
(4-7)

Accordingly, aerobic bacteria also gained the ability to grow quickly and are expected to have grown widely and thickly in the primitive sea.

Photosynthetic bacteria such as blue-green algae or cyanobacteria in the present days belong to eubacteria, indicating that they evolved from eubacteria (see Figure 4-2). Now, why was the emergence of photosynthetic bacteria largely delayed, compared with that of the first living organism? A possible reason was that photosynthetic bacteria emerged near the surface of the primitive sea, which was rich in solar light, in sharp contrast to the first living organism which emerged near hydrothermal vents in the deep sea. Thus, for the emergence of photosynthetic bacteria, the concentrations of organic compounds such as sugar, amino acids and nucleotides near the surface of the primitive sea had to increase to sufficiently high values. Organic compounds near the surface of the primitive sea will have been supplied by continuous discharge of compounds from submarine hydrothermal vents. Thus, it is likely that it took a long time for the concentrations of organic compounds near the surface of the primitive sea to become sufficiently high.

Another possible reason is that there were high-energy solar rays, called the solar wind, which had quite a harmful influence on living organisms lying near the surface of the primitive sea. In fact, photosynthetic bacteria emerged just when the terrestrial magnetism, which protected the earth surface from the solar wind, was formed (see Figure 4-1).

How did photosynthetic bacteria emerge? The first living organism emerging near hydrothermal vents established a mechanism for acquiring chemical energy by the fermentation of glucose, as discussed in Section 4.2.1. Thus, similarly, photosynthetic bacteria will have emerged by establishing a mechanism for acquiring chemical energy through solar light-induced photochemical reactions. In contrast to the fermentation of glucose, which occurred in a homogeneous solution, a mechanism for acquiring chemical energy from solar light was constructed by using a bimolecular lipid membrane (BLM) and an ion pump embedded in it. The full oxidation of organic compounds by molecular oxygen in aerobic bacteria was also attained by using a BLM and an ion pump.

Evolution by interaction between unicellular living organisms

The emergence of photosynthetic and aerobic bacteria, which had the ability to grow quickly and thickly, must have made the primitive sea full of primitive unicellular living organisms. Thus, it is expected that not only interaction between living organisms and the environment but also interaction between living organisms themselves became important and affected the evolution of life. The emergence of such a situation is supported by the following considerations.

Chloroplasts and mitochondria are important intracellular particles in eukaryotic cells in the present days. Chloroplasts have the ability to carry out photosynthesis, while mitochondria have the ability to allow high-energy organic compounds to react with molecular oxygen and release large energy. L. Margulis et al. revealed in the 1970's that chloroplasts and mitochondria in eukaryotic cells have their own inherent genetic substances (DNA), the structures of which are different from those of parent eukaryotic cells but similar to those of prokaryotic bacteria such as blue-green algae and aerobic bacteria [2]. Based on this finding, Margulis et al. concluded that the formation of intracellular organelles such as chloroplasts and mitochondria was a result of symbiotic association between nucleated eukaryotic cells and prokaryotic bacteria. This conclusion indicates that there were really strong interaction between primitive living organisms, which exerted a large influence on the evolution of primitive living organisms. Margulis' conclusion is also interesting in that it explains how animals and plants emerged (see Figure 4-1). Theoretically, we can say that symbiotic association between nucleated eukaryotic cells and prokaryotic bacteria occurred because it led to the realization of a lower entropy production rate.

According to the history of the evolution of life shown in Figure 4-1, it appears that eukaryotic unicellular organisms with membrane-bounded genes emerged a little before the above symbiotic association between eukaryotic cells and prokaryotic bacteria. This fact might suggest that the emergence of membrane-bounded genes also happened for protecting genes (RNA or DNA, substances for controlling reactions in a living organism) from strong interactions among primitive bacteria.

The emergence of multi-cellular living organisms

The emergence of multi-cellular living organisms was really an epoch-making event in the evolution of life. In fact, it led to rapid emergence of a huge diversity of multi-cellular living organisms, as will be discussed later. On the other hand, it is important to note also that it took a very long time for multi-cellular living organisms to emerge. The first multi-cellular living organism emerged about one billion years ago[*1] (Figure 4-1). Surprisingly, this indicates that the age of unicellular living organisms lasted for about three billion years or so. The emergence of multi-cellular living organisms took one billion years even after the emergence of photosynthetic and aerobic bacteria. Why was it so delayed?

[*1] Recent studies suggest that the evolution from unicellular eukaryotes to multi-cellular ones occurred in different ways for animals, plants and fungi. It is presumed that the first multi-cellular animal was poliferia (sponge) with no cellular differentiation, which evolved from a colony of choanomodana (choanoflagellate) i.e. a unicellular heterotrophic eukaryote with a flagellum and a collar.

A plausible reason is that primitive living organisms had evolved in the direction of completing self-organization, self-stabilization and self-replication as unicellular living organisms. Therefore, for producing multi-cellular living organisms, they had to construct an essentially new mechanism for self-organization, self-stabilization and self-reproduction. Here was a great barrier. For example, it is said that multi-cellular eukaryotes had an advantage in the effectiveness of capturing nutrients in the primitive sea because they were able to use cooperative work of a large number of living cells. However, for successful cooperative work, a large number of living cells had to behave simultaneously and harmoniously in a synchronized manner. This meant that the formation of a new continuous internal connection of living cells to one another was needed.

Almost nothing is known about how a new continuous internal connection was formed. As discussed earlier, primitive unicellular bacteria existed each as a reaction network in which reactions were in or near equilibrium and internally continuously connected with one another. Such a continuous internal connection of chemical reactions was actually limited within an individual living cell surrounded by a cell membrane. Therefore, for producing a multi-cellular living organism, primitive unicellular bacteria had to gain new ability to communicate with one another, for example, the ability to use signal transport [*2]. Primitive unicellular bacteria may have prepared such ability in the age when they strongly interacted with one another in the primitive sea.

[*2] For producing a multi-cellular living organism, unicellular living organisms first of all needed to gain the ability to adhere to one another. According to recent studies, this problem was overcome by incorporating a new chemical substance called cadherin.

Once multi-cellular living organisms emerged, they evolved rapidly, as mentioned above. In fact, recent fossil studies show that several ten kinds of multi-cellular living organisms with soft bodies, including jellyfish-like ones with two germ layers, lived in the Ediacaran period (0.640 − 0.542 billion years ago) just after a great glacial time called Snowball Earth (see Figure 4-1). In the following Cambrian period (0.542 − 0.510 billion years ago), a huge diversity (several ten thousand kinds) of multi-cellular living organisms emerged. This sudden emergence of a huge diversity of multi-cellular organisms is called Cambrian explosion (see Figure 4-1).

An expected great merit of the formation of a multi-cellular living organism is that it can have a large-sized body and hence can have a diversity of high functions without heavily increasing the entropy production rate because metabolism basically proceeds within small individual cells. In fact, a large-sized body can produce a diversity of high functions because it can allot various individual functions to parts of a large-sized body and combine them so that they as a whole produce high functions. In this way, multi-cellular living organisms had a new great possibility of producing a diversity of high functions.

The emergence of the ability to behave intentionally with an aim

The most characteristic ability of multi-cellular living organisms is that they can behave intentionally with an aim. For example, animals move here and there to look for foods. Plants also grow in the direction of sun light. It was mentioned above that a multi-cellular living organism was able to produce a diversity of high functions by allotting various individual functions to parts of a large-sized body and combining them. The ability to behave intentionally with an aim can be regarded as a typical example of such high functions.

Now, how was a multi-cellular living organism able to have such ability? This problem can be solved if we assume that a multi-cellular living organism acquired the ability to gain and store information about the outer world. As mentioned in Section 4.2, a unicellular living organism (or a reaction network in which reactions were in or near equilibrium) had the internal ability to organize itself so that it as a whole realized a minimum entropy production rate or a well-organized harmonious stable state with the lowest resistance. Therefore, if a multi-cellular living organism acquired the ability to gain and store information about the outer world, such a multi-cellular living organism was able to organize itself so that "a multi-cellular living organism and information about the outer world" as a whole realized a minimum entropy production rate or a well-organized harmonious stable state with the lowest resistance. This means that a multi-cellular living organism was able to behave intentionally with an aim.

It is to be mentioned here that the emergence of intellectual ability of a human being can also be explained by exactly the same mechanism as the above. In fact, the human brain works so that "a human body and information about the outer world" as a whole realize a harmonious stable state, as will be discussed in Chapter 5. This fact gives strong support to the above mechanism for the emergence of the ability to behave intentionally with an aim.

The emergence of the ability to behave intentionally with an aim in multi-cellular living organisms was really an epoch-making event. As discussed in Section 4.2, a living thing sustains its activity by utilizing the energy released from high-energy organic compounds such as sugar. Multi-cellular living organisms acquired the ability to catch high-energy organic compounds intentionally by utilizing the energy released from such high-energy organic compounds. Then, they gained high-level ability to sustain themselves. By the way, human beings have acquired the ability to produce high-energy organic compounds intentionally by utilizing the energy released from high-energy organic compounds.

4.4 The Principle of Evolution

In the preceding sections we have considered the origin and the evolution of life based mainly on the internal ability of a reaction system placed under a constant thermodynamic force. Now, the internal ability of a reaction system represents a "result" of activities of the grand internal ability, as argued in Section 3.2.1. Therefore, in this section we consider the origin and the evolution of life based on the grand internal ability itself. By this consideration we can gain a general view of the origin and the evolution of life.

The evolution of life occurs via the formation of a new continuous internal connection through interaction between a living thing and the environment

The discussions of the origin and the evolution of life in the preceding sections indicate that the area of internally continuously connected substances and processes became wider and wider as living things evolved to a higher level. Let us again look back on the foregoing discussions. Firstly, a reaction system placed under a nearly constant thermodynamic force was produced in a colloidal drop in a small inorganic compartment near a hydrothermal vent, such as illustrated in Figure 4-3. At this stage, chemical reactions in a colloidal drop were slow and far from equilibrium. Therefore, only microscopic processes other than chemical reactions such as intra- and inter-molecular energy transfer in translational, rotational and vibrational motion of solute and solvent molecules, spatial movement of them, and so on were in or near equilibrium and internally continuously connected with one another. Then, continuous inflow of various chemical substances into a colloidal drop led to increases in the kinds of chemical substances and reactions in it and hence led to increases in reaction rates, finally leading to the formation of a reaction network in which almost all reactions were in or near equilibrium and were internally continuously connected with one another.

A continuous internal connection of chemical reactions in a reaction network thus formed became wider and wider later through the emergence of coupling between RNA and polypeptide formation and the formation of a cell membrane. In this way, a unicellular living organism was born. A continuous internal connection in a unicellular living organism became further wider later through the achievement of symbiotic association between eukaryotic cells and prokaryotic bacteria and through the formation of a multi-cellular living organism. Finally, a multi-cellular living organism acquired the ability to gain and store information about the outer world and came to be able to behave intentionally with an aim. This meant that a multi-cellular living organism acquired the ability to form a continuous internal connection between a multi-cellular living organism and information about the outer world. The intellectual ability of a human being emerged from here.

In this way, the area of internally continuously connected substances and processes in a reaction system or a living organism became wider and wider as it evolved to a higher level. Interestingly, this is a natural result if we remember that all natural things are governed by the grand internal ability introduced and discussed in Section 1.4.1, 2.1.3, 2.2 and 3.2.1. This is because the grand internal ability works so as to make various things interact with one another according to their inherent properties and motion, form as widely-spreading continuous internal connections of them as possible under given conditions in various quality levels and aspects, and complete the formation of such continuous internal connections.

In fact, we can clearly see the activity of the grand internal ability in the discussions of the origin and the evolution of life given in the preceding sections. Continual inflow of various organic compounds into a colloidal drop in a small inorganic compartment and their interactions (reactions) in it led to the formation of a reaction network in which almost all reactions were internally continuously connected with one another. Moreover, continual inflow of various organic compounds into such a reaction network and their interactions (reactions) in it led to the emergence of coupling between RNA and polypeptide production and the formation of a cell membrane. Interactions between primitive cells growing thick in the primitive sea led to the formation of intracellular particles such as chloroplasts and mitochondria by symbiotic association on the one hand and led to the emergence of multi-cellular living organisms on the other hand. Interactions of multi-cellular living organisms with the outer world led to the emergence of the ability to gain and store information about the outer world and that of the ability to behave intentionally with an aim.

From these considerations, we are led to the conclusion that the origin and the evolution of life were really brought about by the grand internal ability. This means that the origin and the evolution of life occurred via the formation of a new continuous internal connection through interaction between a reaction network or a living thing and the environment, though we cannot be conscious of such a continuous internal connection itself. Theoretically, it is reasonable to consider that interactions between a variety of things lead to the formation of a new continuous internal connection of them through the overlapping of their original continuous internal connections or by other advanced mechanisms because in the very truth everything exists as a continuous internal connection of all other things (see Section 1.4.1 and 2.1.2). In fact, the attainment of *Tai-toku* indicates that thorough exercise, i.e. thorough interactions between constituent elements lead to the formation of a new continuous internal connection of them (see Section 1.3.2). Quantum mechanics also indicates that interaction between things leads to the superposition of their wave functions, i.e. the formation of a new continuous internal connection of them (see Section 2.1.2).

How have evolution and heredity achieved harmony?

The above mechanism, however, involves an important problem to be considered carefully. It is now well known that structures and processes of a living thing are completely controlled by genes that are kept stable. Therefore, the traditional theory of evolution such as Central Dogma definitely denies the concept of the inheritance of acquired traits and says that evolution can occur only via a mutation in genes (see Section 4.1.1). According to this idea, the above mechanism that the evolution of life occurs via the formation of a new continuous internal connection should be denied because it includes the concept of the inheritance of acquired traits.

It should be noted here that the origin and the evolution of life cannot be explained by Darwin's theory of mutation and natural selection alone, or to be accurate, by the Darwinism established in the 20th century alone[*1]. This is because a system following this theory is in an entirely passive state against an attack of the outer world and has no internal ability to organize itself so that it realizes a free independent harmonious stable state. Therefore, this theory cannot explain the emergence of autonomous dynamic self-organizing ability or free independent spirit and wisdom of living things. Thus we have to say that the Darwinism established in the 20th century, including Central Dogma, also has a serious problem.

[*1] It is said that Darwin himself had a different idea from the Darwinism established in the 20th century, as already mentioned in a footnote #1 of Section 4.1.1. It appears that Darwin's original idea included the thought that a mutation occurred so

that it improved adaptation, similar to Lamarck.

Now, how can we overcome the above problem? Originally, evolution and heredity or evolution and self-stabilization are mutually conflicting concepts of the same type as discussed in Section 1.2.3. Therefore, they are impossible to stand together under word-based understanding. Thus, the traditional theory of evolution such as Central Dogma, which did not take into account the existence of the internal invisible world, dealt with the mutual conflict between evolution and heredity by adopting the concept of a mutation (an *accidental* alteration) in a gene. However, in such a traditional idea, evolution is unfairly suppressed, heredity or self-stabilization being too emphasized. In the very truth, even mutually conflicting concepts realize dynamic harmony, as discussed in Section 1.4.1. Therefore, it is important to look for another possibility of overcoming the above problem.

Indeed, there is no doubt that living organisms have given the top priority to the stabilization of genes (DNA) because it is absolutely important for multi-cellular living organisms to keep genes in each cell completely stable for accurately controlling a tremendously large number of complex chemical reactions. However, it should be noted that this argument indicates that genes are not the most basic element for controlling structures, properties and functions of a living organism. A living organism keeps genes stable so that stable genes keep a living organism stable. Namely, a living organism uses genes as a tool for stabilizing itself. If we use a metaphor, genes in a living organism are like the law in human society. The law indeed controls our life and is transmitted from generation to generation. However, this is because we human beings use the law as a tool for keeping society stable. This metaphor indicates that the most essential element that determines the evolution of living organisms is not genes but living organisms themselves or the internal ability of living organisms to organize them themselves. Thus, we can expect that living organisms have a certain proper means of controlling their genes including those in gametes.

In harmony with this expectation, recent advancement of molecular biology has revealed that the traditional theory of evolution or the traditional concept of genes should be greatly changed, as briefly surveyed at the end of Section 4.1.1. An important fact is that DNA chains form gene adjustment networks consisting of gene switching and adjustment genes as well as genes themselves. Accordingly, evolutionary developmental biology has now concluded that the diversity of the phenotype of living things in most cases arises not from differences in genes but from differences in gene adjustment networks or gene switching. This means that the evolution of living things has mostly occurred by changes in gene switching rather than changes in genes. In this way, recent molecular biology has come to show a possibility that the evolution of life occurred via the formation of a new continuous internal connection through interactions between living things and the environment.

Some events giving support to the occurrence of evolution via the formation of a new continuous internal connection

Moreover, we can show a number of events that give support to the occurrence of evolution via the formation of a new continuous internal connection. In the first place, all plants and animals on the present earth have *completely* harmonious structures and functions *with no exception*. For example, a giraffe indeed has a long neck but simultaneously has a body and legs with completely harmonious structures with a long neck. The same holds for all plants and animals. Such facts are entirely difficult to explain by "mutations in genes and natural selection" alone, without taking into account the formation of a new continuous internal connection by interaction between living things and the environment. The concepts of mutations and natural selection involve no power to achieve complete harmonization in a living body.

Secondly, the emergence of multi-cellular living organisms was soon followed by the appearance of a large diversity of multi-cellular living organisms in the Ediacaran period and the following Cambrian period, as mentioned in Section 4.3 (see also Figure 4-1). Such an event is also quite difficult to explain as due to lucky mutations in genes because it is implausible that a large diversity of favorable lucky mutations happened by chance for a relatively short period of time. Such an event can be best explained as due to the formation of a large variety of new continuous internal connections through interactions between multi-cellular organisms and the environment.

In fact, as discussed in Section 4.3, a multi-cellular living organism acquired the ability to organize itself so that "a multi-cellular living organism and information about the outer world" as a whole realized a well-organized free independent harmonious stable state with the lowest resistance. Namely, a multi-cellular living organism acquired the ability to behave intentionally with an aim. In addition, a multi-cellular living organism acquired the ability to produce a diversity of high functions by allotting various individual functions to parts of a large-sized body and combining them so that they as a whole produced high functions. Therefore, it is quite reasonable to assume that multi-cellular living organisms were able to form a large variety of new continuous internal connections in the direction of producing desirable high functions through interactions between them and the environment. Such new continuous internal connections will have affected gene adjustment networks including those in gametes.

Thirdly, it is known that a variety of Homo species such as Homo habilis, Homo erectus and Homo neanderthalensis emerged before Homo sapiens emerged. In addition, such various Homo species emerged in the order of increasing brain size, suggesting that they emerged according to a certain rule. Thus, this fact is also very difficult to explain as due to accidental alterations in genes. It is probable that ancient monkeys noticed through their life (or through their interaction with the environment) that common properties of natural things and imaginary things with such properties were useful for their life and then enlarged their brain for catching as many such properties or such things as possible [*2], as mentioned in Section 1.2.1. This trial finally led to the emergence of Homo sapiens.

[*2]Common or repeatedly observed properties are expected to have been relatively easy to catch because such properties existed as unchanging nervous patterns in the brain (see Chapter 5).

Fourthly, the rate of evolution in the evolution of life became higher and higher as the evolution went up to an advanced stage though structures and functions of living organisms produced by the evolution became more and more complex. For example, it took a very long time of about 2 to 3 billion years for unicellular living organisms to evolve to multi-cellular living organisms while it took only 0.35 billion years for the first multi-cellular living organism to evolve to high animals, e.g. vertebrate (see Figure 4-1). Moreover, it took only about several million years for ancient monkeys (the ancestor of human beings) to become human beings though ancient monkeys had to produce a high-quality big brain.

This fact is also quite difficult to explain in terms of lucky mutations in genes. On the other hand, this fact can be reasonably explained if we take into account "a positive feedback mechanism" for the improvement of the ability to form a new continuous internal connection, which will be discussed in Section 6.3. To be brief, a living organism in an advanced stage of evolution consists of a complexly-entangled multi-fold multi-dimensional network of a huge number of continuous internal connections and thus has the ability to form a diversity of high-quality new continuous internal connections.

An information-based connection has played an important role in the evolution of life

There is another important argument that supports the thought that living organisms have evolved via the formation of a new continuous internal connection through their interactions with the environment. A glance of the evolution of life indicates that the quality of the mechanism for forming a new continuous internal connection became higher and higher as the evolution went up to an advanced stage. At the initial stage, direct physicochemical interactions (i.e. chemical reactions) between chemical substances led to the formation of a new continuous internal connection of them. The emergence of the first living organism on the primitive earth can be explained by this mechanism. However, the emergence of multi-cellular living organisms is impossible to explain by this mechanism. As mentioned in Section 4.3, a continuous internal connection of chemical reactions was limited within an individual cell of primitive bacteria surrounded by a cell membrane. Therefore, for the emergence of multi-cellular living organisms, a new continuous internal connection of individual cells had to be formed.

In my opinion, such a new continuous internal connection in a multi-cellular living organism is formed by utilizing an information-based connection between continuous internal connections in individual cells, as illustrated in Figure 4-6. Here, an information-based connection refers to an indirect connection between things by the use of signal transmission or analogs, in sharp contrast to direct physicochemical connections between chemical substances. An important point is that the

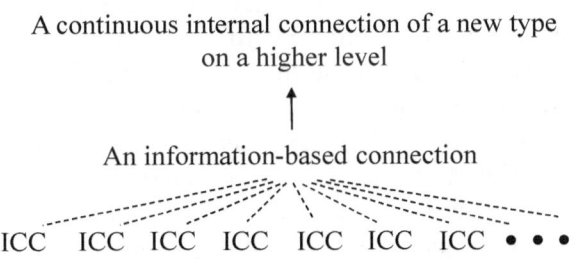

ICC = a continuous internal connection

Figure 4-6 The formation of a continuous internal connection of a new type, by using an information-based connection.

formation of an information-based connection between continuous internal connections leads to the formation of a continuous internal connection of a higher quality or on a higher level.

Careful investigations indicate that there are various kinds of information-based connections. It appears that even a reaction network at the initial stage already had an information-based connection. For example, collective catalytic actions in a reaction network such as proposed by Kauffman and his group may be regarded as an information-based connection formed between internally continuously connected chemical reactions. Coupling between RNA and polypeptide formation and the formation of a cell membrane in a reaction network may also be regarded as an information-based connection of a similar type.

An information-based connection is also formed in a hierarchical way. A high animal such as a human being has a continuous internal connection of various organs such as the head, the heart, the lung, and so on, because they have to work so that they as a whole realize a harmonious stable state with the lowest resistance. It is expected that such a continuous internal connection is obtained by making an information-based connection between continuous internal connections formed in various organs, which are in turn each obtained by making an information-based connection between continuous internal connections in individual cells.

Moreover, an information-based connection is formed between a living organism and things in the outer world. A shellfish produces a shell and bees produce a honeycomb outside their bodies. We human beings produce various machines such as bicycles, cars and robots outside our body. An important point is that a shell is produced so that "a shellfish and a shell" as a whole realizes a well-organized harmonious state. The same holds for a honeycomb for bees and a machine for a human being. These facts indicate that a continuous internal connection is formed between a living organism and things in the outer world. In fact, it was argued in Section 1.3.2 that a person who has attained *Tai-toku* about riding a bicycle has formed a continuous internal connection of his or her body and a bicycle. Now, it is certain that such a continuous internal connection is formed by making an information-based connection between a living organism and things in the outer world because there is no direct contact between them.

The above argument for an information-based connection between a living organism and things in the outer world can be generalized to an information-based connection between two or more living organisms.

Namely, an information-based connection can be formed between various living organisms. The emergence of the ecological system can be explained by this principle. In this way, a large diversity of continuous internal connections of different qualities is formed by the use of an information-based connection.

The origin of an information-based connection between continuous internal connections lies in continuous internal connections themselves

Now, how is an information-based connection formed between continuous internal connections? A prominent characteristic of an information-based connection is that it is visible, in contrast to a continuous internal connection which is invisible. In fact, relations between individual cells in a multi-cellular living thing, those between various organs in high animals, and those between a living organism and things in the outer world are all visible. Collective catalytic actions, a cell membrane, and coupling between RNA and polypeptide formation in a precursor of the primitive life are also visible and can be understood within the realm of word-based understanding. This means that an information-based connection represents an "external" connection between truly real things existing each as a continuous internal connection.

The argument that an information-based connection is visible means that it is of a mechanical character, because visible things have individual unchanging properties and have a mechanical character (see Section 1.1.1 and 3.2.3). Thus, we can obtain the following interesting conclusion.

> *An information-based connection of a mechanical character can be formed between continuous internal connections of a non-mechanical character and this leads to the formation of a continuous internal connection of a new type on a higher level, which is of a non-mechanical character.*

In fact, living organisms incorporate a variety of information-based connections of a mechanical character, without losing their basic qualities of a non-mechanical character such as autonomous dynamic self-organizing ability or free independent harmonious spirit and wisdom. For example, high animals make use of dissipative structures such as nervous impulses and brain waves, which are information-based connections of a mechanical character (see Section 3.2.3).

How can we understand the above argument? According to Tanaka's book [4], only signals are transmitted in an information-based connection. Signals are formed by agreed rules between a signal sender and a signal receiver. Therefore, signals have meanings only between a signal sender and a signal receiver. In addition, the energy and mechanisms for creating signals and those for receiving signals are also provided by sites of a signal sender and a signal receiver, respectively. Therefore, signal transmission (or an information-based connection) merely represents "results" of activities of continuous internal connections (or truly real things existing as continuous internal connections) acting as a signal sender and a signal receiver. Therefore, it is natural that continuous internal connections of a non-mechanical character can form an information-based connection of a mechanical character without losing their basic qualities of a non-mechanical character. The origin of an information-based connection between continuous internal connections is within continuous internal connections themselves[*3].

[*3] It is interesting to note here that the above argument has an important philosophical meaning. It was argued in Section 1.2.3 and 1.4.2 that scientific laws only represent "results" of activities of truly real things. According to the above argument, this means that scientific laws only represent information-based connections between continuous internal connections or truly real things, underlying continuous internal connections being completely overlooked. Quantum mechanics gives us the same view because it says that all things in the external visible world represent "results" of activities of an invisible state described by a wave function (see Section 2.1.1 and 2.1.2).

Continuous internal connections can produce a diversity of high-level abilities by the use of an information-based connection

Let us consider the foregoing arguments in more detail. As mentioned earlier, the formation of an information-based connection between a large number of continuous internal connections leads to the formation of a continuous internal connection on a higher level (see Figure 4-6). This means that a large number of continuous internal connections can produce a diversity of high-level abilities by forming an information-based connection between them.

In fact, for example, a multi-cellular living organism, produced by forming an information-based connection between a large number of individual living cells, demonstrates a diversity of high-level abilities such as adaptive behavior and intentional motion with an aim, which are far beyond abilities of constituent individual cells. Similarly, the human brain, produced by forming an information-based connection between a large number of neurons, demonstrates a diversity of high-level abilities such as thinking, creating and judging, which are far beyond abilities of constituent individual neurons (see Chapter 5). Another interesting example is human society, which is formed by an information-based connection between a large number of individual persons. Human society demonstrates a huge diversity of high-level properties and functions in various fields, which are far beyond abilities of constituent individual persons.

There are examples of another type. A watermelon produces delicious fruits with hard seeds in such a way that animals eat fruits together with seeds and excretes seeds somewhere in a distant place with no digestion. It is clear that this is a very effective way for multiplication of a watermelon. Surprisingly, a watermelon behaves as if it were aware that an animal digests only a fruit and excretes seeds somewhere without digestion [*4]. It is highly probable that such excellent ability of a watermelon comes from an information-based connection between continuous internal connections in water melons and those in animals. Not only a watermelon but also almost all fruits have seeds of the same character as water melons. It is also known that blooming plants and insects show similar cooperative behavior.

[*4] This fact becomes easier to understand if we regard water melons and animals together as one living body. By experiences we can understand different parts of one living body to work in a cooperative manner.

Relatively short lifetimes of high plants and animals in the present days can also be regarded as a result of evolution based on information-based connections between them. Originally, living organisms can have much longer lifetimes if they want, because unicellular bacteria in the present days have continued to survive for about four billion years though through cell division. Therefore, it is very likely that living organisms chose to have relatively short lifetimes at the stage at which sexual reproduction started so that they were able to evolve more effectively.

From the above examples, it is clear that continuous internal connections can produce a diversity of high-level abilities by forming an information-based connection between them though it is unknown how an information-based connection is formed. The evolution of life is a representative field of science, which should be understood from a deep level by taking into account the internal invisible world.

References

(1) W. Hiromatsu et al. (editors), *Encyclopedia of Philosophy and Thought* (in Japanese), Iwanami-shoten, Tokyo, 1998.

(2) M. Kimura, *A Consideration of the Evolution of Living Things* (in Japanese), Iwanami-shinsho, Iwanami-shoten, Tokyo, 1988.
(3) M. Mitchell, *"Complexity: A Guided Tour"*, Oxford University Press, Inc., Oxford, 2009: The Japanese edition translated by H. Takahashi, Kinokuniya Shoten, Tokyo, 2011.
(4) H. Tanaka, *Life and Complex Systems* (in Japanese), Baihukan, Tokyo, 2002.
(5) J. Isemura, et al. (editors), *Modern Biology 2, Living Bodies and Energy,* Iwanami-koza, Iwanami-shoten, Tokyo,1966.
(6) A. I. Oparin, *The Origin of Life – The Emergence of Life and Its Development at the Initial Stage,* Moscow, 1966: The Japanese edition translated by M. Ishimoto, Iwanami-shoten, Tokyo, 1969.
(7) M. Calvin, *Chemical Evolution – Molecular Evolution towards the Origin of Living Systems on the Earth and Elsewhere,* Oxford University Press, Inc., Oxford, 1969: The Japanese edition translated by F. Egami, Y. Kuwano, T. Kimura, and K. Nakamura, Tokyo-kagaku-dojin, Tokyo, 1970.
(8) P. L. Luisi, *The Emergence of Life From Chemical Origins to Synthetic Biology,* Cambridge University Press, Cambridge, 2006: The Japanese edition translated by T. Shirakawa, P. Y. Gunji, NTT Shuppan, Tokyo, 2009.
(9) J. L. Monod, *Le Hasard et la Nécessité, Essai sur la philosophie naturelle de la biologie moderne,* Alfred A. Knopf, Inc., Paris, 1971: The Japanese edition translated by I. Watanabe and M. Murakami, Misuzu-shobo, Tokyo, 1972.
(10) S. Kauffman, *At Home in the Universe: The Search for Laws of Self-organization and Complexity*, Oxford University Press, Inc., Oxford, 1995: The Japanese edition translated by F. Yonezawa, Nihon-keizai-shinbun-sha, Tokyo, 1999.
(11) For example, E. V. Koonin and W. Martin, *Trends in Genetics*, **21** (12), 647 (2005).
(12) D. Braun, et al., *Proc. Natl. Acad. Sci. USA,* **104** (22), 9346 (2007).

5. The Origin of Human Ego and Consciousness

Human ego and consciousness are familiar concepts to us. However, surprisingly, their true meanings have not been clearly understood to date. In this chapter we consider where the origin of human ego and consciousness is, based on a new theory of the origin and the evolution of life proposed in the preceding chapter.

The conventional understanding of human ego and consciousness

At first, let us examine how human ego and consciousness have been understood to date. The concept of human ego was for the first time established by Descartes, as mentioned in Section 1.3.2 (see also Section A3.1.1). According to Descartes, human ego is completely free from anything, only depending on pure thinking. It is the subject that looks at the outer world including his or her own ego itself, obtains the correct objective knowledge, and rationally judges based on it. Thus, a person who has established his or her ego can live a free independent and rational life.

In psychology, human ego is regarded as the subject of conscious activities of an individual person. Consciousness is regarded as the perception of the ego's activities. A glance at our daily life indicates that our ego does a great variety of things. Our ego is in contact with the outer world through the senses and recognizes what exist there and how they exist there. Our ego is also in contact with inner bodily or mental activities and recognizes in what state they are. In addition, our ego keeps information about the outer and the inner worlds as memories and freely evaluates, judges and plans. Furthermore, our ego unifies all recognized things and keeps them in harmony with one another in order to protect it itself from falling into confusion.

Now, such conventional understanding of human ego may look reasonable but in reality it only describes observed properties and functions of human ego as they are, by regarding them as given a priori, without clarifying anything about the origin of them. Such a way of understanding is characteristic of word-based scientific understanding, as emphasized in Section 1.2.3. Therefore, the fundamental problem of where the origin of human ego is has still been left unclear. In fact, for this reason, the dualism of body and mind has not been overcome to date (see Section 1.1.2, 1.2.3 and 1.4.2).

The origin of human ego

Now, where is the origin of human ego? The best way to solve this problem is to investigate how it emerged in the course of the evolution of life. It was argued in Section 4.2 that continual inflow of various organic compounds into a colloidal drop in a small inorganic compartment and their interactions (chemical reactions) in it led to the formation of a reaction network in which almost all reactions were in or near equilibrium and internally continuously connected with one another. Such a reaction network had the internal ability to organize itself so that it as a whole realized a well-organized free independent harmonious stable state with the lowest resistance. The emergence of the first living organism on the primitive earth was successfully explained based on the formation of such a reaction network. Thus, the "internal ability" of such a reaction network can be regarded as the most primitive form of human ego.

The internal ability of a reaction network mentioned above advanced to that of a unicellular living organism and further advanced to that of a multi-cellular living organism. A multi-cellular living organism acquired the ability to gain and keep information about the outer world and thus acquired the ability to behave so that "a multi-cellular living organism and information about the outer world" as a whole realized a well-organized free independent harmonious stable state. This meant that a multi-cellular living organism acquired new ability to move toward the outer world intentionally with an aim (see Section 4.3). Intellectual ability of human beings can be explained as an advanced form of such new ability.

Based on these arguments, we can safely say that human ego is the internal ability of an individual person to organize him- or herself so that he or she realizes a well-organized free independent harmonious stable state with the lowest resistance. Such internal ability comes from infinitely-spreading fully-continuous internal connections of constituent elements which a person has formed.

We can also say in the following way. As argued in Section 1.4.1, 2.1.3, 3.2.1, etc., all things in nature and the world including human beings are controlled by the grand internal ability to make various things interact with one another according to their inherent properties and motion, form as widely-spreading continuous internal connections of them as possible under given conditions in diverse quality levels and aspects, and complete the formation of such continuous internal connections. Therefore, various ideas and behavior of a human being can be regarded as "results" of activities of the grand internal ability. According to this view, human ego is nothing else but the grand internal ability a person catches within him- or herself[*1].

> [*1] If we say in further detail, the grand internal ability comes from the basic qualities of microscopic particles such as electrons and atomic nuclei, as argued in Section 2.1.3. Namely, it comes from the wave nature of electrons and atomic nuclei (or the internal power to make them exist simultaneously in all possible spatiotemporal positions under given conditions) and their various kinds of interactions (potential energy) and free motion (kinetic energy).

A living organism survives by taking energy and chemical substances from the outer world and utilizing them as a driving force (see Section 4.2 to 4.4). In this sense, we should say that the power of human ego comes from such energy and chemical substances.

How is a visual image formed?

Is the above explanation of human ego correct? In order to examine this question, let us consider how a sensory image is formed. As mentioned earlier, human ego is in contact with the outer world through the senses and recognizes what exist there. Therefore, it is certain that a sensory image is formed by human ego. Fortunately, we can now understand in fair detail how a visual image is formed in our brain. Brain science has recently made great progress through the invention of a new technique of detecting an active area in the brain, called fMRI (functional magnetic resonance imaging). According to Sakai's book (2008) [1], a visual image is formed roughly as follows.

The human brain consists of the occipital lobe, the temporal lobe, the parietal lobe, the frontal lobe, and so on. The occipital lobe lying at the back of the head consists of various visual areas. The first, the second, and the third visual areas take charge of the presence or the absence of objects as well as their spatial arrangements in the outer world, the fourth visual area has charge of color, the fifth visual area has charge of movement, and so on. The temporal lobe at the right and the left sides of the head also consists of a variety of visual areas such as the face-perception area (the area which becomes active when a person looks at a human face), the building-perception area (the area which becomes active when a person looks at the scenery such as a building, a street, and a mountain), and so on. Similarly, the parietal lobe at the upper part of the head and the frontal lobe at the front part of the head also consist of various areas.

Visual information from the outer world is thus treated as follows. When a person looks at certain objects in the outer world, visual areas located in the occipital lobe become active. Namely, information from the outer world is first received in visual areas in the occipital lobe. To be accurate, information is first received by the first visual area in the occipital lobe and then transmitted to the second visual area, the third one, and so on. The visual information is further transmitted to the parie- tal lobe and to the frontal lobe. Visual information from the outer world is also transmitted to the temporal lobe. Namely, it is transmitted to the face-perception area, the building-perception area, and so on, and then transmitted to the frontal lobe. In addition, visual information is memorized in the face perception area, the building perception area, and so on by the act of the hippocampus and also recalled by the act of the hippocampus.

Furthermore, visual information is transmitted in the reverse direction. When a person pays attention to a certain particular object in the outer world, an active part first emerges in the frontal lobe and it is transmitted to the parietal lobe and to visual areas in the occipital lobe or in the temporal lobe. The activity in the frontal lobe gives a weight to information from a certain particular object (for example, an object which is located in a particular position or has a particular quality). This fact indicates that the frontal lobe accepts information from the mind and works in harmony with it.

In this way, a visual image is created after active areas have traveled forward and backward in the brain. No consciousness of a visual image emerges without having the backward movement of active areas from the frontal lobe to the occipital lobe and/or the temporal lobe. This result indicates that the whole area of the brain works to create a visual image. This result also indicates that a person is not conscious of all nervous processes. Thus, it often happens that a person is conscious of nothing even when active parts emerge and travel the brain.

Furthermore, the above result indicates that the human brain not only receives information from the outer world but also positively acts on it. For example, the frontal lobe gives a weight to information from a certain particular object, as mentioned above. Nevertheless, interestingly, recent brain science has revealed that there is no special nervous cell or site in the brain, which acts as the controlling center or commandant for nervous processes. When a person looks at something in the outer world, various areas of the brain become active and active areas move here and there, and this is all. Nothing else happens in the process of producing a visual image.

There is another important fact. Recent studies in brain science have disclosed that the human brain only produces a harmonious meaningful image. Let us assume that a person looks at different pictures by the right and the left eyes. For example, let us assume that a person looks at a picture of a human face by the right eye and looks at a picture of a building by the left eye. Such a person never has visual images of both pictures simultaneously. Such a person also never has randomly mixed visual images of these pictures. Such a person only has an image of either a human face or a building, though such an image is replaced with each other alternately in a short period of time of several or several ten seconds. There are a number of experiments of different kinds which lead to the same conclusion. For instance, when a person looks at a human face resembling Marilyn Monroe, the person recognizes it as Marilyn Monroe even if it is exactly not Marilyn Monroe.

Now, the above-mentioned results of recent studies of brain science can be summarized as follows. (1) A visual image is created spontaneously after active areas traveled forward and backward in the brain. (2) A human being is only conscious of a "result" of various nervous processes. A human being is not conscious of all nervous processes. (3) The human brain not only receives information from the outer world but also positively acts on it. (4) Nevertheless, there is no special nervous cell or site which acts as the control center or commandant to nervous processes in the brain. (5) The whole area of the brain, including the occipital lobe, the temporal lobe, the parietal lobe and the frontal lobe, together with the hippocampus treating memorized information, works to create a visual image. (6) The brain produces only a meaningful harmonious image as a whole.

How can we explain the above results? They can be explained reasonably if we consider that a visual image is produced by human ego, i.e. the grand internal ability we catch within ourselves, as mentioned earlier. We should rather say that the above results can be explained only by taking into account the grand internal ability. First of all, a visual image is spontaneously produced without using any controlling center or commandant for nervous processes in the brain. This fact can only be explained by taking into account the grand internal ability we catch within ourselves. In fact, the grand internal ability works so as to make various things (various nervous patterns due to information coming from a diversity of sources) interact with one another, form as widely-spreading a continuous internal connection of them as possible under given conditions, and complete the formation of such a continuous internal connection. Thus, the grand internal ability allows the spontaneous production of a visual image.

A visual image is also produced so that all nervous patterns due to information coming from the outer and the inner worlds as well as memorized information take a harmonious meaningful form. This fact can again be explained only by taking into account the grand internal ability we catch within ourselves. In fact, the grand internal ability produces a thing which realizes a free independent harmonious stable state, as argued in Section 1.4.1, 2.1.3, 3.2.1, etc.

Furthermore, the human brain not only receives information from the outer and the inner worlds but also positively acts on it. Nervous patterns due to information coming from various sources exist individually and separately and besides have a widely-spreading continuous shape with no definite boundary. Therefore, no visual image with a definite meaning can be produced if there is no positive act on such nervous patterns. In other words, no visual image is produced by physiological or physicochemical processes alone. Such a positive act comes from the grand internal ability we catch within ourselves.

In general, it has been believed that nervous cells (or neurons) are responsible for brain activities such as perception, thinking, feelings, and so on. However, when we only look at activities of neurons, i.e. when we only look at the motion of active parts in the brain, we can hardly understand how a visual image is formed. In fact, a visual image is spontaneously and suddenly produced after active areas move here and there, as mentioned earlier. This indicates that a visual image is actually produced in the internal world we cannot be conscious of, in agreement with the above argument.

How are ideas created?

Next, let us consider how ideas (corresponding to words, concepts and laws) are created. The creation of ideas can be explained in a similar way to the production of a visual image.

When a person looks at an object in the outer world, a visual image is created, as mentioned above. Such a visual image exists as a continuous internal connection of various nervous patterns due to information coming from various sources. Let us now assume that a person looks at various objects in the outer world, which have "common properties". In such a case, various visual images (i.e. various continuous internal connections of nervous patterns) are produced in the brain and besides they should overlap one another [*2] in the area of "common properties", just as dumped newspapers overlap one another. Thus, if such an overlapping area of visual images (or continuous internal connections) is picked out and formed as a new continuous internal connection, such a new continuous internal connection gives a new idea representing "common properties" or an imaginary thing with such common properties.

> *²The word of "overlap" is here used in a broad sense. A similar argument was given in Section 4.4. Namely, it was argued there that interactions between a variety of things lead to the formation of a new continuous internal connection of them through the overlapping of their original continuous internal connections or by other advanced mechanisms.

It should be noted again that it is by a positive act of human ego that an overlapping area of visual images (continuous internal connections) is picked out and formed as a new continuous internal connection. The creation of a new idea with a definite meaning cannot be explained without taking into account such a positive act (i.e. the act of the grand internal ability we catch within ourselves) because visual images in an overlapping area still exist individually and separately and in addition an overlapping area has a widely-spreading continuous shape with no definite boundary.

The above mechanism for creating a new idea can be generalized to the creation of highly abstract ideas in science and philosophy. When a person creates a large number of sensory images and ideas (i.e. continuous internal connections of nervous patterns) and keeps them as memories, various overlapping areas are created between them and moreover overlapping areas further overlap one another, resulting in the formation of doubly, triply, or more-time overlapping areas in sensory images and ideas (continuous internal connections of nervous patterns). Similar processes occur when a person learns various pieces of knowledge and think of them. Thus, if such a multi-fold overlapping area is picked out and formed as a new continuous internal connection, it gives a new idea with a highly abstract meaning. It is interesting to note that this mechanism explains why the formation of a large-sized brain was necessary for ancient monkeys to evolve to human beings. This mechanism also explains why words with more and more abstract meanings have been produced in human history with the advancement of human understanding.

Our ego and the true self

Our ego does various things, as mentioned at the beginning of this chapter. It creates various sensory images and ideas, as discussed above. In addition, our ego keeps sensory images and ideas as memories and freely evaluates, judges, plans and leads us to action. Furthermore, our ego unifies all recognized things and keeps them in harmony with one another. Such various acts of our ego can all be clearly explained if we consider that our ego is the grand internal ability we catch within ourselves.

Now, our ego plays an absolutely important role in our life, as mentioned above, but has a serious limit. This is because the origin of our ego, i.e. the grand internal ability works in multiple ways, as discussed in Section 1.4.1. Namely, the grand internal ability produces our ego and lets it show lively activity. Such activities of our ego, i.e. interactions between our ego and the outer world lead to the formation of a new continuous internal connection of them. This means that our ego comes to have a limit because it does not include a new continuous internal connection thus formed. Then, our ego comes to notice that it has a limit and tries to overcome it. After all, the grand internal ability produces our ego, lets it show lively activity, produces a limit in it, makes it overcome such a limit, and advances its ability. The grand internal ability working in such a way can be called the "true self". Thus, we human beings are controlled by two kinds of the subject: our ego and the true self.

C. Jung developed the theory of psychoanalysis in depth psychology, started by S. Freud, and proposed a new concept of the collective unconscious [2]. According to Jung, the personal unconscious, with which Freud was mainly concerned, originates from particular experiences of an individual person in the past, while the collective unconscious is irrespective of such personal experiences in the past. It is shared by all people. Jung explains the collective unconscious based on archetypes (original patterns) and archetype images. Thus, Jung's collective unconscious may have a meaning similar to the above "true self". However, depth psychology is developed only by inference based on experiences of patients with mental illness and is not necessarily clear enough. On the other hand, the concept of the "true self" in this book is developed based on real experiences of all people and has a firm basis. Accordingly, the meanings of Jung's collective unconscious are expected to be better understood when they are investigated based on the "true self" in this book.

References

(1) K. Sakai, *Brain Science of the Mind* (in Japanese), Chuko-shinsho, Chuo-koron-shinsha, Inc. Tokyo, 2008.
(2) H. Kawai, *The Structure of the Unconscious* (in Japanese), Chuko-shinsho, Chuo-koron-shinsha, Inc. Tokyo, 1977.

6. How Is Creation Achieved?

In the preceding chapter we considered how we human beings create sensory images and ideas. Based on this consideration, in this chapter we investigate how we achieve creation. Interestingly, the achievement of creation happens via the formation of a new continuous internal connection, in the same way as the evolution of life discussed in Section 4.4. Accordingly, the clarification of the principle of creation is effective for understanding the evolution of life.

6.1 How Has Creation Been Understood to Date?

Creation has been regarded as one of the most valuable human activities

To begin with, let us consider how creation has been understood to date. In general, we human beings have our own jobs and display activity with particular purposes in daily life. In such a situation, creation emerges as giving a fundamental solution to a difficult problem that has long been left unsolved. Creation also appears as a manifestation of a new significant fact or possibility which nobody has ever imagined or expected. Thus, creation makes a breakthrough in human civilization and brings about great progress in societies. Therefore, creation has been regarded as one of the most valuable human activities. In fact, persons who achieved great creation in the past, such as Aristotle, Descartes, Newton, Hegel, *Shaka-muni*, Confucius, *Shin-ran*, and so on, have left their names behind. Not only in science and philosophy but also in fields of art, technology, business and politics, a large number of persons who achieved great creation in the past have handed down their names to posterity.

Apart from such great creation, there are various sorts of daily creation such as completing a good piece of work and hitting on a new splendid idea. Such daily creation also makes important contributions to the progress of societies. First of all, such daily creation is important for our own personal life.

Naturally, creation does not always exert a beneficial effect on our life or societies. Some of created things cause harmful or destructive effects on them. Therefore, it is necessary to make appropriate selection about created things. Note that the achievement of such selection is also regarded as creation. How created things are selected in societies will be discussed in Section A3.3.

Creation has been regarded as happening by chance

In this way, creation is a very valuable thing. However, surprisingly, it seems that nothing essential has been clarified about how creation is achieved and how we can improve creativity. This is most probably because creation has been really a strange event. In fact, a person hits on a new idea suddenly and unexpectedly in a flash of inspiration. Thus, even a person who has hit on a new idea cannot know why he or she has been able to do so. Thus, creation is quite difficult to understand. Strangeness is a basic characteristic of creation.

Accordingly, creation has in general been understood to happen by chance. Creation is just accidental good luck. Therefore, it is probably a common opinion at present that it is far beyond scientific understanding to reveal how creation is achieved. Some of people may even say that creativity is the innate ability of individuals and impossible to control a posteriori. In such a situation, it may sound a silly talk to say that we can achieve creation in the desired direction. However, this is roughly the case, as will be discussed in the next section. In fact, living things have evolved in their desired direction. Human beings have also constructed their culture in their desired direction.

The origin of creation and creativity has not been clarified yet

Now, why has creation been regarded as a strange event? To my understanding, a main reason is that no origin of creation and creativity has been clarified yet. In fact, there are many facts which support this opinion. Let us examine some of them in order to know what limit traditional understanding has possessed.

In the first place, several theories have been proposed about generation and growth, which have close relation to creation and creativity. However, no origin or cause of these events has been clarified yet. For example, G. Hegel, famous German philosopher in the 19th century, dealt with generation and growth theoretically, based on his unique philosophy called dialectic. He says [1], "Everything is always within itself and simultaneously in relation to the others and thus everything exists in the internal unification of mutually opposing concepts." According to Hegel's philosophy, the internal unification of mutually opposing concepts leads to generation and growth.

Hegel's philosophy has another prominent characteristic in that it pays attention to properties arising from a whole consisting of widely interacting things. Hegel says [1], "Causes and results in causality only express a few moments of entire universal connections of a huge variety of events in the world and also only represent a few links of entire chains of developments of things. Causality expresses the entire universal relations in the world only fragmentarily and in a one-sided manner". Interestingly, this opinion of Hegel is very similar to an opinion of researchers in complexity science in the present days.

In this way, indeed Hegel succeeded in constructing a great theoretical system about generation and growth, which was linked with the highest human idea. However, he had still not achieved the clarification of the origin of generation and growth. A serious problem in his philosophy is that he considered that all things existed in the form of clearly defined concepts and thus the internal unification of mutually opposing concepts was described by logic. Therefore, problems of how mutually opposing concepts are internally unified and by what mechanism the internal unification of mutually opposing concepts brings about generation and growth have been left uncertain. We have to say that Hegel's philosophy only gave a "logical or external explanation" to generation and growth and provided almost nothing about the origin of them. In other words, Hegel's philosophy was still within word-based logical understanding and failed to overcome the intrinsic limits of it (see Section 1.2.3). In general, dialectic has been regarded as a philosophy that allows us to surmount mechanism such as seen in Descartes' philosophy. However, dialectic still has a mechanical character because it is based on word-based logical understanding.

Theoretical consideration of creation was also given by M. Polanyi, famous physical chemist and philosopher in the middle of the 20th century [2]. Based on several psychological experiments, Polanyi proposed the concept of "tacit knowledge", which refers to non-linguistic comprehensive knowledge as compared with linguistic analytical knowledge. He then argued that it was by tacit knowledge that a human being was able to distinguish a variety of human faces, exchange opinions, and perform sports and other technical skills. Moreover, he said that creativity in science came from tacit knowledge, namely, a scientist was able to become aware of a problem and discover a new fact by tacit knowledge.

I agree with Polanyi's opinion on some basic points. Tacit knowledge may have the same meaning as "understanding without using words" discussed in Section 1.3.1. However, Polanyi did not clarify where tacit knowledge comes from, why it has a tacit character, why it can act as creativity, how we can obtain it, and how we can improve it. Polanyi's discussions also lack the clarification of the origin of creation and creativity, in a similar way to Hegel's philosophy.

Recently, complexity science has constructed some fundamental models which allow us to investigate mechanisms for the adaptive and evolutional behavior of living things. However, it still has difficulty in completely explaining them, as pointed out in Section 1.1.2.

Creativity is impossible to control intentionally

Next, I would like to tell my personal experience, which led me to the conclusion that the origin of creation and creativity had not been clarified yet. I had interest in creativity in my youth and since then I have occasionally considered what the origin of creativity is and how we can improve it.

In the former half of the 1960's, when I was a university student, Japan slowly escaped from the damage of defeat in the second world war and started to enter into the age of industrial growth. In harmony with this trend, some of the informed people emphasized the importance of creativity for promoting science and technology. In general, they told examples of great creation in the history of science and taught us what way of thinking is important for achieving creation. For example, they recommended the following way of thinking.

Have a great dream and purpose. Have a wide view of the world. Have the spirit of exploration and adventure. Have intellectual curiosity. Have accurate knowledge. Have a flexible head. Do not follow prevailing ideas. Endure loneliness. Have the eye to intuit the essence of things. Consider thoroughly.

Indeed, most of creative persons had such abilities, indicating that these catchphrases were certainly important for improving creativity. Therefore I was initially attracted by such catchphrases and made effort to have abilities expressed by them.

However, I gradually came to feel that such catchphrases were not necessarily effective. The most important problem was that knowing and understanding such catchphrases were not equal to having corresponding abilities. For example, even if I had completely understood the importance of having a flexible head, this did not mean that I had such a head. The same held for all catchphrases. Originally, the above catchphrases represented abilities of creative persons, i.e. "results" of improvement of creativity. Therefore, it was natural that persons who had not yet improved creativity were unable to have such abilities. The above catchphrases only represented a "target" for improving creativity, not a "way" to improve creativity. In other words, they only represented the external appearance of creativity, not the origin of creativity.

In this way, I was finally led to the conclusion that creativity was impossible to control intentionally. There was sharp contrast between creativity and knowledge, because knowledge was possible to obtain intentionally. A. Einstein told a similar idea.

We are able to do what we desire to do, but we are unable to decide freely what we desire to do.

Thus, even though various valuable catchphrases were told, they all simply fell on us like water off a duck's back.

Various techniques of producing a good idea have been proposed

There is another example which shows that the origin of creation and creativity has not been clarified yet. In the field of industrial technology, various techniques of producing a good idea have been proposed for winning in mutual competition. The techniques have been successfully adopted in many companies in the world, indicating that they are indeed effective. At first, let us briefly survey some representative techniques proposed thus far, according to a book (1983) edited by Japan Creativity Society [3].

Brainstorming

Brainstorming was proposed by A. F. Osborne in 1939. In this technique, a group of people thinks about one subject at the same time so that they can create good ideas and solve a problem. The aim of this technique is to prepare artificial conditions under which people can easily hit on a new idea. Thus, in actual practice, a number of people freely talk of their opinions without any criticism and restriction and simultaneously hear unknown or strange opinions of others, which may induce keen interest or excitation. Such a free and stimulating atmosphere is expected to offer many chances of people hitting on new ideas. This technique is practically very useful because it can be easily adopted with a fruitful result. The technique is also important in that everybody can participate in and contribute to activities of a group. Therefore, this technique is now widely used in many companies in the world.

In general, a human being cannot think without having anything given. In other words, human thinking is initiated by a certain seed for thinking. A strange or unusual seed is more effective. In addition, personal thinking is usually limited to a certain particular area. Moreover, the human brain has a tendency to treat information in a particular fixed way. Thus, it is necessary to escape from such a customary way of thinking for working out a new idea. In brainstorming, these requirements are met by hearing a chain of ideas in a group of people. Each person can think of a subject in a diversity of new ideas, hypotheses or imaginations in a group.

Synectics

Synectics was started by W. J. J. Gordon in 1944 and completed by him and G. M. Prince in 1960. A prominent characteristic of this technique is to pay attention to psychological processes in creative activity and aim at constructing a technique for creative thinking based on the clarification of them. It is said that "syn" in synectics means not only "a number of people doing together" but also "combining apparently indifferent things". Namely, synectics is based on the belief that creation is to make a new combination of apparently indifferent matters and produce something with superior functions. Creation is not the production of something from nothing but the production of a new combination of already existing things. Thus, this technique makes much of looking at usual things from unusual or abnormal points of view and also looking at strange or curious things from usual or normal standpoints.

In synectics, these requirements are met by a chain of "associations" initiated by analogy, based on collective thinking by a group of people. It is believed that association plays a key role in connecting ideas with no logical relations. Naturally, it is desirable that a chain of associations extends as widely as possible. For this purpose, synectics utilizes the psychology of Freud as a theoretical basis.

The KJ method

The KJ method was worked out by J. Kawakita in around 1965 as a method of producing a new idea from a large number of qualitative data. In an ordinary method, collected data are classified into groups according to a rule prepared in advance. On the other hand, in the KJ method, collected data are classified according to properties of data themselves with no rule prepared beforehand. Kawakita says, "Let data talk". This method begins with collecting a large number of qualitative data by fieldwork or others. The data are then labeled with key subjects and classified according to the distance between the subjects. This procedure is repeated and several sets of data are produced. Next, such sets of data are spatially arranged so as to have a reasonable meaning. Finally, the content of spatially arranged data is described in the form of sentences.

The techniques surveyed above have a common characteristic in that they emphasize the importance of *divergent thinking*, i.e. free thinking of a subject from various points of view with a flexible head. The techniques have been successfully adopted in companies etc., as mentioned earlier, indicating that they are really effective. However, they all are just techniques. Therefore, we cannot be aware of where the origin of creation is and how we can improve creativity, based on them.

From considerations described thus far, we can clearly see that the origin of creation and creativity has not been clarified yet, as mentioned earlier. A main reason is that creation and creativity have been considered within the realm of word-based understanding, without taking into account the internal invisible world. Creation is to reveal what has been left unknown. Therefore, it is in principle impossible to understand within the realm of word-based understanding.

6.2 The Principle of Creation

Now, let us consider systematically how creation is achieved, by taking into account the internal invisible world. Here we deal with only creation achieved by consideration or exercise. Accidental discovery will be discussed in the next section.

Arriving at the very truth is equivalent to achieving creation and vice versa

Note first that arriving at the very truth is equivalent to achieving creation and vice versa. For example, when we met a problem and have overcome it by thorough consideration and pursuit, we achieve free thinking together with a wholly clear state of mind full of delight and simultaneously create a new idea, as mentioned in Section 1.4.1. The achievement of free thinking and a wholly clear state of mind indicates that ideas (nervous patterns) in our brain and things in the real world are fully continuously connected with one another and have completely agreed with each other, i.e. we have arrived at the very truth. On the other hand, the creation of a new idea means that we have achieved creation. Thus, this example indicates that arriving at the very truth is equivalent to achieving creation and vice versa.

If we add some words, the achievement of free thinking and a wholly clear state of mind implies that we have formed a continuous internal connection of ideas in our brain and things in the real world. A new idea we create upon arrival at the very truth comes from such a newly-formed continuous internal connection. Namely, a new idea we create upon arrival at the very truth represents common or repeatedly observed parts of such a newly-formed continuous internal connection, as argued in Section 1.4.3.

The same argument holds for the attainment of *Tai-toku*. It was mentioned in Section 1.3.1 and 1.3.2 that we arrive at the very truth when we have attained *Tai-toku*. In fact, at this time we achieve true freedom and a wholly clear state of mind full of delight, indicating that we have really arrived at the very truth. Moreover, at this time we simultaneously acquire new abilities and create a new self of us. Thus, this example also indicates that arriving at the very truth is equivalent to achieving creation and vice versa. It is highly probable that strong interaction between constituent elements of "the ability gained by the attainment of *Tai-toku*" during thorough exercise leads to the formation of a continuous internal connection of them. New abilities and a new self we create upon the attainment of *Tai-toku* emerge from such a newly-formed continuous internal connection.

The above considerations indicate that the essence of creation is to form a *new* continuous internal connection of ideas (nervous patterns) in our brain and things in the real world, or in other words, to form a *new* continuous internal connection of ideas in the brain, which is in agreement with a continuous internal connection of things in the real world. This conclusion explains why thorough exercise or thorough consideration and pursuit are necessary for the achievement of creation and why creation can give a solution to a problem that has been left unsolved till then. Moreover, the conclusion explains why a wholly clear state of mind full of delight is realized upon the achievement of creation. These are all due to the formation of a *new* continuous internal connection of ideas in our brain and things in the real world.

Strikingly, the origin of creation and creativity which has been entirely difficult to understand within the realm of word-based understanding can be easily understood if the internal invisible world is taken into account. Traditional understanding has only looked at creative behavior of a human being in the external visible world.

Two types of creation: a discovery type and a harmonization type

Let us consider how creation is achieved in more detail. For this purpose, we divide, for convenience, creation into two types: a discovery type and a harmonization type.

In general, creation is achieved when we met a difficulty and have overcome it, as mentioned above. Here, we have two kinds of difficulties: a difficulty arising from the lack of a necessary thing or knowledge and a difficulty arising from ill qualities of existing things or knowledge. Corresponding to such two kinds of difficulties, there are two kinds of creation. When a difficulty arises from the lack of something necessary, it is overcome by discovering a necessary thing. This creation is called a discovery type. On the other hand, when a difficulty arises from ill qualities of existing things or knowledge, it is surmounted by achieving improvement in qualities of existing things or knowledge and realizing a peaceful harmonious state. This creation is called a harmonization type or a quality-improvement type.

Which type of creation becomes important depends on personal or social conditions. Note here that the division of creation into the above two types is not absolute. Actual creation has characteristics of both types in different weights.

How is the creation of a discovery type achieved?

It was argued in Section 1.4.1 that we can arrive at the very truth only in a particular quality level and aspect, depending on what problem we have overcome. Arriving at the very truth is equivalent to achieving creation, as mentioned above, and thus this means that we can achieve creation only in a particular quality level and aspect. This conclusion is in harmony with the above argument that we have two types of creation. Now, it was also argued in Section 1.4.1 that we can form an internal continuous connection only in a particular quality level and aspect, depending on what problem we have overcome. This implies that we form a *new* continuous internal connection of ideas in different quality levels and aspects when we achieve two types of creation.

At first, let us consider what *new* continuous internal connection of ideas we form when we achieve the creation of a discovery type. Theoretical creation in science is a typical example of the creation of this type. As already discussed in Section 1.2.1, the discovery of a new concept, law or theory in science is achieved by picking out common or repeatedly observed properties of natural things. Now, how can we pick out common or repeatedly observed properties of natural things? This problem was already discussed in Chapter 5.

When a person looks at various objects in the outer world, various visual images (i.e. various continuous internal connections of nervous patterns) are produced in the brain. Then, when various objects have some "common properties", various visual images overlap one another in the area of "common properties" just as dumped newspapers overlap one another. Thus, if such an overlapping area of visual images is picked out and formed as a new continuous internal connection, such a new continuous internal connection gives a new idea representing "common properties" or an imaginary thing with common properties.

Furthermore, when a person creates a large number of sensory images and ideas (i.e. continuous internal connections of nervous patterns) and keeps them as memories, various overlapping areas are created between them and moreover overlapping areas further overlap one another, resulting in the formation of doubly, triply or more-time overlapping areas. Similar processes occur when a person learns various words or concepts and think of them. Thus, if such a multi-fold overlapping area is picked out and formed as a new continuous internal connection, it gives a new idea (or a new concept, law or theory) with a highly abstract meaning.

As emphasized in Chapter 5, it is by a positive act of our ego (the grand internal ability we catch within ourselves) that an overlapping area is picked out and formed as a new continuous internal connection. For the creation of a discovery type, a difficulty arises from the lack of a necessary thing or knowledge, as mentioned earlier. Therefore, it is likely that thorough consideration and pursuit are unawares carried out so that they form an overlapping area in a place where a necessary thing or knowledge exists[*1].

[*1] This means that creation is achieved in the desired direction, as mentioned in Section 6.1. This conclusion is supported by the following consideration. As argued in Section 4.4, a multi-cellular living organism such as a human being has the internal ability to organize itself so that "a multi-cellular living organism and information about the outer world" as a whole realize a free independent harmonious stable state. Therefore, it is likely that a person who feels a difficulty arising from the lack of a necessary thing or knowledge carries out thorough consideration and pursuit so that an overlapping area is formed in a place where such a necessary thing or knowledge exists.

Somebody may say that the achievement of theoretical creation in science can be explained in terms of nonlinear effect such as studied in complexity science, without assuming the formation of a continuous internal connection of ideas. However, I cannot agree with such an opinion, as already mentioned in Section 1.3.2. Such a way of understanding cannot explain the emergence of free independent harmonious spirit and wisdom upon the achievement of theoretical creation because a system in nonlinear kinetics is of a mechanical character (see Section 3.2.3).

How is the creation of a harmonization type achieved?

The attainment of *Tai-toku* discussed in Section 1.3.1 is a typical example of the creation of a harmonization type. In fact, the attainment of *Tai-toku* allows a person to overcome his or her awkward behavior and realize a free independent harmonious state. The attainment of *Satori* in Buddhism, which will be discussed in detail in Appendix, is another typical example of the creation of a harmonization type. For example, *Shaka-muni*, originator of Buddhism, felt intense mental pain to see the death of a person or to notice a temporary character of human life and overcame it after thorough consideration and pursuit. This is called *Shaka-muni*'s *Satori*. This meant that *Shaka-muni* noticed that word-based understanding had severe intrinsic limits and transcended it and realized a peaceful harmonious state.

Now, what *new* continuous internal connection of ideas do we form when we achieve the creation of a harmonization type? The creation of this type appears when we surmount a difficulty arising from ill qualities of existing things or knowledge and realize a peaceful harmonious state, as mentioned earlier. Such a situation is realized when various ideas (i.e. various continuous internal connections of nervous patterns) in our brain overlap one another just as fish scales or roof tiles overlap and the whole area of such an overlap is picked out and formed as a new continuous internal connection. This is because the formation of such a new continuous internal connection leads to the fusion of various ideas covering a diversity of quality levels and aspects and the achievement of free skillful and unconscious behavior or free independent harmonious spirit and wisdom. In this case, it is likely that thorough exercise or thorough consideration and pursuit are unawares carried out so that ill qualities of existing things or knowledge are removed, i.e. various ideas overlap one another just as fish scales or roof tiles overlap (see a footnote #1 of this section).

There is another important mode of overlapping of ideas, which leads to the attainment of the creation of a harmonization type. Details of this mode will be explained in Section A2.4.

Creation is achieved via three steps of "learning and activity, doubt-pursuit, and the jump"

Creation has another important characteristic. It is achieved via three steps of *"learning and activity, doubt-pursuit, and the jump"*. We here consider theoretical creation in science as an example, because *Tai-toku* was discussed in detail in Section 1.3 and 1.4 and *Satori* will be discussed in detail in Appendix.

Let us look at daily activities of a scientist. A scientist usually sets up a research subject, accumulates information about related past work, does experiments and thinks of results of them. While a scientist displays such activity, he or she meets a problem that has been left unsolved till then. If it is an important problem, a scientist starts to carry out thorough consideration and pursuit about it. Then, surprisingly, a scientist usually falls into an absolutely difficult situation. When a scientist has overcome such a situation, he or she achieves creation.

Thorough consideration and pursuit should improve understanding and thus lead to a clear state of mind. However, the opposite is usually the case. By thorough consideration and pursuit, a scientist is in most cases driven into an absolutely difficult situation and then suddenly and unexpectedly reaches a solution to a problem.

Let us look at an example. In 1934, H. Yukawa, Nobel laureate, theoretically predicted the presence of a meson as the key particle for generating the binding force between a proton and a neutron in an atomic nucleus. He later wrote an autobiographical book entitled Tabibito (The Traveler) [4] and described how he had been thinking of his problem before he hit on the idea of a meson. Let us cite some words in it.

> *The period from 1932 autumn to 1934 autumn was the hardest time for me. However, to be in worry was simultaneously to be in pleasure.* ·····
>
> *I left Kyoto University and became a full-time lecturer of Osaka University and started my course of lecture on electromagnetism, at which I was not very much good. However, my head was always full of the problem of a force within an atomic nucleus.* ···
>
> *No bright idea crossed my head while I studied in day time. On the other hand, various ideas stroke me while I lay down in bed at night. They were developing freely without any obstacles of a series of mathematical equations. In the meantime I became tired and fell asleep. In the next day morning I reconsidered again what I thought in bed last night, but they were all quite senseless things. Just as sunlight in the morning dispersed, so my hope dispersed as if it were a nightmare. How many times I had such an experience!* ·····

These words indicate that Yukawa continued consideration of his problem in quite a thoroughgoing way. These words also indicate that Yukawa had fallen into an absolutely difficult situation just before he achieved creation. He then hit on the idea of a meson suddenly and unexpectedly.

"Learning and activity" lead to the formation of a new continuous

internal connection

How can we explain the above characteristic of creation? It was argued earlier that creation is achieved by forming a new continuous internal connection of ideas in the brain, which is in agreement with a continuous internal connection of things in the real world. Therefore, the above characteristic of creation is expected to appear in the process of forming such a new continuous internal connection of ideas. Thus, let us consider this process in more detail. For simplicity, "continuous internal connection" is hereafter abbreviated to CIC.

Note first that carrying out thorough consideration and pursuit means making a variety of ideas interact with one another. Naturally, ideas are made interact with one another also when a scientist shows other research activities. Another important point to be noted is that it is *truly real ideas* that are made interact with one another in a scientist's brain though a scientist is conscious of only ideas with common or repeatedly observed parts of them, i.e. ideas with individual unchanging meanings (see Section 1.4.3 for truly real ideas). Now, truly real ideas exist each as a CIC of all other truly real ideas (see Section 1.4.1, see also Chapter 5). Thus, it is highly probable that thorough consideration and pursuit of a scientist, i.e. interactions among various truly real ideas in a scientist's brain lead to the formation of a new CIC through the overlapping of their CIC's or by other mechanisms (see a footnote #2 of Chapter 5). Interestingly, this argument indicates that the achievement of creation happens by essentially the same mechanism as the evolution of life discussed in Section 4.4. This is natural if we consider that both processes are controlled by the grand internal ability.

The formation of a new CIC is reflected in the mind as feelings of interest, unclearness, unease and doubt

Now, what happens when a new CIC of ideas has been formed? Creation, in particular, great creation is in general achieved via the formation of a new CIC of an enormous number of truly real ideas lying in a huge diversity of quality levels and aspects. Therefore, it is expected that such a new CIC is formed in the following way. A number of new *partial* CIC's are formed here and there while a scientist shows research activities. Then, they grow and join one by one and finally all of them are united, resulting in the formation of a new CIC of a completed form. At this final stage, creation is achieved and a new idea is created. Now, if creation is achieved in this way, a scientist has a set of new partial CIC's, i.e. a new CIC of a non-completed form before he or she has achieved creation. Feelings of interest, unclearness, unease, doubt, etc. appear from this situation.

A set of new partial CIC's, i.e. a new CIC of a non-completed form is finally converted to a new CIC of a completed form[*1] i.e. a new idea, as mentioned above. Therefore, the formation of a set of new partial CIC's is nothing else but the formation of *a new internal possibility*[*2]. Accordingly, it is plausible that the formation of a set of new partial CIC's is reflected in the mind as interest. On the other hand, a set of new partial CIC's is formed by a scientist's ego (i.e. the grand internal ability a scientist catches within him- or herself), as argued in Chapter 5. This means that a set of new partial CIC's is controlled by a scientist's ego so that it grows and forms a new CIC of a completed form. Therefore, at the stage at which it has remained a new CIC of a non-completed form or a new internal possibility, feelings of unclearness, unease, doubt, etc. appear in the mind.

[*1] The difference between a new CIC of a non-completed form and a new CIC of a completed form is explained as follows. A new CIC of a completed form has realized a free independent harmonious state and hence demonstrates a thing with individual unchanging properties (such as a new concept, law or theory) in a common or averaged form (see Section 1.4.2). On the other hand, a new CIC of a non-completed form has not yet reached such a stage. Therefore, it demonstrates nothing with individual unchanging properties even in a common or averaged form. It is completely hidden. However, the formation of it is reflected in the mind as feelings of interest, unclearness, unease, doubt, etc., as mentioned above.

[*2] A new *internal* possibility, which we cannot be conscious of, should be clearly distinguished from a new possibility we can be conscious of. The latter new possibility is in general formed by logical thinking based on existing knowledge.

The fact that the formation of a new CIC of a non-completed form is reflected in the mind as feelings of interest, unclearness, unease, doubt, etc. is very important because it gives direct evidence for the argument that the achievement of creation happens via the formation of a new CIC.

Creation is achieved via the jump

In the final stage of creation, a scientist is driven into an absolutely difficult situation and then suddenly and unexpectedly reaches a solution to a problem, as mentioned earlier. This stage can be explained as follows. A scientist's activities, in particular, thorough consideration and pursuit lead to large growth of a set of new partial CIC's or a new internal possibility. However, the formation of a new CIC of a completed form is not easily achieved because it needs the completion of an infinitely-spreading fully-continuous connection of truly real ideas in a huge diversity of quality levels and aspects, as mentioned earlier. Here is severe difficulty arising from the complexity of nature and the world.

Therefore, just before the achievement of creation, there appears a situation in which a set of largely growing new partial CIC's, which is very close to a new CIC of a completed form, is formed but still remains a non-completed form. Such a situation is reflected in the mind as an absolutely difficult situation. Then, further growth of a set of largely growing new partial CIC's leads to sudden formation of a new CIC of a completed form.

In this way, we can reasonably explain why creation is achieved via the three steps of *"learning and activity, doubt-pursuit, and the jump"* if we take into account the internal invisible world. In general, partial CIC's in the brain are spontaneously and unconsciously formed by daily activities in our life. Therefore, we suddenly and unexpectedly start to feel interest, unclearness, doubt, etc. The growth of partial CIC's also occurs spontaneously and unconsciously. Therefore, we suddenly and unexpectedly hit on a new idea and reach a wholly clear state of mind[*3].

[*3] In the attainment of *Tai-toku*, the ability acquired by exercise goes up nonlinearly with the amount of it (see Figure 1-2). This fact can be explained as follows: The growth of new partial CIC's happens linearly with the amount of exercise on an unconscious level but the ability in the external visible world appears nonlinearly when new partial CIC's have reached a critical level.

Logical thinking and trial-and-error thinking

The fact that creation is achieved by the three steps of *"learning and activity, doubt-pursuit, and the leap"* can be explained in a different way. As mentioned in Section 1.3.1 and 1.4.3, there is a great difference between carrying out thorough consideration and pursuit and accumulating a large amount of knowledge. Thorough consideration and pursuit lead to the achievement of creation while the accumulation of a large amount of knowledge does not lead to such an event. Why does such a difference appear?

When we accumulate knowledge, the main aim is to accept and understand new knowledge. Therefore, our head only follows logical connections in accumulated knowledge. Namely, we only do logical thinking, with the meanings of individual words in knowledge kept unchanged. On the other hand, when we carry out thorough consideration and pursuit, we in general contend with a problem. Therefore, in this case, our thinking is mainly controlled by feelings of unclearness, unease, doubt, and so on. Knowledge or individual words in it are only used as a tool for finding a solution to a problem. Therefore, their meanings are left able to change. In short, we rely on knowledge when we accumulate knowledge while we critically utilize knowledge when we carry out thorough consideration and pursuit.

It should be noted here that it has long been believed that words have unchanging meanings (see Section 1.1.1 and 1.1.2). However, in reality, the meanings of ideas (truly real ideas) and hence those of words change during consideration and pursuit, as discussed in Section 1.3.2, 1.4.3, 2.1.2, and so on. If we take into account this point, it becomes clear that there is a great difference between carrying out thorough consideration and pursuit and accumulating knowledge. Thorough consideration and pursuit leads to changes in the meanings of words and thus leads to the achievement of creation, while the accumulation of knowledge leads to no change in the meanings of words. Actually, when we carry out thorough consideration and pursuit, we expect that the meanings of words will change appropriately so as to give a solution to a problem.

Note also that we in general do trial-and-error thinking when we carry out thorough consideration and pursuit. We have no way but to do trial-and-error thinking when we meet a problem. Trial-and-error thinking is a method of carrying out thorough consideration and pursuit. Thus, trial-and-error thinking is also of a very creative character. It appears that such an important fact has been overlooked in traditional understanding because it has been believed that words have unchanging meanings, as mentioned above. If no change happened in the meanings of words, trial-and-error thinking would only be a method of waiting for accidental good luck.

The difference between logical thinking and trial-and-error thinking can be understood in the following way as well. When we have continued trial-and-error thinking about a problem and reached a solution to it, we describe the solution in an ordered form by using words with clearly defined meanings. A solution thus described is logical knowledge. Namely, logical knowledge and hence logical thinking express a "result" of trial-and-error thinking. This consideration indicates that a person who has continued trial-and-error thinking about a problem and reached a solution to it has much deeper understanding than a person who only learns a solution (a result of trial-and-error thinking) by words.

Some of people may think that a person who always does logical thinking is strong in logical thinking. However, this is not the case. Just a person who always feels doubt and continues trial-and-error thinking is strong in logical thinking because such a person has a large number of continuous internal connections of ideas, which constitute a basis for logical thinking. A person who always does logical thinking has no such continuous internal connections.

Unconscious and conscious trial-and-error thinking

In this way, trial-and-error thinking plays an important role in our life. However, we have to note that there is no absolute difference between trial-and-error thinking and logical thinking. Even logical thinking can act as *unconscious* trial-and-error thinking and unawares brings about a change in the meanings of ideas or words. Certainly, we do logical thinking according to definite rules. However, logical thinking makes various ideas interact with one another and thus in some cases leads to the formation of partial CIC's of a non-completed form. This is a natural result of the fact that truly real ideas exist each as a CIC.

Accordingly, we have two kinds of trial-and-error thinking; conscious and unconscious. We do conscious trial-and-error thinking when we feel interest, unclearness, doubt, and so on. Trial-and-error thinking of this type is effective because it is done with a focus placed on the clarification of the origin of such feelings. However, this trial-and-error thinking is possible only when we feel interest, unclearness, doubt, and so on. On the other hand, unconscious trial-and-error thinking (i.e. ordinary logical thinking) is done even when we have no such feeling and thus plays a key role when we have no such feeling though it is not effective. In summary, creation is in general achieved in the following way.

Logical thinking (unconscious trial-and-error thinking) based on learned knowledge → the formation of partial CIC's and the appearance of feelings of interest, unclearness, doubt, and so on → conscious trial and error thinking → the completion of a new CIC and the creation of a new idea accompanied with a wholly clear state of mind

This summary again indicates that creation is achieved via three steps of *"learning and activity, doubt-pursuit, and the jump"*.

6.3 How Does the Internal Ability Work in Processes of Achieving Creation?

Considerations in the preceding sections indicate that in processes of achieving creation we human beings are wholly controlled by internal processes we cannot be conscious of. For getting better understanding of this situation, we in this section examine how the internal ability works in processes of achieving creation.

Strangely, the correct answer emerges suddenly and unexpectedly in an absolutely difficult situation

In creation, a person suddenly and unexpectedly hits on a new idea (or a new concept, law or theory) when he or she is driven in an absolutely difficult situation, as discussed earlier. Nevertheless, strangely, such a new idea gives the correct answer to a question which a person has long been tackling till then.

For example, Yukawa carried out thorough consideration and pursuit about the problem of what the binding force between a proton and a neutron in an atomic nucleus was. He was finally driven into an absolutely difficult situation and then suddenly and unexpectedly hit on the idea of a meson. Strangely, the idea gave him the correct answer to the problem he had long been considered till then.

According to my understanding, there are many similar examples in the history of Buddhism. As will be explained in detail in Section A1.2.1, Shaka-muni, originator of Buddhism, felt intense mental pain to see the death of a person or to notice a temporary character of human life. Then, he entered into a mountain for overcoming the mental pain and carried out hard practice for about six years. Finally, he was driven into an absolutely difficult situation and suddenly and unexpectedly noticed that everything incessantly changes and has no individual unchanging quality. Strangely, this idea gave him the correct answer to his problem. Similarly, Shin-ran, Japanese Buddhist monk, had also long been asking where Buddha was (see Section A2.2). Then, he was finally driven into an absolutely difficult situation and suddenly and unexpectedly noticed that he had been entirely guided by Buddha from the beginning to the end. This idea gave him the correct answer to his question.

The sudden and unexpected appearance of a new idea when a person is in an absolutely difficult situation clearly indicates that it is not a direct result of logical thinking. Nevertheless, such a new idea gives the correct answer to a question a person has long been tackling till then. This is really a strange fact. Such a fact can be explained only by assuming that a new idea appears based on the formation of a set of new partial CIC's or a new internal possibility, as discussed in Section 6.2.

Opposite feelings of interest and difficulty appear simultaneously just before the achievement of theoretical creation

In Section 6.2, some words of H. Yukawa were cited from his autobiographical book, Tabibito (The Traveler) [4]. The words indicate that he was strongly attracted by his problem just before he achieved creation though he felt severe difficulty in it. For example, he says, "*The period from 1932 autumn to 1934 autumn was the hardest time for me. However, to be in worry was simultaneously to be in pleasure.*" He also says, "*I became a full-time lecturer of Osaka University and started my course of lecture on electromagnetism. However, my head was always full of the problem of a force within an atomic nucleus.* Yukawa was driven in a state in which he was unable to stop thinking of his problem though he heavily suffered from severe difficulty in it.

Simultaneous emergence of opposite feelings of severe difficulty and strong interest is quite a strange state. Such a state can never be achieved intentionally. Nevertheless, such a state is really realized just before the achievement of creation. It was argued in Section 6.2 that various feelings of opposite characters such as interest, unclearness, unease, doubt, etc. emerge before creation is achieved. The above-mentioned simultaneous emergence of opposite feelings of severe difficulty and strong interest is a typical case of such a situation. We cannot explain it without taking into account the formation of a set of new partial CIC's or a new internal possibility.

Some people may consider the emergence of a feeling of difficulty is a troublesome matter. However, we should note that the emergence of such a feeling has a positive meaning. The above argument indicates that a feeling of difficulty emerges from the formation of a set of new partial CIC's, i.e. a new internal possibility. Therefore, the emergence of a feeling of difficulty suggests the formation of a new internal possibility. Therefore, we should rather be delighted to have such a feeling.

The above argument also indicates that there is no difficulty that cannot be overcome because a feeling of difficulty comes from the formation of a new internal possibility. Therefore, any severe difficulty can necessarily be overcome [*1]. In case no new internal possibility is formed, a feeling of difficulty itself does not emerge.

[*1] This does not mean that a difficulty is always overcome in the correct way or in a favorable way. In fact, we human beings often went wrong in the past.

It is not easy to feel interest, unclearness, doubt, and so on and carry out thorough consideration and pursuit

The importance of the formation of a set of new partial CIC's or a new internal possibility can be understood from the following consideration as well. In general, it may be believed that everybody can feel interest or doubt about existing knowledge. However, this is a big mistake. The arguments in Section 6.2 indicate that *only* a person who has formed a set of new partial CIC's or a new internal possibility can have such a feeling. With no new internal possibility, everything looks natural and reasonable.

Moreover, with no new internal possibility, it is impossible to carry out thorough consideration and pursuit. We cannot carry out thorough consideration and pursuit about a problem only for a reason that it is important. We need to have something internal, which strongly attracts our mind. In other words, we have to reach a state in which we cannot stop thinking, such as Yukawa reached. Similar arguments were given in Section 1.4.3.

Let us look at some examples. As mentioned earlier, Shaka-muni, originator of Buddhism, felt intense mental pain to see the death of a person or to notice the temporariness of human life. Then, he entered into a mountain and carried out hard practice for about six years. Surprisingly, he gave up his family and palace to continue his consideration and pursuit. Indeed, many of people may feel intense mental pain to see the death of a person and consider whether they can escape from it. However, most of people will soon reach a conclusion that a human being necessarily dies and this is a "clear fact" or something like this and give up thinking further. They may rather think that it is foolish to continue meaningless consideration for a long time. However, Shaka-muni continued thorough consideration and pursuit and finally reached the very truth about human life, which nobody had ever reached. Shaka-muni must have had deep insight into the essence of human life and felt doubt about it on quite a deep level. For this reason, he was able to continue thorough consideration and pursuit about the origin of his mental pain [*2]. The limits of existing knowledge Shaka-muni noticed were only able to be noticed by Shaka-muni.

[*2] When we cannot reach a solution to a difficult problem after thorough consideration and pursuit, we have to decide whether we further continue consideration or not. In such a case, logic is usually powerless. Thus, after all, we decide to continue consideration when we have something internal, which strongly attracts our mind.

Similarly, a Japanese Buddhist monk, Shin-ran, who discovered the correct way to arrive at the very truth in the early 13th century, as mentioned in Section 1.4.3, carried out extraordinarily thorough consideration and pursuit and reached the very truth, which nobody had ever arrived at. He must also have had deep insight into the essence of human life and felt doubt about the traditional teachings of Buddhism on quite a deep level (see Section A2.2 and A2.3).

What person has high creativity?

The above examples indicate that a person who has deep insight into the essence of human life can feel interest or doubt about existing things from a deep level and can carry out thorough consideration and pursuit. This fact can be explained as follows.

It was argued in Section 1.4.3 (see also Figure 1-6) that we can efficiently form a new CIC of ideas in the area in which we can carry out smooth thinking, i.e. we have a large number of CIC's of ideas. A person who has deep insight into the essence of human life is one who has a large number of CIC's of ideas and can carry out smooth thinking over wide and deep areas of knowledge. Therefore, such a person can efficiently form a set of new partial CIC's and hence can effectively feel interest or doubt about existing things and effectively carry out thorough consideration and pursuit.

Creation, in particular, great creation is achieved by forming a new CIC of an enormous number of ideas lying in a huge diversity of quality levels and aspects, as mentioned in Section 6.2. Therefore, for achieving great creation, it is necessary that thorough consideration and pursuit proceed smoothly and efficiently over wide and deep areas of knowledge in a huge diversity of quality levels and aspects. This means that we in advance have to have a large number of CIC's of ideas in a huge diversity of quality levels and aspects.

Both the ability to feel interest or doubt about existing things or knowledge and the ability to carry out thorough consideration and pursuit are the ability to achieve creation. Therefore, they can be called creativity. The above consideration indicates that we can have high creativity when we have formed CIC's of ideas over wide and deep areas of knowledge in a huge diversity of quality levels and aspects. The same conclusion was obtained in Section 1.4.3.

New internal possibilities are formed in society

We have thus far considered creation and creativity in personal life. Next, let us consider creation and creativity in society. In traditional understanding and philosophies, in which the internal invisible world has not been taken into account, it appears that human creativity has been simply understood to be some special skill. However, the foregoing arguments in this chapter indicate that human creativity has a much deeper meaning. Namely, human creation is achieved by the grand internal ability a person catches within him- or herself. Accordingly, the arguments given in Section 6.2 can be safely generalized to creation and creativity in society.

In harmony with the above expectation, we can show that a new internal possibility is formed one after another in society, in the same way as in personal life. We usually do various things with particular aims in society. However, such activities of us act as unconscious trial-and-error processes in society and lead to the formation of a set of new partial CIC's or a new internal possibility. Let us look at some examples.

Example 1
The history of science demonstrates that science has advanced in a characteristic way. Namely, individual facts were first caught separately and then they were understood together in a connected form later. For example, in the Aristotelian idea in ancient Greece, the motion of earthly bodies and that of heavenly ones were understood individually and separately. Namely, earthly bodies were understood to have vulgar properties of falling down while heavenly bodies were understood to have noble properties of moving in circles. However, in Newtonian mechanics in the Renaissance age, both earthly and heavenly bodies were understood together to have the same properties.

There are a large number of similar examples in the history of science. In the 18th century, heat and atomic motion were understood separately. However, in the 19th century, it was revealed that heat originates from atomic motion and thus they only represent different faces of one thing. Inorganic substances such as stones, water, and iron metal and organic substances such as uric acid were initially regarded as having different natures. However, in the 19th century, it was proved that they are of the same nature. Darwin's theory of evolution in the 19th century disclosed that human beings and other animals are of essentially the same nature though they had long been believed to be of a different character.

Quantum mechanics established in the early 20th century revealed a variety of new connections. For example, Heisenberg's uncertainty principle says that the position and the momentum of a particle, together with the energy of a quantum state and its lifetime, which have been regarded as essentially different things till then, cannot be understood separately (see Section 2.1.2). Quantum mechanics also says that particles of the same kind cannot be understood separately. Furthermore, Einstein's theory of relativity has revealed that "time and space" and "mass and energy" cannot be understood separately.

Why does science show such a characteristic way of advancing? This is not because human beings wanted to first understand things individually and separately. Human beings always wanted to catch common properties of truly real things, as discussed in Section 1.2.1. Accordingly, science's characteristic way of advancing is just because a set of new partial CIC's or a new internal possibility was formed in the history of society. When new knowledge is created based on a new CIC or a new internal possibility, which was not included in old knowledge, the old knowledge looks as if it were created by looking at things individually and separately. Thus, science's characteristic way of advancing indicates that a new internal possibility has been formed in society one after another.

Example 2

The history of science also shows that it has advanced in the following way: people's activities brought about the development of technology and industry, which led to the production of a new area of human activity and then led to the creation of a new science and a new view of the world. This fact also indicates that a new internal possibility was formed one after another in society.

In the age of four ancient civilizations of Mesopotamia, Egypt, Indus and China, farming was the major calling. Thus, people paid attention to plants and animals together with the great power of nature which brought people various blessings such as solar light, rain and winds. Probably for this reason, people in those years understood natural phenomena by personification and established a view that gods governed nature. It seems that such a view of nature was the first theoretical (systematic) recognition of nature which human beings acquired.

The developments of farming technologies in the age of four ancient civilizations, brought by the discovery of bronze and iron implements, led to the accumulation of wealth, which in turn brought about the advancement of commerce and trade on the one hand and continuing battles on the other hand. The advancement of commerce and trade really became important in ancient Greece, while continuing battles became a severe problem in East Asia. Thus, people in ancient Greece came to live a free independent and exploratory life such as marine trade. Such a way of living led to the advancement of an objective (scientific) way of understanding and an objective view of nature (see Section A3.1.1). On the other hand, people in East Asia had to make strenuous efforts to work out a way to avoid battles and as a result created unique philosophies (see Section A3.1.2).

The development of commerce and trade next brought about the advancement of machines such as handcarts, ships, clocks and water mills for production and transport of goods. Thus, people came to pay attention to the motion of bodies and physical forces acting on bodies. In this way, mechanical motion was studied in detail. This led to the construction of Newtonian mechanics and a geometrical mechanistic view of nature in the Renaissance age.

After the Renaissance age, the development of machines led to the invention of a steam engine in the latter half of the 18th century, which made people direct their eyes to the "driving force" of machines. Thus, people came to pay attention to the power included in the interior of substances such as heat, electricity, magnetism and vitality. This trend led to the establishment of electromagnetism and thermodynamics on the one hand and the clarification of inner structures of substances on the other hand in the 19th and 20th centuries. The clarification of inner structures of substances led to the disclosure of the vast microscopic world. Modern advanced science and technology are based on these discoveries.

Now, what new field appears in the present days? The development of study of the microscopic world in modern science has brought us various high technologies such as electronic, molecular, biological and informational technologies. Our views and activities have really spread into the whole world and even into the universe. Thus, it appears that many of people have come to direct their eyes to understanding of complex systems such as living organisms, the human brain, artificial life, artificial intelligence, the ecological system, information societies, globalized economic and political systems, and so on (see Section 1.1.2). A new theory proposed in this book will be helpful for promoting the understanding of such complex systems.

Example 3

It is often said that scientific discovery in most cases happens by chance. The word of "serendipity" is sometimes used to emphasize the importance of accidental discovery. Indeed, C. Columbus discovered new islands near the continent of America by chance. The discovery of X-ray by W. Röntgen also happened accidentally during his experiments about a cathode ray. The synthesis of a new organic electrical conductor, polyacetylene, in Shirakawa's laboratory was also achieved by chance. The laboratory had been searching for a suitable condition for polymerization of acetylene. One day, a student made a mistake of using a catalyst of an amount one thousand times larger than instructed, by taking gram for milligram. Interestingly, such a mistake led to the discovery of a condition for efficient polymerization.

However, further consideration teaches us that there is no essential difference between an accidental discovery and, for example, theoretical creation achieved by thorough consideration and pursuit. It was argued in Section 6.2 that creation is achieved via the formation of a set of new partial ICC's or a new internal possibility. Similarly, an accidental discovery is also made on the basis of the formation of a set of new partial ICC's or a new internal possibility in society. For example, Columbus's discovery of new islands was made in the Age of the Grand Voyage. Namely, Columbus's discovery was made in nearly the same age as Copernicus's proposal of the heliocentric system and Galilei's discovery of the fundamental laws of motion. All these discoveries appeared on the basis of a new revolutionary view of nature (or a new internal possibility) formed in society in those years. Similarly, Roentgen's discovery of X-ray was made in nearly the same age as discoveries of electrons, atomic spectra, and quantum mechanics. These examples indicate that various kinds of creation on the same level happened in nearly the same age based on the formation of a new internal possibility in society, irrespective of whether they are of a theoretical type or of an accidental-discovery type.

It is not easy to become aware of a new internal possibility formed in society

Now, let us again return to the consideration of creation and creativeity in personal life. If a new internal possibility is formed in society, we can achieve great creation by becoming aware of such a new internal possibility. In fact, it is no exaggeration to say that great creation in human history was in most cases achieved in this way.

It should be noted, however, that this does not mean that to become aware of a new internal possibility formed in society is the best way to achieve great creation. This is because it is not easy to become aware of a new internal possibility formed in society. As discussed in Section 1.3.2, 2.1.2 and so on, we can never be conscious of a CIC. Therefore, a set of new partial CIC's or a new internal possibility formed in society is also completely hidden. Only a person of high creativity, who can form a new internal possibility by his or her own consideration and pursuit, can become aware of a new internal possibility formed in society.

The above argument can be explained as follows. The formation of a set of new partial CIC's or a new internal possibility in society is completely hidden, as mentioned above. This means that such formation causes no change in structures and properties of individual things in the external visible world. However, such formation brings about a change in the meaning of individual things in the external visible world. In fact, for example, no change happened in the motion of earthly and heavenly bodies in the Renaissance age. However, the meaning of the motion of earthly and heavenly bodies has changed in the Renaissance age. A person of high creativity such as Galillei, Descartes and Newton sharply noticed such a change in the meaning.

A person of high creativity has a great many experiences of arriving at the very truth over wide and deep areas of knowledge in a huge diversity of quality levels and aspects, as mentioned earlier. Therefore, such a person can form new partical CIC's of ideas and hence can feel interest, unclearness, unease, doubt, etc. about phenomena in nature and the world. This means that such a person can notice a change in the meaning of individual things.

On the other hand, a person who only has "understanding by words" can merely look at things individually and separately. Therefore, such a person can notice nothing because almost no change happens in structures and properties of individual things in the external visible world, as mentioned earlier.

How can we improve creativity?

Finally, let us consider how we can improve creativity. This problem has also been left unsolved to date, as discussed in Section 6.1. The clarification of the origin of creation and creativity has made it possible to solve this problem.

As mentioned above, a person who has a great many experiences of arriving at the very truth has high creativity. Therefore, we can say that we can improve creativity by having an experience of arriving at the very truth. Namely, we can improve creativity by achieving creation, *Tai-toku* and *Satori*.

As discussed in Section 6.1, creativity is impossible to freely obtain intentionally, in contrast to knowledge. However, this does not mean that we cannot improve creativity a posteriori. Creativity is indeed the ability beyond word-based understanding, i.e. the ability coming from the internal invisible world. However, we can improve it by controlling the internal invisible world. Strikingly, the clarification of the origin of creation and creativity teaches us how we can control the internal invisible world.

There is another interesting aspect about the improvement of creativeity. The above argument indicates that the process of improving creativity has a positive feedback mechanism. Namely, the achievement of creation, *Tai-toku* and *Satori* leads to improvement in creativity, which in turn makes creation, *Tai-toku* and *Satori* easier to achieve.

In this connection, we have to note also that the process of improving creativity induces a negative feedback mechanism under an ill condition. Namely, a person who only has "understanding by words" has little chance of achieving creation and hence little chance of improving creativity. Therefore, such a person all the more comes to rely on "understanding by words". To make matters worse, creativity is the ability we cannot be conscious of. Therefore, it is difficult to notice a decrease in creativity, in contrast to the lack of knowledge. Everybody can know whether or not he or she has enough knowledge. However, nobody can know whether or not he or she has sufficient creativity. Accordingly, as to creativity, a person easily falls into the worst situation in which he or she is ignorant of his or her ignorance.

Lastly, it should be emphasized that the above positive feedback mechanism in the process of improving creativity can be applied to events in nature and the world. It was mentioned in Section 4.4 that the rate of evolution in the evolution of life becomes higher and higher as the evolution goes up to an advanced stage. This fact indicates that a positive feedback mechanism really works in creativity in nature. Similarly, the rate of the development of human civilization becomes higher and higher as the civilization goes up to an advanced stage. For example, human beings emerged about several million years ago. The use of letters started about 5,000 years ago. Great progress of science and technology started only 200 years ago. These facts clearly indicate that a positive feedback mechanism also works in creativity in society.

References

(1) V. Lenin, *Notes on "Science of Logic" by G. Hegel*, 1914-16: The Japanese edition translated by K. Matsumura, Iwanami-bunko, Iwanami-shoten, 1956.
(2) M. Polanyi, *The Tacit Dimension*, Routledge & Kegan Paul Ltd., London, 1966: The Japanese edition translated by K. Sato, Kinokuniya-shoten, Tokyo, 1980.
(3) Japan Creativity Society (ed.), S. Ito (editor-in-chief), *Theories and Methods of Creativity* (in Japanese), Journal of Japan Creativity Society No. 1, Kyoritsu Shuppan, Tokyo, 1983.
(4) H. Yukawa, *"Tabibito" (The Traveler) – A Reminiscence of a Physicist* (in Japanese), Kadokawa-shoten, Tokyo, 1960.

Appendix

A1. The Philosophy of Buddhism (I)
– The Disclosure of the Internal World and the Internal Ability –

The purpose of this book has been fundamentally achieved by the discussions in Chapter 1 to 6. In this appendix we consider a remaining problem, which was pointed out near the end of Preface.

Indeed, a new interpretation of modern science by taking into account the internal invisible world has given convincing solutions to many important problems that have been left unsolved to date, as discussed in the preceding chapters. However, to be strict, there remains a serious problem to be considered further. Namely, the internal invisible world is in principle impossible to express by words. The internal invisible world expressed by words is not the internal invisible world itself. Therefore we have not yet obtained *the truly correct solutions* to problems that have been left unsolved to date. Word-based scientific understanding is in essence of an approximate character and only represents the external appearance of the very truth even if we take into account the internal invisible world. Accordingly, we have to work out a way to directly catch the very truth lying in the internal invisible world for gaining the truly correct solutions.

In this Appendix we consider how we can deal with this problem for readers who are interested in it. Fortunately, this problem can be successfully handled based on the philosophy of Buddhism, which teaches us that we can transcend word-based understanding and arrive at the very truth. Just arrival at the very truth provides us with the truly correct understanding. Namely, it allows us to jump to the world far beyond "understanding by words" and grasp the true meanings of individual words. It should be noted here that only the philosophy of Buddhism can deal with arrival at the very truth. This is because all other philosophies including science deal with only things we can be conscious of, i.e. things with common or repeatedly observed properties. Therefore, the philosophy of Buddhism plays the key role in dealing with the above problem.

Though the philosophy of Buddhism has constructed a great theoretical system of a unique character, we in this Appendix do not aim at the full understanding of it. We only make use of it for understanding the origin and the evolution of life and related topics from a deeper level. Namely, we only aim at clarifying what it means to transcend word-based understanding and arrive at the very truth. The final goal is to disclose where the very truth on the highest level is and how we can arrive at it. To date, the philosophy of Buddhism has not necessarily been discussed in this direction. Therefore, many original interpretations are added in this book according to my understanding. Especially, the following point is to be mentioned here. In general, the Buddhist teachings have been taught without giving any detailed explanation to why they are regarded as correct. Such a way of teaching is not desirable if the Buddhist teachings are taken as a philosophy. Therefore, this book develops many discussions for making the correctness of the Buddhist teachings clear.

Buddhism started from Shaka-muni's *Satori*. Then it has advanced historically in the same way as science. Therefore we consider it along its history. I believe that it is the best way to understand Buddhism along its history because in general the process of historical advancement agrees with the process of human understanding. The long history of Buddhism can be divided into three stages. The first stage, Shaka-muni's *Satori* and the disclosure of the very truth about human life, will be explained in Section A1.2. The second stage, the rise of a new thought of Buddhism, called Mahayana Buddhism, in ancient India and the con- struction of a new profound Buddhist view of nature, will be explained in Section A1.3. The third stage, the transmission of Mahayana Buddhism to North East Asia such as China, Korea and Japan and its original progress in this region, will be explained in Section A1.4. The third stage involves the progress of the philosophy of Buddhism to a completed stage but this part will be explained in the next chapter.

A1.1 Fundamental Characteristics of the Philosophy of Buddhism

Buddhism only says that you should transcend word-based understanding and arrive at the very truth

At first, let us explain fundamental characteristics of the philosophy of Buddhism. It should be emphasized first that Buddhism now has an appearance of religion but the essence of Buddhism is, to my knowledge, a philosophy that has been revealed by a great number of seekers after the truth. In fact, Buddhism has no a-priori proposition or dogma. It only says that you should attain *Satori*, i.e. you should transcend word-based understanding and arrive at the very truth. In Buddhism, attaining *Satori* is equal to transcending word-based understanding and arriving at the very truth, to be accurate, arriving at the very truth about human life or about nature and the world. Thus, a Buddha simply refers to a person who has arrived at the very truth about human life or about nature and the world, not a god with supernatural power.

Such a characteristic of the Buddhist teachings originates from Shaka-muni's *Satori*. Shaka-muni, originator of Buddhism, felt intense mental pain to see the death of a person or to notice the temporariness of human life. Then, he carried out hard ascetic practice and finally overcame the mental pain. He achieved a wholly clear state of mind completely released from any worry and agony and full of free independent spirit and wisdom, as will be explained in the next section. Now, an important point is that Shaka-muni met a problem (such as the fear of death) that was impossible to solve within the realm of word-based understanding and nevertheless overcame it and achieved a wholly clear state of mind. This meant that Shaka-muni overcame a problem by transcending word-based understanding and arriving at the very truth. In addition, this also meant that Shaka-muni discovered that word-based understanding has intrinsic limits and transcending it is necessary for arriving at the very truth. Thus, Buddhism emphasizes that you should transcend word-based understanding and arrive at the very truth, as mentioned above.

The importance of Shaka-muni's Satori

The importance of Shaka-muni's *Satori* lies in the fact that we human beings really have a great many problems that are impossible to solve within the realm of word-based understanding. The removal of the fear of death is a typical example of such problems. Also, we can never gain the ability to ride a bicycle based on word-based understanding alone, as argued in Section 1.3.1. The same argument applies to various kinds of ability gained by the attainment of *Tai-toku*.

To tell the truth, *every* phenomenon or event in nature and the world is impossible to correctly understand within the realm of word-based understanding. This is because words only represent common or repeatedly observed parts or individual unchanging parts of natural things and fail to catch natural things themselves. An important example is that we can never correctly understand the basic qualities of living things such as free independent harmonious spirit and wisdom by words (see Section 1.1.1, 1.1.2, 3.2.2 and so on). We can also never correctly understand creation and creativity based on word-based understanding, as discussed in Section 1.4.3 and Chapter 6.

In general, for overcoming a problem, it is necessary to gain the correct knowledge about nature and the world. Now, many of people usually believe that the correct knowledge is obtained by a scientific way of understanding and this is the only way to do so. However, Buddhism has revealed that word-based scientific understanding has severe intrinsic limits. Therefore, it asserts that any problem can be overcome only by transcending word-based understanding and arriving at the very truth. The importance of Buddhism is that it has not only revealed that word-based understanding has severe intrinsic limits but also disclosed that we human beings have the ability to transcend them and arrive at the very truth.

What ability or wisdom can we obtain by overcoming the intrinsic limits of word-based understanding?

Now, what ability or wisdom can we obtain by transcending word-based understanding and arriving at the very truth? As already discussed in Section 1.3 and 1.4, we have many experiences of transcending word-based understanding and arriving at the very truth in daily life. The philosophy of Buddhism is constructed based on experiences of the same type as those discussed in these sections. Therefore, the fundamental meanings of transcending word-based understanding and arriving at the very truth were already described in these sections.

Theoretically, the very truth (the way that truly real things or natural things themselves exist) is in the internal world we cannot be conscious of. Therefore, we cannot catch it by words, as already emphasized earlier. We can only catch it as "internal" wisdom. Thus, arrival at the very truth means that we take the very truth within ourselves in the form of the internal wisdom that works spontaneously and unconsciously when it is necessary as if it were instinctive ability (see also Figure 1-6 in Section 1.4.3). If we acquire such internal wisdom, we can think and behave spontaneously and unconsciously in accord with the very truth. This certainly means that we have arrived at the very truth.

In general, when we have arrived at the very truth, truly real ideas in our brain and truly real things in nature and the world are fully continuously connected with one another and completely agree with each other, as discussed in Section 1.3.2 (see also a footnote #2 of Section 2.1.1). Therefore, the way that truly real things exist in nature and the world is exactly reflected on our head and besides the grand internal ability to govern all things in nature and the world is directly embodied in our mind. Accordingly, we gain "a big sphere of internal wisdom" in our mind, words only floating on its surface. At this stage, we can acquire the true meaning of words.

We can also say as follows. When we have arrived at the very truth, our understanding extends deep into the area we cannot be conscious of though word-based understanding is always restricted within the known area. Thus, we gain special ability, which may be called "ability be- yond word-based understanding", "understanding without using words" or "understanding without relying on words". Also, we gain high-level ability to freely control individual words.

It is important to note also that it is only arrival at the very truth that provides us with free independent harmonious spirit and wisdom (creativity). Indeed, word-based scientific understanding is useful but it is of an approximate mechanical character. Therefore, it has a tendency to deny the existence of the spirit or mind. In fact, it is said that recent great progress of brain science and computer science has come to show a possibility of denying the existence of the free will of a human being, as mentioned in Section 1.4.2. Word-based scientific understanding has a danger of making a human being like a machine.

The internal wisdom we acquire by arriving at the very truth is important also in that it provides us with the firm basis for living an undisturbed peaceful harmonious life. When we remain within the realm of word-based understanding of an approximate mechanical character, we are forced to live a superficial technical life and have to suffer from limitless unclearness, unease and loneliness at the most basic level even though we can live apparently a fruitful and satisfied life.

Further details of the importance of arriving at the very truth will be discussed in Section A2.4, A3.2 and A3.3.

How can we arrive at the very truth?

We already discussed this problem in Section 1.3 and 1.4. As emphasized in Section 1.4.3, strangely, this problem has been left unclear in the long history of Buddhism though it appears that many Buddhist monks had attained *Satori*. In close relation to this fact, it is interesting to note that there is a similar strange situation in the field of creation as well. Certainly, many people have achieved creation up to now but how we can achieve creation has been left uncertain to date, as discussed in Section 6.1.

In my opinion, this problem was clarified by a Japanese Buddhist monk, Shin-ran, in the early 13th century, as already mentioned in Section 1.4.3 and will be explained in detail in Chapter A2. However, this fact has not been clearly recognized to date. Some discussions of how we can arrive at the very truth were given in Section 1.4.3.

The Buddhist teachings are difficult to understand

It is often said that Buddhist teachings are difficult to understand. Certainly, there are some reasons for such an opinion emerging. Firstly, Buddhism teaches us to transcend word-based understanding and arrive at the very truth, as mentioned earlier. Now, there is a non-removable gap between the very truth and word-based understanding, as discussed in Section 1.2.1. This means that truly real things lying in the very truth have a strange way of existing, which is impossible to correctly understand by human ideas and words. Accordingly, the Buddhist teachings have special difficulty in understanding. There is no such difficulty in other philosophies, which are constructed based on things we are conscious of and do not deal with truly real things.

Secondly, the very truth or truly real things are in the internal invisible world and impossible for us human beings to be conscious of. Therefore, when Buddhist monks attained *Satori* and arrived at the very truth, they were unable to be conscious of the very truth they reached and thus were unable to express it by words. Buddhist monks were only able to express what they looked at in the very truth they reached, i.e. the external appearance of the very truth they reached. The same situation exists for a person who has attained *Tai-toku*. It was mentioned in Section 1.3.3 that a person who has attained *Tai-toku* about riding a bicycle cannot express the substance of his or her ability by words. Therefore, for correctly understanding the Buddhist teachings, after all we have to have essentially the same experience as Buddhist monks. Buddhist monks' words only express things we shall be able to look at when we have reached the very truth.

Thirdly, Buddhism has historically advanced in the same way as science, as mentioned earlier. Therefore, in my opinion, Buddhism in the initial stage taught people some incomplete teachings. The most important problem was that the correct way to attain *Satori* had long been left uncertain and for this reason incomplete ideas of the very truth had been taught for a long time, as will be surveyed in Section A2.1. Naturally, this was an unavoidable matter but the incompleteness had caused severe confusion in various aspects (see Section A2.1).

Science and the philosophy of Buddhism have made progress with the same event as the base

As mentioned thus far, there is a large difference between word-based scientific understanding and what Buddhism teaches. However, interestingly, they have a close relation to each other. Science and the philosophy of Buddhism have made progress with the same event as the base.

As mentioned in Section 1.4.1, 1.4.3 and 6.2, when a person met a problem and has overcome it by thorough consideration and pursuit, he or she creates "a new idea" and simultaneously achieves a wholly clear state of mind full of delight. The achievement of a wholly clear state of mind means that a person has formed a new continuous internal connection of truly real ideas in his or her brain and truly real things in nature and the world. This also means that a person has produced *a new truly real idea* consisting of such a newly formed continuous internal connection. Now, a person cannot be conscious of such *a new truly real idea*. A person can only be conscious of a common or repeatedly observed part or an individual unchanging part of it. This means that "a new idea" a person creates when he or she has overcome a problem represents a common or individual unchanging part or an individual unchanging part of *a new truly real idea* he or she has produced.

Now, science has made progress by paying attention to "new ideas" people created and organizing them systematically in a consistent way, based on experiments or experiences. On the other hand, the philosophy of Buddhism has made progress by paying attention to a wholly clear state of mind people achieved and pursuing the profound meanings of it in a thoroughgoing way. Science has looked at human creative behavior from the outside and collected "results" of it, while the philosophy of Buddhism has looked at human creative behavior from the inside and pursued the origin of it. After all, science and the philosophy of Buddhism have looked at the same event from different sides.

A1.2 The Birth of Buddhism

A1.2.1 Shaka-muni Opened the Door to the Very Truth

A brief survey of Shaka-muni's life

Buddhism started from Shaka-muni's *Satori*, as mentioned earlier. Therefore, let us briefly survey Shaka-muni's life and consider how he attained *Satori*. According to literature [1-5], Gautama Siddhartha (the original name of Shaka-muni) was born in 463 B.C. (or 566 B.C. according to another method of estimation) as a prince of the Shaka clan, who dwelt in a southern foothill of the Himalayas. Shaka-muni is an honorific title given to Gautama Siddhartha after he attained *Satori*, where muni means saint. His mother suddenly died several days after his birth and he was brought up by a younger sister of his mother.

In the childhood, Gautama Siddhartha lived a happy life and was well educated as a prince. He married at the age of sixteen. However, when he was in the twenties, he saw the death of a person and came to feel unendurable unease and think about the temporariness of human life. It is said that he continued thinking about problems such as "I have luxuries of a palace, a healthy body and a happy family. However, what significance do they have if I shall die someday?" and "What is the true meaning of human life?" The spiritual struggle went on in his mind limitlessly. Therefore, when he became twenty nine years old, just when he had the first child, he made up his mind to give up his palace and family in order to look for a solution to his problem.

Why did Gautama Siddhartha take such a bold action? Presumably, he was a man of deep thinking. Besides, the loss of his mother so soon after his birth might have deeply affected his mind. Another important reason was that the state of affairs in ancient India in those years strongly influenced his mind. Since the 6th century B.C., farming technology developed in ancient India and this led to prosperity in commerce and industry and advancement in a monetary system. A number of new countries were established and many royal families rose. Such a new energetic atmosphere led to continual outbreak of battles among countries. Thus, it is very likely that Gautama Siddhartha felt intense unease to see such an unstable social situation and this acted as the true origin of his spiritual struggle mentioned above. On the other hand, such a new energetic atmosphere in ancient India also gave rise to a lot of fresh thinkers, called Samana. They left home and gave up the world and tried to create their own novel thoughts through hard ascetic practice. Gautama Siddhartha looked at activities of such fresh thinkers and made up his mind to join them.

He first went to the south and visited two hermits near the River Ganges. Then, he moved to Magadha Land, the most prosperous area in ancient India at that time, and started practicing asceticism in a mountain. He continued practice for a long period of time of six years. However, he was unable to get any solution to his problem. He noticed that continuing hard ascetic practice alone was not necessarily effective. Then, he came down from a mountain and washed himself in a river and ate a cup of milky rice gruel, kindly given by a girl in a village. He got refreshed and sat down at the foot of a linden tree and fell into deep contemplation. His five friends, with whom he had continued asceticism together, thought him to be corrupted and left him.

While Gautama Siddhartha continued contemplation under a linden tree, he suddenly and unexpectedly achieved a wholly clear state of mind completely released from any worry and agony and full of free independent spirit and wisdom. He was able to transcend all individual things in his life such as living, existing, aging, dying, and so on and stand firmly without relying on anything in the external visible world. At this moment, he discovered the very truth, later called the principle of *Engi*, and became a Buddha (a person who has arrived at the very truth about human life, a spiritually awakening person). This famous historical event is called Shaka-muni's *Satori*[*1].

> [*1]Shaka-muni's *Satori* is not a mysterious miracle. We can reasonably explain it if we take into account the internal invisible world (see the last part of Section A1.2.2).

After attaining *Satori*, Gautama Siddhartha or Shaka-muni went around in high spirits to tell people what he disclosed. At first, he taught it to his five friends. They initially thought Shaka-muni to be corrupted and shunned him. However, as his attitude was full of confidence and delight, they talked with him and believed him and became his first followers. Shaka-muni then went to Rajagriha Castle in Magadha Land and won the belief of King Bimbisara. After then, Shaka-muni continued going about in countries to teach people his way of living until he died at the age of eighty.

Shaka-muni paid primary attention to the achievement of a wholly clear state of mind

Now, let us consider characteristics and meanings of Shaka-muni's *Satori*. The most important point in Shaka-muni's *Satori* is that he paid primary attention to the achievement of a wholly clear state of mind, not to the acquisition of individual words (knowledge). Here was a unique attitude of Shaka-muni and simultaneously here was a great revolution in the way of understanding. It was for this reason that Shaka-muni was able to disclose the very truth lying in the internal world. As mentioned in Chapter 1, in traditional understanding and philosophies including science, people have paid attention to individual ideas and words. Shaka-muni's attitude was quite different from such a traditional way of understanding.

Why did Shaka-muni pay primary attention to the achievement of a wholly clear state of mind? While he continued hard ascetic practice in a mountain, he probably recognized that his mental pain, i.e. the fear of death or the fear of the temporariness of human life was impossible to remove by logic. In fact, it is a clear fact that a human being necessarily dies. Therefore, the fear of death can never be removed by any elegant theory as far as such a clear fact is present. Shaka-muni encountered a problem that was impossible to solve within the realm of word-based understanding, as mentioned earlier. Nevertheless, strangely, he was suddenly and unexpectedly able to remove his mental pain and achieve a wholly clear state of mind. Probably Shaka-muni himself was surprised to look at such a result. Here was certainly truth beyond logical understanding. Presumably for this reason, he paid primary attention to the achievement of a wholly clear state of mind.

Another possible reason is that Shaka-muni was strongly affected by social conditions in ancient India. Namely, he lived in a social situ- ation in which the removal of confusion and the achievement of harmony were more important than the acquisition of new knowledge, as will be discussed in Section A3.1.2. For this reason, he paid attention to the achievement of a wholly clear state of mind in which harmony was realized. Also, for the same reason, his idea was willingly accepted by people in ancient India.

Shaka-muni discovered that a person can transcend word-based understanding and arrive at the very truth

Next, what conclusion did Shaka-muni reach by paying attention to the achievement of a wholly clear state of mind? Before he attained *Satori*, he must have understood various things in his life such as living, existing, aging, dying and so on or wealth, health, knowledge, fame, and so on individually and separately and relied on them, in the same way as we now do. For this reason, he felt intense mental pain to notice that such individual things were of a temporary character.

However, the situation absolutely changed when he attained *Satori*. He achieved a wholly clear state of mind completely released from any worry and agony and full of free independent spirit and wisdom, as mentioned earlier. He came to be able to transcend all individual things in his life and stand firmly without relying on anything in the external visible world. At this time, Shaka-muni will have noticed that the achievement of a wholly clear state of mind meant that his understanding agreed with the real world, i.e. he reached the very truth. Then, he will have recognized that his understanding before the attainment of *Satori* gave him an illusion (a wrong view) and for this reason he had to suffer from intense mental pain. The very truth was where such an understanding was transcended. Furthermore, he will have become aware that everything in the very truth was in a fully dynamic free independent peaceful harmonious state. This is because he achieved a wholly clear state of mind full of free independent spirit and wisdom when he attained *Satori*, i.e. he arrived at the very truth.

In this way, Shaka-muni must have reached the following conclusions: (1) understanding by individual ideas and words, i.e. traditional word-based understanding gives a person a wrong view of nature and the world and hence brings him or her many worries and agonies, (2) a person can transcend such word-based understanding and arrive at the very truth at which he or she can gain the correct view, and (3) a person realizes a free independent peaceful harmonious life when he or she has arrived at the very truth, i.e. everything in the very truth is in a free independent peaceful harmonious state. After all, Shaka-muni discovered that there exists the internal invisible world beyond word-based understanding and a person can live a peaceful harmonious life if he or she stands with the internal invisible world as the base.

This was really quite an important discovery. We human beings can only be conscious of individual words and things, such as wealth, health, knowledge, love, fame, and so on, and hence we can only rely on them on a conscious level, as discussed in Section 1.2.1. In addition, such individual words and things are very useful and indispensable for human life. Therefore, we in general want to maintain a peaceful happy life based on such individual words and things. However, Shaka-muni made a revolutionary discovery such as mentioned above. Thus, based on it, he asserted that "visible shapes and color" were only of an imaginary character and relying on them led to limitless worries and agonies. He also asserted that a person was able to escape from all worries and agonies and attain a peaceful harmonious life if he or she had transcended individual words and things. In fact, Shaka-muni demonstrated by his own experience that he was able to escape from the fear of death by transcending individual things. Naturally, nobody in ancient India was aware of such an amazing fact. Thus, a great many people, who suffered from intense mental pain in confused society, were strongly impressed with *Shaka-muni*'s achievements and followed his teachings. Then Buddhism rose.

Here, it will be interesting to consider why Shaka-muni was able to escape from the fear of death. Individual ideas or words such as life and death are what human beings have artificially created by picking out common or repeatedly observed properties of truly real things in human life (see Section 1.2.1). Therefore, a person can transcend such ideas or words and hence feelings coming from them if he or she really arrives at the very truth and looks at truly real things as they are. There are no ideas or words of life and death in the very truth (see Section 1.4.1). Thus, the fear of death appears solely because a person remains within the realm of word-based understanding.

Somebody may ask whether Shaka-muni's *Satori* really existed in the actual human history. At least, it is clear that Buddhism rose from Shaka-muni's *Satori*. In addition, we have a similar experience to Shaka-muni's *Satori* in daily life. As mentioned earlier, when we met a difficult problem and have overcome it by thorough consideration and pursuit, we create a new idea and simultaneously achieve a wholly clear state of mind full of delight. Shaka-muni's *Satori* can be regarded as a typical case of such an experience in our daily life.

A1.2.2 Initial Buddhism

The rise of Initial Buddhism

After attaining *Satori*, Shaka-muni continued going about in countries to teach people his way of living, as mentioned earlier. Unfortunately, there is no literature which Shaka-muni himself directly wrote. Words he told during his life were transmitted among people by words of mouth and later given shapes by his followers. Historically, an association of Shaka-muni's followers, called "Samgha", was formed in a while after his death [4]. It split into two branches about one hundred years after then. Buddhism before the split is called Initial Buddhism or Primitive Buddhism.

Initial Buddhism has a set of sutras, called "Agama" [1,2,4]. Suttanipata [6] is one of the oldest sutras and is known as a collection of Shaka-muni's words spreading in people after his death. Initial Buddhism has also several fundamental principles such as the principle of *Engi* or *Engi*, *Juni-in-en*, and *Shi-tai* [1-4], which Shaka-muni's successors formulated for explaining Shaka-muni's *Satori* based on transmitted words of him.

Shaka-muni's Satori in sutras of Initial Buddhism

How is Shaka-muni's *Satori* explained in sutras of Initial Buddhism? As to the problem of what state Shaka-muni reached when he attained *Satori*, the following words are recorded in Suttanipata [6].

> *He (an awakened person, a person who has arrived at the very truth, a person who has attained Satori, a Buddha) does not addict himself to any pleasure coming from passions and desires. His mind is not impure. He has transcended all illusions (incorrect views arising from understanding by words). An awakened person has the eye to see through a variety of things*

clearly. (#161)

The mind of a sage (an awakened person) is well embodied in deeds and words. It is natural that you are willing to follow the sage who is endowed with clear intelligence and behavior. (#163b)

A person who lives in peacefulness, a person who is rid of judgment of good and evil, a person who is extricated from dirty, a person who knows this world and that world, and a person who has transcended life and death; such a person is called one who seeks after the truth. (#520)

He (an awakened person) entirely controls and governs all things which he has looked at, learned and considered. How would it be possible to guide such a person who behaves without any covering to the area of discernment and illusion (the area of understanding by words)? (#793)

A (true) Brahman (an awakened person) has transcended Bon-noh (a person's desires and attachments for individual things such as living, existing, wealth, and so on). Even if he sees or knows anything, he is never attached to it. He is neither eager for desires nor eager for anti-desires. He has nothing to attach himself to as the best in this world. (#795)

Mr. Nanda! It is not by opinions, learning or knowledge that a person reaching the very truth is called a sage. It is a sage that excludes all Bon-noh and behaves without any desires and sufferings. (#1078)

Words in parentheses are added by the author of this book (Y. N.).

The above words clearly indicate that to attain *Satori* in Shaka-muni's *Satori* is to transcend individual ideas and things in human life and achieve a wholly clear state of mind completely released from any worry and agony and full of free independent spirit and wisdom, as mentioned in Section A1.2.1.

The very truth is where word-based understanding is transcended

Suttanipata also emphasizes that the very truth Shaka-muni teaches has quite a different meaning from the truth in traditional understanding. This is because the very truth Shaka-muni teaches is where word-based understanding is transcended, though the truth in traditional understanding is based on word-based understanding. Thus, a person who has arrived at the very truth no longer relies on ideas and words. Such a person only look at what he or she looks at as they are. On the other hand, a person who has the truth in traditional understanding strongly relies on individual ideas and words and thus has various (wrong) opinions and hence often disputes. In fact, the following words are recorded in Suttanipata.

> *I (Sakya-muni) do not have what I teach. I only know clearly that attachments are attachments and errors are errors and as a consequence I merely see the peacefulness of mind. (#837)*

> *About opinions which one person says to be correct and true, another person says that they are false and unreliable. In this way, they have different opinions and dispute with each other. Why do persons who walk on different paths not tell the same thing? (#883)*

> *Only one truth exists and no second truth is present. Therefore, a person who has reached this truth has no dispute with others. (On the other hand) Those (persons who have not yet reached this truth or persons who have understanding by words) praise their own truths, which are different from others. Therefore, they do not speak of the same thing. (#884)*

> *Depending on traditional opinions, learning, commandments, pledges and thoughts, and despising other thoughts, those persons (who have not yet reached the very truth, who have understanding by words) take pleasures in their own theories and say that opponents are fools and do not reach the truth. (#887)*

Shaka-muni emphasizes that the very truth is where a person has transcended word-based understanding.

For the same reason as the above, Shaka-muni also stressed the importance of paying attention to real things rather than knowledge. It is recorded in a sutra, named *Nehan-kyo*, compiled in the age of Initial Buddhism [1-4] that Shaka-muni, just before he died, taught to his followers as follows.

> *After I shall die, make you a light, rely on you yourself, and do not depend on anyone else. Make Dharma (constituent elements of the real world) the ground, rely upon them, and do not depend on anything else.*

Where is the mind of a person who has attained Satori?

Word-based understanding is the only perceptive ability of human beings (see Section 1.2.1). Therefore, if a person has transcended word-based understanding, this means that he or she has leapt over this world and moved to another world lying beyond the bounds of human consciousness. Accordingly, *Satori* is called *Mezame* (spiritual awakening), *Gedatsu* (extrication from the human bondage), *Higan ni wataru* (jumping to a peaceful ideal land) and *Nehan* (nirvana, a state of tranquility).

Now, if a person who has attained *Satori* leaps to another world, where is the mind of such a person? About this question, interesting dialogues are recorded in Suttanipata.

> *Dear Saint! If he (an awakened person, a person who has attained Satori) does not return to the area of discernment and illusions (the area of understanding by words) and remains a place free from the human bondage, does he still exist? Also, does his ability of discernment still exist? (#1073)*

> *For example, a fire, which has been blown away by strong wind, is destroyed and can no longer be recognized as a fire. In a similar way, a saint, who is liberated from his name and body, is destroyed and can no longer be recognized as a being. (#1074)*

> *Does a man who is destroyed not exist, or does he permanently exist and is he not lost? (#1075)*

> *About a man who is destroyed, there is no criterion for judging whether he exists or not. About such a man, there is no reminder for discussion. When everything is entirely eradicated, any room for discussion is also completely exterminated. (#1076)*

The questions cited above are of much interest. Such questions are intrinsic to the philosophy of Buddhism, which deals with the very truth or truly real things that lie beyond the bounds of human consciousness. This issue will be discussed again in later sections.

The principle of Engi

We have thus far looked at Shaka-muni's words recorded in Suttanipata. Next, let us consider the fundamental principles of Initial Buddhism, which Shaka-muni's followers formulated for explaining Shaka-muni's *Satori*. At first, let us consider a fundamental principle called the principle of *Engi* or *Engi*. As already mentioned, it is said that Shaka-muni noticed this principle when he suddenly and unexpectedly reached a wholly clear state of mind. Therefore, this principle constitutes the very basis of the philosophy of Buddhism.

The principle of *Engi* deals with relations among a variety of constituent elements of human life such as existence, living, aging, dying, knowing, desires, affection, and so on, which are called Dharma in Buddhism. Then, the principle says as follows [1-4].

> *In the very truth, constituent elements of human life such as existence, living, aging, dying, knowing, and so on emerge through dependence on others.*

Engi refers to "emerging through dependence on others" or "existing in dependence on others". By attaining *Satori*, Shaka-muni transcended individual ideas and things and stood firmly without relying on anything in the external visible world. He also achieved a wholly clear state of mind completely released from any worry and agony and full of free independent spirit and wisdom. Thus, he must have seen that in the very truth everything realized a fully dynamic free independent peaceful harmonious state, as mentioned in Section A1.2.1. The principle of *Engi* is an expression of such a state in the very truth [*1].

[*1] It was mentioned in Section A1.1 that whan a person has arrived at the very truth, he or she gains a big sphere of internal wisdom in his or her mind, words simply floating on its surface. We may be able to say that the principle of *Engi* expresses such a situation.

Juni-in-en (Twelve causes and conditions)

The principle of *Engi* was formalized in various ways in Initial Buddhism. The simplest expression is given as follows [3].

> *When this exists, that exists. If this is born, that is born. When this is absent, that is absent. If this is destroyed, that is destroyed.*

Here "this" or "that" refers to one of constituting elements of human life. Certainly, these words indicate that everything emerges through dependence on others.

The most famous expression of the principle of *Engi* is *Juni-in-en*, where *juni* means twelve and *in-en* means causes and conditions. This is composed of twelve constituent elements of human life arranged in an appropriate order [1-4],

> (1) *unclearness (the ignorance of the very truth)* (2) *intentional*

behavior and acts (3) *knowing (composed of seeing, hearing, smelling, tasting, toughing, and discernment)* (4) *name and shape (mind and body)* (5) *six sense organs (including a thinking organ)* (6) *contact (of sense organs with objects)* (7) *reception (of feelings and emotions)* (8) *desires* (9) *attachments* (10) *existence (living)* (11) *birth* (12) *aging and death.*

It is said that the number of "twelve" has no significant meaning. An important point is that various constituent elements of human life emerge through dependence on others and form a series of causes and conditions of worries and agonies in human life.

Thus, *Juni-in-en* is in general interpreted as follows: Worries and agonies come from thinking of "aging and death", which come from thinking of "birth", which in turn come from thinking of "existence (living)", and so forth. The word of "unclearness" is placed at the first position. This means that all constituent elements of human life and hence all worries and agonies come from "unclearness" (the ignorance of the very truth).

Juni-in-en is also interpreted in the reverse direction. If "unclearness" is removed (i.e. if we have arrived at the very truth or if we have transcended individual ideas and things), constituent elements such as (2) "intentional behavior and acts", (3) "knowing", ··· disappear one after another and thus all worries and agonies also disappear. Note again that the first word of "unclearness" does not mean having no knowledge. It means that a person may have much knowledge but only understands it by words.

The true meaning of the principle of Engi cannot be understood without having essentially the same experience as Shaka-muni

The principle of *Engi* (or *Juni-in-en*) expresses the very truth about human life, which Shaka-muni reached. However, the principle of *Engi* described by words only expresses what Shaka-muni looked at when he had attained *Satori*, as mentioned in Section A1.1. Namely, it only expresses the external appearance of the very truth Shaka-muni reached. "The very truth he reached" itself is in principle impossible to express by words.

A similar argument was given in Section 1.3.1. By looking at a person who has attained *Tai-toku* about riding a bicycle, i.e. who has arrived at the very truth about riding a bicycle, we say that to ride a bicycle is to carry out a variety of things such as to take the balance of a body, to exert an appropriate force on left and right legs alternately, and so on simultaneously in harmony with one another. However, a beginner cannot ride a bicycle even if he or she has completely understood such words. The very truth about riding a bicycle, expressed by words, does not correctly represent "the very truth about riding a bicycle" itself.

A famous theoretician in ancient India in the 3rd century, Nagarjuna (Ryuju in Japanese), asserted that the principle of *Engi* expressed by words was not correct in a strict sense and was only valid from a conventional point of view [2]. According to his opinion, every constituent element of human life emerges through dependence on others and is of a freely changing dynamic character and thus has no immutable quality. Therefore, such a thing is in principle impossible to express by words. Therefore, he says that words such as "this" and "that" or existence, living, aging and dying never correctly express constituent elements of human life.

These arguments imply that we can by no means gain the truly correct understanding of the principle of *Engi* by words. We have to make much effort after we have understood it by words. Actually, this effort occupies the main part of processes of gaining the truly correct understanding of the principle of *Engi*. Here is a unique feature of the Buddhist teachings. To tell the truth, we can obtain the truly correct understanding of the principle of *Engi* (the very truth Shaka-muni reached) only when we have had essentially the same experience as Shaka-muni.

Accordingly, the phrase of *correctly understanding the principle of Engi* or *clearly looking at the principle of Engi* has quite a significant meaning. It has the same meaning as arriving at essentially the same very truth as Shaka-muni reached and having obtained the truly correct understanding of the principle of *Engi*.

Shi-tai (Four Truths)

Another fundamental principle of Initial Buddhism, called *Shi-tai*, also explains Shaka-muni's *Satori* in a simple form. It consists of the following four truths [1-4].

1. *Human life is agony.*

2. *Desires and attachments only lead to the collection of agony.*

3. *The extinction of desires and attachments leads to Satori.*

4. *A way of attaining Satori is Hassho-do (eight correct ways of living).*

It seems that *Shi-tai* teaches us how Shaka-muni attained *Satori*, as explained below.

The first item of *Shi-tai* says that human life is agony if a person remains within the realm of word-based understanding. It is probable that this item expresses the fact that Shaka-muni felt unendurable unease to look at the death of a person or to notice a temporary nature of human life before he attained *Satori* (see Section A1.2.1). In fact, the following words are recorded in Suttanipata.

> *Any place in the world is not stable. Everything is agitated and in anxiety. I have sought for a stable place on which I can rely but I have been unable to find any place which is free from the bondage of human life.* (#937)

Shaka-muni will at first have sought for a stable place on which he was able to rely. However, he was unable to find such a place. The recognition that human life was of an unstable temporary nature must have made Shaka-muni's sufferings entirely serious. This is because this recognition showed that things such as existence (living), wealth, health, fame, love, and so on, which he strongly desired to maintain for a long time, were all gone someday.

The most severe problem was that anybody was unable to avoid aging and death. Everybody strongly desired to be young and live long but was destined to age and die someday. There are many words about the death of a human being in Suttanipata. Let us cite some of them.

> *Human life in this world has no fixed length. Nobody knows how long he or she can live. Human life is pitiful, short, and linked to agony.* (#574)

> *What are born in this world cannot escape death. They age and die. The fate of living things is as such.* (#575)

> *A young person and a person in the prime of life, and a foolish person and a wise person all surrender themselves to death. Everybody necessarily goes to death.* (#578)

> *Look! Persons are enticed away one by one just like sheep which are conducted to a slaughterhouse though their relatives gazing at them are overcome with sorrow.* (#580)

The agony of death is the most fundamental in human life. Any happiness, pleasure and ability become valueless before death. In this way, human life is full of agony.

The second item of *Shi-tai* explains why human life becomes full of agony. This is because people are attached to individual words and things, which are necessarily lost someday. In general, people want to maintain a peaceful happy life by relying on individual things such as existence (living), wealth, health, knowledge, love, fame, and so on, as mentioned earlier. However, Initial Buddhism says that having desires and attachments for individual things is equivalent to accumulating worry and agony because they are necessarily gone someday.

The third item of *Shi-tai* explains how Shaka-muni escaped from his agony. He succeeded in extinguishing desires and attachments for individual things, which were the causes of his agony. Note here that extinguishing desires and attachments never means giving up having desires and attachments. It also never means disregarding them. When Shaka-muni attained *Satori*, he achieved a wholly clear state of mind completely released from any worry and agony and full of free independent spirit and wisdom, as mentioned earlier. He also stood firmly without relying on anything in the external visible world. If Shaka-muni had extinguished desires and attachments by giving up them or disregarding them, he could never have realized such a lively state of mind. Extinguishing desires and attachments in Shaka-muni's *Satori* meant transcending them and acquiring high-level ability to freely control them, as will be explained later.

Now, how was Shaka-muni able to transcend desires and attachments for individual things? There are some important points to be noted here. Firstly, he faced a problem that was impossible to solve by logical thinking, as mentioned in Section A1.2.1. In fact, it is entirely impossible to overcome the fear of death by logical thinking because it is a clear fact that a human being dies someday. Logical thinking is powerless in overcoming a problem arising from a clear fact. Human intellect is not free from restraint by inevitability. Secondly, Shaka-muni did not rely on any absolute or supernatural power such as a god in order to solve his problem. This was most probably because he paid primary attention to the removal of *actual* agonies in his mind. Presumably, he will have clearly recognized that any elegant idea or theory, including the concept of a god, was powerless in removing his *actual* agonies. Relying on supernatural power is equal to escaping from a difficulty.

Shaka-muni's aim was to remove actual agony in his mind, as mentioned above. His mental pain emerged from the fact that he looked at the death of a person or he noticed the temporariness of human life or the unstable nature of the world. Therefore, he first looked for a stable place on which he was able to rely, as mentioned earlier. However, he was unable to find such a place anywhere. Then, he will have come to try to pursue the origin of his mental pain probably in order to find a way to remove it. Fortunately, this choice of Shaka-muni was correct. After thorough consideration and pursuit, he suddenly and unexpectedly reached a wholly clear state of mind. This was indeed a strange event but really occurred. How Shaka-muni gained such a success will be explained at the end of this section.

Interestingly, Shaka-muni's achievement of a wholly clear state of mind indicated that his agony did not come from the fact that a human being necessarily dies someday. Otherwise, he could not have achieved a wholly clear state of mind. His agony came from his incorrect view of human life. Before Shaka-muni attained *Satori*, he looked at individual things in his life such as living, aging, dying, and so on separately and had intense desires and excessive attachments for them. For this reason, he felt the fear of death to see the death of a person. When he attained *Satori* and transcended individual things, desires and attachments for them disappeared and hence worries and agonies also disappeared.

The above argument constitutes the kernel of Shaka-muni's *Satori*. In fact, a large number of words about it are recorded in Suttanipata. Let us cite some examples.

> *Various sufferings in the world come from constituent elements of personal life (such as existence, wealth, fame, desires, attachments, etc.). Non-awakened persons unawares rely on constituent elements of their life and undergo sufferings repeatedly. Therefore, look at causes of sufferings by the clear eye and make effort not to depend on them. (#728)*
>
> *A person who is not subordinated to individual things does not stagger. However, a person who is subordinated to individual things is attached to this or that and cannot overcome the circulation of sufferings. (#752)*
>
> *Look! (Non-awakened) people think "things with no entities" to be "things with entities" and are excessively attached to them. These people think that this is the truth. (#756)*
>
> *Sages (awakened persons) say that things, which other people regard as pleasures, are agonies. On the other hand, sages know that things, which other people regard as agonies, are pleasures. Look at the truth which is difficult to understand. Non-awakened persons are at a loss here. (#762)*

The words in the last item indicate that Shaka-muni himself recognized that the very truth he disclosed was difficult to understand.

In Buddhism, a person's desires and attachments for individual things such as existence, living, wealth, love, social position, fame, and so on are called *Bon-noh*. Then, Buddhism says that a main cause of worries and agonies is *Bon-noh*, which in turn arise from illusions (an incorrect view of nature and the world arising from word-based understanding) or unclearness (the ignorance of the very truth). Thus, Initial Buddhism teaches people to transcend illusions and *Bon-noh* and reach the very truth. For example, a sutra of Initial Buddhism entitled *Yuikyo-kyo* teaches as follows [3].

> *All things (all shapes and color) in the world are only of a temporary imaginary character. They are like dew on leaves or bubbles in a river. You lose them someday with necessity. This is the truth of the world. Do not have any worry and agony with this matter. Make effort and reach the very truth and destroy illusions and Bon-noh by the light of wisdom.*

Note that Initial Buddhism emphasizes the importance of arriving at the very truth. The extinction of illusions and *Bon-noh* is a result of arriving at the very truth.

Now, people in the present days will feel more or less resistance to the above conclusion of Initial Buddhism. Roughly speaking, *Bon-noh* and illusions in Buddhism refer to acts of a person's ego. Therefore, it appears that Initial Buddhism denies a positive role of a person's ego. Here is why many of people in the present days feel resistance to the conclusion of Initial Buddhism. Somebody may say, "I want to have existence, wealth, love, social position, fame, and so on even though I cannot escape from worries and agonies." Certainly, we human beings usually want to have existence, wealth, social position, and so on. Science has advanced based on such desires and attachments of human beings.

According to my understanding, Initial Buddhism does not say that we should throw desires and attachments for individual things away. It only says that we should transcend desires and attachments for individual things. When we have transcended desires and attachments for individual things and reached the very truth, we acquire high-level internal wisdom to freely control them so that they work rightly. Therefore, we can gain a free independent peaceful harmonious life. It was argued in Chapter 5 that we human beings have two kinds of the subject: our ego and the true self. It seems that initial Buddhism teaches us to rely not on their ego but on the true self.

However, even if we understand the conclusion of Initial Buddhism in this way, there will still remain resistance to it. A problem is what the ideal state of human life is. Initial Buddhism says that a state in which desires and attachments are thoroughly eradicated or a state in which complete peace and harmony are realized is the ideal state of human life. However, it is possible to say that a state full of desires and attachments, which leads us to lively activities, is the ideal state of human life. Many people in the present days will agree with the latter opinion. Certainly, the realization of complete peace and harmony is important for our life. However, if our life is always in the state of complete peace and harmony, we shall feel it boring. Such a problem in Initial Buddhism arose from the fact that it had not yet reached a completed stage of the philosophy of Buddhism. In fact, many issues were left unsolved at this stage, as will be discussed in Section A1.3.1, A2.1 and A2.4.

The fourth item of *Shi-tai* expresses the way to attain *Satori*, which Initial Buddhism teaches. The way consists of eight correct ways of living and is called *Hassho-do*[1-4].

(1) Correct view (2) correct intention (3) correct words (4) correct behavior (5) correct occupation (6) correct efforts (7) correct thought (8) correct meditation with a concentrated mind.

Hassho-do teaches people to approach a peaceful harmonious state Shaka-muni attained through the control of personal life by their own effort.

Regretfully, in my opinion, *Hassho-do* does not express the correct way to attain *Satori*, as argued in detail in Section A2.1. It is only what Shaka-muni's followers artificially invented based on transmitted words of Shaka-muni. Namely, it merely offers a superficial phenomenological way to attain *Satori*. As already mentioned in Section 1.4.3 and A1.1, the revelation of how to attain *Satori* was a very difficult problem in the long history of Buddhism. In Initial Buddhism, it was well recognized that Shaka-muni transcended desires and attachments for individual things and attained *Satori* but it remained unknown how he transcended them. Accordingly, all discussions of how Shaka-muni attained *Satori*, described in this book, are solely based on my understanding.

San-po-in (Three Theorems)

Buddhism has a famous fundamental principle called *San-po-in* (Three Theorems). It is said that *San-po-in* was formulated in the age of Mahayana Buddhism[2]. However, this principle best expresses the main conclusion of Shaka-muni's *Satori* in a simple form. *San-po-in* consists of the following three theorems [1-4].

1. *Everything incessantly changes.*

2. *Everything has no inherent unchanging quality.*

3. *The attainment of Satori leads to peacefulness and tranquility*

The first and second theorems emphasize that in the very truth things have an entirely different way of existing from those in the external visible world. Then, the third theorem teaches people that the truly correct understanding of these theorems (i.e. the arrival at the very truth) leads us to a peaceful harmonious life.

We human beings usually understand things in nature and the world by the use of logical connections. However, if we have arrived at the very truth, words disappear and hence logical connections also disappear. Instead, we have a network of continuous internal connections (see Section 1.4.1) or a network of continuous internal connections and information-based connections (see Section 4.4). Thus, it is likely that *San-po-in* expresses a situation in which we have understood things in nature and the world by the use of such a network of continuous internal connections.

A theoretical consideration of how Shaka-muni attained Satori

Finally, it will be important to make it clear that Shaka-muni's *Satori* is not a mysterious miracle but a real event. Indeed, Shaka-muni suddenly and unexpectedly reached a wholly clear state of mind by thorough consideration and pursuit, as mentioned earlier. This was really a strange fact. However, we can successfully explain it if we take into account the internal invisible world.

As mentioned earlier, Shaka-muni noticed the very truth, later called the principle of *Engi*, when he suddenly and unexpectedly achieved a wholly clear state of mind. Therefore, let us consider how he had reached the principle of *Engi*. Shaka-muni's mental pain emerged from the fact that he saw the death of a person, i.e. he noticed the temporariness of human life or the unstable nature of the world. Therefore, he at first searched for a stable place but was unable to find such a place anywhere, as mentioned earlier. Then he will next have tried to investigate the origin of his mental pain probably with the aim of finding a way to remove it. Fortunately, this choice led him to *Satori*.

Let us conjecture how he investigated the origin of his mental pain. It is expected that he first asked as follows.

"Where does my mental pain come from?"

This was an important question, from which Shaka-muni's *Satori* lastly emerged. He must soon have found an answer.

"I desire to live long but I know that I shall die someday. The death of a person reminded me of this fact. Thus I feel agony to see death."

This is a reasonable answer but Shaka-muni was unable to remove his mental pain by this answer. Then, he had to continue asking further.

"Why do I desire to live long?"

He may have answered as follow.

"I have keen affection for my existence and thus I desire to live long."

This is also a reasonable answer. However, Shaka-muni was still unable to remove his mental pain. In this way, similar asking and answering will have continued one after another.

"Where does my keen affection for my existence come from?"

"I see that I live here and I perceive that my life is here. Therefore I am attached to my existence and have keen affection for it."

This is because any answer did not allow him to remove his mental pain.

By such continuation of asking and answering, Shaka-muni probably became aware that various aspects (constituent elements) of his life such as existence, living, desires, attachments, seeing, and so on emerged one after another as causes of his agony (mental pain). In addition, he will also have recognized that his agony emerged from his desires and attachments for various things in his life.

Accordingly, it is possible to assume that Shaka-muni discovered the principle of *Engi* at this stage and attained *Satori*. However I do not adopt this interpretation. If this interpretation is the case, it follows that Shaka-muni attained *Satori* straightforwardly by logical thinking alone. Such a conclusion is in disagreement with the aforementioned argument that *Satori* is attained by transcending word-based understanding.

Actually, any answer did not allow him to remove his agony, as mentioned above. Therefore, he had to continue asking and answering limitlessly and had finally fallen into an absolutely difficult situation in which his asking and answering went round and round in circles and continued endlessly. Certainly, he faced a problem that was impossible to solve by logical thinking. Nevertheless, Shaka-muni did not stop asking and answering. He never gave up removing his agony. Here we can see extraordinarily superior ability of *Shaka-muni* (see also the discussion of Section 6.3). In fact, such thorough consideration and pursuit brought about the jump in Shaka-muni's understanding. When Shaka-muni was under deep contemplation, he suddenly and unexpectedly reached a wholly clear state of mind, as mentioned earlier.

What jump happened in Shaka-muni's head? Before attaining *Satori*, he had intense desires and excessive attachments for his existence or living, from which his severe worries and agonies appeared. Such intense desires and excessive attachments emerged from the fact that he looked at constituent elements of his life individually and separately. *Before attaining Satori, Shaka-muni was only able to look at constituent elements individually and separately.* He was never able to escape from such a state. This was because truly real ideas (nervous patterns) in his brain existed individually and separately at this stage and were impossible to control intentionally, as argued in Section 1.4.3.

However, the situation greatly changed by continuing thorough consideration and pursuit about the origin of his agony. The jump happened in Shaka-muni's understanding and he came to be able to look at individual ideas and things in a continuous way with no separation. As a result, he transcended individual ideas and things and achieved a wholly clear state of mind. The principle of *Engi* arose from here. Shaka-muni's thorough consideration and pursuit led to the formation of a continuous internal connection of constituent elements of his life.

In this way, we can reasonably explain how Shaka-muni attained *Satori* and reached the principle of *Engi*. Interestingly, the attainment of Shaka-muni's *Satori* happened in essentially the same way as the attainment of *Tai-toku* discussed in Section 1.3.2 and the achievement of theoretical creation in science explained in Section 6.2. Thus, we can say that *Satori*, *Tai-toku* and theoretical creation are analogous events and have similar characteristics to one another. In either of these events, we reach the very truth though there are large differences in the quality level and aspect (see Section 1.4.1 and 6.2).

The above consideration also indicates that Shaka-muni transcended word-based understanding by forming a continuous internal connection of individual ideas and things. The formation of a continuous internal connection gave him the ability beyond logical understanding.

A1.3 Mahayana Buddhism

In the early 1st century B.C., about 250 years after Shaka-muni's death, a new movement in Buddhism rose in ancient India. The new movement was supported by a number of groups of people and continued over a period of about ten centuries. A certain group called the new movement "Mahayana" (which means "a large vehicle" for attaining *Satori*). Thus, new Buddhism brought by the new movement is now called Mahayana Buddhism.

A1.3.1 A Brief Survey of Mahayana Buddhism

Mahayana Buddhism has a large number of sutras, theoretical books, explanatory books and commentary books but they all deal with particular subjects with focused aims. Therefore, before proceeding to detailed explanations of them, let us briefly survey why and how a new movement rose in Buddhism and what thought it reached, according to my understandings of sutras and introductory books [1-4,7-9].

The motive and features of a new movement

Mahayana Buddhism is clearly characterized by the concept of Bodhisattva, where a Bodhisattva refers to a person who carries out hard Buddhist practice with the aim of accomplishing the ability to conduct the common people as well as him- or herself to *Satori*. The accomplishment of such ability is called the attainment of *complete Satori* and the way that a Bodhisattva conducts the common people as well as him- or herself to *Satori* is called *Bosatsu-jo* or *Butsu-jo*. Here *Bosatsu* and *Butsu* refer to Bodhisattva and Buddha, respectively. As will be explained later, people who started new Buddhism (Mahayana Buddhism) were in general not monks in monasteries but general people living in towns or villages. For convenience, we hereafter call people who started new Buddhism *new Buddhists in Mahayana Buddhism*. Strikingly, they primarily aimed at conducting the common people to *Satori*. Namely, their final aim was to establish *Bosatsu-jo* or *Butsu-jo*. On the other hand, Buddhist monks following the traditional thought aimed at conducting them themselves to *Satori*.

Theoretically, the rise of a new movement was an inevitable result of historical progress of the philosophy of Buddhism. Shaka-muni carried out thorough consideration and pursuit about the origin of his mental pain and attained *Satori* by transcending individual ideas and things, as discussed in Section A1.2.1. However, Shaka-muni's *Satori* had a personal character in that the object of his thinking was actually restricted within constituent elements of his personal life, as can be seen from the items of *Juni-in-en* in Section A1.2.2. Namely, his thinking did not necessarily extend widely to constituent elements of other persons in the outer world. This was an unavoidable result because Shaka-muni for the first time achieved *Satori*. However, such insufficiency had to be overcome sooner or later by his successors. New Buddhists in Mahayana Buddhism, i.e. people who started new Buddhism (Mahayana Buddhism) really tackled this problem by using the concept of Bodhisattva.

We can safely say that the philosophy of Buddhism has advanced by overcoming various sorts of separation in human ideas, which arise from the intrinsic limit of word-based understanding. For example, Shaka-muni encountered a problem arising from separation between life and death. He then overcame the problem by discovering that in the very truth everything emerged through dependence on others, i.e. everything was in a fully dynamic peaceful harmonious state. On the other hand, new Buddhists in Mahayana Buddhism started activity as a Bodhisattva and encountered a problem arising from separation between new Buddhists and the common people or separation between "I" and "you" or between the subject and the object. By encountering such a problem, new Buddhists in Mahayana Buddhism were forced to clarify how in the very truth people and things in the outer world existed. Shaka-muni overcame the problem of "what am I?" while new Buddhists in Mahayana Buddhism were requested to surmount the problem of "what are you?"

In this way, new Buddhists in Mahayana Buddhism greatly widened the range of vision from personal life to the whole world. As a result, they succeeded in disclosing a profound Buddhist view that everything in nature and the world, including a living thing and a human being, exists in a fully continuous dynamic and harmonious state though it demonstrates individual unchanging properties in the external visible world.

How did a new movement start?

The association of Shaka-muni's followers (called Samgha) split into two branches because of a difference in opinions about commandments about 100 years after his death, as already mentioned in Section A1.2.2. The branches later further split into about twenty branches. Buddhism before the split is called Initial or Primitive Buddhism, while Buddhism after the split is called Branch Buddhism.

Buddhist monks in Branch Buddhism remained within the traditional thought based on the *Satori* of a personal character. It is said [1,2] that they thought that only a limited number of persons such as Shaka-muni was able to achieve *Satori* and become a Buddha. Thus, they gave up becoming a Buddha and simply aimed at becoming an Arahant (a person who has completely destroyed *Bon-noh*, a person who deserves esteem). They were financially supported by rich citizens and were absorbed in establishing theoretical systems (called "Abhidharma") in order to make the legitimacy of branches' thoughts clear [4]. Another important feature of

Buddhist monks in Branch Buddhism is that they stopped going about in countries for teaching the common people their ideas, in contrast to Shaka-muni. They were confined to monasteries separated from the common people [4,7].

On the other hand, people who started new Buddhism, i.e. new Buddhists in Mahayana Buddhism were in general not monks in monasteries but general people who believed Shaka-muni's teachings, as already mentioned earlier. It is said that new Buddhists in Mahayana Buddhism were active, rich, and truly fascinating persons and livelily worked for the common people with high ideals and deep affection [7]. Presumably, they played an important role in teaching Shaka-muni's words to the common people because Buddhist monks in Branch Buddhism were confined to monasteries.

In the early 2nd century B.C., Maurya Dynasty, which had been governing the whole India and giving support to Buddhism, was ruined. Since then, short-term dynasties rose one after another in North West India and people in this district were severely tormented by plunder and tyranny till around the 3rd century A.D. In such a miserable situation, new Buddhists in Mahayana Buddhism in this district probably played more and more important roles in helping the common people to relieve sufferings. Through such activities, new Buddhists may also have started to feel doubt on the attitude of Buddhist monks in Branch Buddhism who were confined to monasteries.

In the beginning of the 1st century B.C., another important event happened. Buddhist monks in the most influential branch of Branch Buddhism completed a theoretical system (Abhidharma) that largely deviated from Shaka-muni's teachings. They categorized constituent elements of human life and concluded that both physical and mental constituent elements had immutable entities on the essential level though they had no unchanging quality on a subjective level [1,2,4,7]. This event will have made the doubt of new Buddhists in Mahayana Buddhism about the attitude of Buddhist monks in Branch Buddhism definite. In fact, new Buddhists in Mahayana Buddhism started marked activity around the beginning of the 1st century B.C.

In this way, new Buddhists in Mahayana Buddhism started activity for constructing new Buddhism mainly in the region of North West India, in parallel to activities of Buddhist monks in Branch Buddhism. It is said that Branch Buddhism had still occupied the dominant place in Buddhism of the whole India even when Mahayana Buddhism became prosperous.

The establishment of a Bodhisattva of a new image

When new Buddhists in Mahayana Buddhism started systematic activity, they must have faced several important problems. At first, they were not Buddhist monks in monasteries but general people and therefore they had to make their position or authority clear. Another more important problem is that they had to make it clear how they were able to conduct the common people to *Satori*.

According to the Buddhist literature [1,2,4], it seems that the first problem was solved by creating a Bodhisattva of a new image. In the age when new Buddhists in Mahayana Buddhism started activity, Buddhism was already widely accepted in ancient India and Shaka-muni was the object of worship. He was believed to have accumulated worthy deeds sufficient to help the common people. A number of grand Buddhist pagodas (or stupas) were built and a number of (fictitious) biographical stories of Shaka-muni were produced. Such stories highly praised Shaka-muni with words familiar to the common people. Therefore, they largely contributed to the wide spread of Buddhism among general people.

Interestingly, the concept of a Bodhisattva was already told in such biographical stories. Namely, Shaka-muni before attaining *Satori* displayed activity as a Bodhisattva in such stories. In addition, the concept of past life (life which a person had before he was born in this world) as well as the concepts of a past Buddha and a vow (*Sei-gan* in Japanese) also entered in Shaka-muni's biographical stories. The concept of past life came from the thought of *Rin-ne* in ancient India, where *Rin-ne* refers to the belief that all living things repeat birth and death many times. The concepts of a Bodhisattva, past Buddha and a vow, which played important roles in Mahayana Buddhism, were already prevalent when new Buddhists in Mahayana Buddhism started activity.

One of famous biographical stories was told roughly as follows [2]. Gautama Siddhartha (i.e. Shaka-muni before attaining *Satori*) in his past life met a past Buddha, named Dipamkara. Then, Gautama had *Bodai-shin* (a mind to seek for the very truth), took a vow of necessarily becoming a Buddha in the future, and held a service. Thus, he was given a prediction and a guarantee of becoming a Buddha in the future by the past Buddha. By this prediction and guarantee, Shaka-muni became a Bodhisattva, i.e. a person who was certain to become a Buddha in the future. Then, Shaka-muni practiced asceticism to accomplish the ability to realize his vow for a long time over his past and present life. He conquered a huge number of hardship and temptation and lastly became a Buddha in his present life.

Probably, new Buddhists in Mahayana Buddhism thought that such stories were useful as a model of their activity. In particular, they paid attention to Gautama Siddhartha (Shaka-muni) at the stage at which he practiced asceticism as a Bodhisattva before attaining *Satori*. Thus, they made up their mind to regard them themselves as a Bodhisattva. However, they changed the meaning of a Bodhisattva in two respects so that it was in harmony with their aim.

In the first place, a Bodhisattva in Shaka-muni's biographical stories referred to a person who was given a prediction and a guarantee of becoming a Buddha in the future by a past Buddha. New Buddhists in Mahayana Buddhism had no such prediction and guarantee. However, they thought that they were able to become a Bodhisattva if they had sincere *Bodai-shin* (a mind to seek for the very truth). By this change, everybody was able to become a Bodhisattva if he or she had sincere *Bodai-shin*.

Secondly, a Bodhisattva in Shaka-muni's biographical stories took a vow of conducting him- or herself to *Satori*. However, new Buddhists in Mahayana Buddhism regarded a Bodhisattva as a person who took a vow of conducting the common people as well as him- or herself to *Satori*. Certainly, new Buddhists in Mahayana Buddhism primarily aimed at conducting the common people to *Satori*, as mentioned earlier.

In general, a vow of a new Buddhist in Mahayana Buddhism (or a vow of a Bodhisattva) involved an item(s) in which he or she declared that he or she never became a Buddha until he or she conducted *all* people to *Satori*[1,2]. New Buddhists dared to choose a hard way toward *complete Satori* though it may have made their own *Satori* impossible. We can see here a very high ideal and noble spirit of new Buddhists in Mahayana Buddhism.

The generalization of the principle of Engi

Next, how had new Buddhists in Mahayana Buddhism solved the second problem? They must have faced a problem of how they were able to conduct the common people to *Satori* or what ability they had to accomplish for conducting the common people to *Satori*. Furthermore, prior to solving this problem, they had to solve a problem of whether on earth it was really possible for them to conduct the common people to *Satori*.

Branch Buddhism had two ideas about the way to attain *Satori*, called *Shou-mon* (attaining *Satori* by hearing *Shaka-muni*'s teachings) and *En-gaku* or *Doku-gaku* (attaining *Satori* by a person's own consideration)[1,2]. Both the ways were based on an individual person's own effort.

In fact, Shaka-muni attained *Satori* by his own effort. On the other hand, new Buddhists in Mahayana Buddhism aimed at conducting the common people to *Satori*. This meant that they proposed the third new idea about the way to attain *Satori* in which the common people were able to attain *Satori* by the ability of a new Buddhist in Mahayana Buddhism or by the ability of a Bodhisattva or a Buddha. This third new idea about the way to attain *Satori* was called *Bosatsu-jo* or *Butsu-jo*, as mentioned earlier. Now, is such a new idea supported by real facts? Also, how is it justified theoretically?

Unfortunately, it is unknown how new Buddhists in Mahayana Buddhism overcame these problems. Presumably, they carefully considered the meaning of Shaka-muni's *Satori*. As stated in Section A1.2.1, Shaka-muni removed his mental pain by arriving at the very truth about his life and revealing that constituent elements of his life followed the principle of *Engi*. According to this argument, for conducting the common people to *Satori*, i.e. for removing worries and agonies of the common people, it is necessary to arrive at the very truth about the common people's life and disclose that constituent elements of their life followed the principle of *Engi*.

Now, when new Buddhists in Mahayana Buddhism carefully considered Shaka-muni's *Satori*, they will have noticed that not only constituent elements of Shaka-muni's life but also those of all people and things in the outer world and besides all people and things in the outer world themselves followed the principle of *Engi*. This is because otherwise Shaka-muni could not have achieved a wholly clear state of mind. A wholly clear state of mind Shaka-muni achieved was just a reflection of the way that all people and things in the outer world and their constituent elements existed, in Shaka-muni's mind.

Thus, new Buddhists in Mahayana Buddhism will have reached the following conclusion.

In the very truth, all people and things in nature and the world and their constituent elements emerge through dependence on others and have no individual unchanging quality. They exist in a fully dynamic free independent harmonious state with no distinction and separation.

Apart from such theoretical consideration, new Buddhists in Mahayana Buddhism will also have tried to conduct actually some of the common people in the neighborhood to *Satori*. Successes in such trials must have given direct evidence for the correctness of the third new way to attain *Satori*, which they proposed. It seems that new Buddhists in

Mahayana Buddhism really succeeded in conducting some of the common people in the neighborhood to *Satori*.

In fact, they brilliantly declared a profound Buddhist view of nature, as will be explained in the next section. They also created a variety of novel concepts such as *In-en* (complex connections among things acting as causes and conditions), *Kuh* (emptiness), *Funi* (being different but not different), *Bodai*, *Hannya*, *Chi-e* (the grand wisdom), *Jihi* (the grand compassion), *Hosshin* (the ultimate essence of Buddha's), *Ohjin* (a manifestation of *Hosshin* in this world), and *Hohjin* (the essence of a Buddha). Such a definite attitude with firm confidence was possible only when they were fully convinced of their new thought with unambiguous evidence.

There are some important points to be noted here. Firstly, the principle of *Engi* generalized above really represented the very truth about all people and things in the outer world and their constituent elements, i.e. the very truth about nature and the world, as discussed above. However, to be accurate, we should say that it only represented the very truth about nature and the world, which new Buddhists in Mahayana Buddhism reached. As argued in Section 1.4.1, we human beings can only reach the very truth in a particular quality level and aspect, depending on what problem we have overcome.

Secondly, the principle of *Engi* generalized above only represented the external appearance of the very truth about nature and the world, in the same way as the principle of *Engi* discussed in Section A1.2.2. "The very truth about nature and the world" itself is in principle impossible to express by words. Accordingly, we cannot attain complete *Satori* even if we have completely understood the principle of *Engi* generalized above by words. We have to make strenuous effort for gaining the truly correct understanding of it, as will be discussed later.

The generalized principle of Engi gave a theoretical basis for Bosatsu-jo or Butsu-jo

The generalization of the principle of *Engi* brought about a big revolution in the philosophy of Buddhism. In the first place, the principle of *Engi* generalized above gave a firm theoretical basis for the third new idea about the way to attain *Satori* (i.e. *Bosatsu-jo* or *Butsu-jo*) which new Buddhists in Mahayana Buddhism proposed. A superior point of the generalized principle of *Engi* compared with the original principle of *Engi* in Initial Buddhism is that the former says that in the very truth all people and things, including Buddha's and Bodhisattva's, exist in a fully dynamic peaceful harmonious state with no distinction and separation even though they look greatly different in the external visible world. This meant that new Buddhists in Mahayana Buddhism were able to acquire the ability to conduct the common people to *Satori*.

The *Satori* of a personal character in Initial and Branch Buddhism was originally far away from complete *Satori*. When new Buddhists in Mahayana Buddhism actually tried to conduct some of the common people in the neighborhood to *Satori*, they must have felt worries to see some of the common people in the neighborhood worried, by sympathy, affection or mental resonance. The *Satori* of a personal character in Initial and Branch Buddhism was powerless for the removal of such worries. This is because such worries were removed only when the common people attained *Satori*. Thus, new Buddhists in Mahayana Buddhism clearly recognized that they were able to attain complete *Satori* only when they had succeeded in accomplishing the ability to conduct *all* people to *Satori*. The *Satori* of a personal character in Initial and Branch Buddhism was only the first step to the achievement of complete *Satori*.

Chi-e (the grand wisdom) and Jihi (the grand compassion)

The generalization of the principle of *Engi* also made it clear that the ultimate goal of new Buddhists in Mahayana Buddhism was to accomplish the internal wisdom (the mind eye) to clearly look at the principle of *Engi* throughout nature and the world [*1]. The principle of *Engi* generalized in Mahayana Buddhism (the very truth about nature and the world Mahayana Buddhism reached) says that in the very truth all people and things in nature and the world and their consistent elements exist in a fully dynamic free independent harmonious state with no distinction and separation. Therefore, if new Buddhists in Mahayana Buddhism gained the truly correct understanding of it, i.e. if they acquired the internal wisdom to look at it clearly, they were able to transcend all individual ideas and things in the external visible world and look at a fully dynamic free independent harmonious state with no distinction and separation throughout nature and the world. This meant that new Buddhists in Mahayana Buddhism achieved the ability to overcome all troubles and worries in nature and the world, i.e. they accomplished the ability to conduct all people to *Satori*.

[*1] It was mentioned in Section A1.2.2 that *clearly* looking at the principle of *Engi* has the same meaning as arriving at essentially the same very truth as that Shaka-muni reached and obtaining the truly correct understanding of the principle of *Engi*. The same argument applies to the principle of *Engi* generalized in Mahayana Buddhism. Clearly looking at the principle of *Engi* throughout nature and the world has the same meaning as arriving at the very truth about nature and

the world and obtaining the truly correct understanding of it.

Actually, new Buddhists in Mahayana Buddhism had not yet accomplished such internal wisdom. They only carried out the generalization of the principle of *Engi* in Initial Buddhism. Also, they only succeeded in conducting some of the common people in the neighborhood to *Satori*. Therefore, they understood the most part of the principle of *Engi* generalized in Mahayana Buddhism only by words. Namely, not only the common people but also new Buddhists themselves in Mahayana Buddhism still had heavy illusions (incorrect views of nature arising from word-based understanding) and intense *Bon-noh* (a person's desires and attachments for individual things such as existence, wealth, social position, social power, fame, love, and so on). They depended on individual ideas and opinions and were only able to look at a variety of things individually and separately.

Thus, new Buddhists in Mahayana Buddhism will have started efforts to gain the truly correct understanding of the principle of *Engi* generalized in Mahayana Buddhism and acquire the internal wisdom to look at it clearly. However, here was another severe problem. It was actually impossible to acquire such internal wisdom because nature and the world were infinite in size and content. In addition, nature and the world incessantly changed and thus illusions and *Bon-noh* emerged one after another. Accordingly, new Buddhists in Mahayana Buddhism must have aimed at approaching the ultimate goal step by step. As mentioned earlier, a new Buddhist in Mahayana Buddhism (or a Bodhisattva) took a vow of conducting the common people as well as him- or herself to *Satori* and carried out hard practice for accomplishing the ability to realize the vow. Therefore, it is very likely that a new Buddhist in Mahayana Buddhism first took a vow on a low level, for example, a vow of conducting a small number of the common people in the neighborhood to *Satori* and accomplished the ability to realize the vow and then took a new vow on a higher level.

In fact, in Mahayana Buddhism, the internal wisdom to look at the principle of *Engi* generalized in Mahayana Buddhism clearly was regarded as the highest wisdom (the grand wisdom) and called *Bodai*, *Hannya* or *Chi-e*. Thus, the ultimate aim of new Buddhists in Mahayana Buddhism was to improve internal wisdom step by step and finally complete *Chi-e* [1-4,7].

Mahayana Buddhism also emphasized the importance of completing the highest compassion (the grand compassion, a hearty wish to conduct all people to *Satori*) called *Jihi* or *Dai-hi* [1-4,7]. The generalized principle of *Engi* in Mahayana Buddhism says that in the very truth all people and things, including a Buddha and a Bodhisattva, exist with no distinction and separation even though they look greatly different in the external visible world, as mentioned earlier. Thus, *Jihi* as well as *Chi-e* was an important conclusion deduced from the generalized principle of *Engi*.

The establishment of practice items necessary for attaining complete Satori

Initial Buddhism proposed *Hassho-do* (eight correct ways of living) as a way of attaining *Satori*, as explained in Section A1.2.2. On the other hand, new Buddhists in Mahayana Buddhism designed a way of attaining complete *Satori*, called *Roku-haramitsu*[1-4,7], where *Roku* means six and *Haramitsu* means completion. Thus, *Roku-haramitsu* means the completion of the following six practice items.

(1) *Fuse (offering wealth, truth and peacefulness to other persons), (2) Jikai (keeping commandments), (3) Nin-niku (enduring difficulties and sufferings), (4) Shojin (making sincere efforts), (5) Zenjo (deep meditation with a concentrated mind), (6) Chi-e (the grand wisdom)*

Roku-haramitsu recommends new Buddhists in Mahayana Buddhism (or Bodhisattva's) to improve the internal wisdom step by step through the control of thoughts and behavior by their own efforts. The completion of the last item, *Chi-e*, was regarded as the final goal.

Roku-haramitsu was described in the first-issued famous sutra of Mahayana Buddhism, *Hannya-Kyo*[4,7], together with the concept of *Kuh* (emptiness), as will be explained in the next section. This indicates that *Roku-haramitsu* was established in the early stage of Mahayana Buddhism. As mentioned earlier, the principle of *Engi* generalized in Mahayana Buddhism provided a theoretical basis for *Bosatsu-jo* or *Butsu-jo*, while *Roku-haramitsu* offered a practice system for it.

The construction of *Roku-haramitsu* was indispensable for attaining complete *Satori* in Mahayana Buddhism because there was a significant difference in the meaning of *Satori* between Initial Buddhism and Mahayana Buddhism. As mentioned in Section A1.2.2, we have to make strenuous effort for gaining the truly correct understanding of the principle of *Engi* after we have understood it by words. Now, Shaka-muni attained *Satori* by carrying out thorough consideration and pursuit about the origin of his mental pain. Thus, the object of his thinking was actually restricted within constituent elements of his personal life. On the other hand, new Buddhists in Mahayana Buddhism (or Bodhisattva's) took a vow of conducting the common people as well as themselves to *Satori* and carried out hard practice to accomplish the ability to realize the vow. An important point is that a new Buddhist (or a Bodhisattva) had to take actual action in the real world for conducting the common people to *Satori*. Otherwise, the mind of the common people would have remained unchanged. Here is why *Roku-haramitsu* included such items as *Fuse* (offering wealth, truth and peacefulness to other persons), *Jikai* (keeping commandments), *Nin-niku* (enduring difficulties and sufferings) and *Shojin* (making sincere efforts).

Interestingly, the attainment of complete *Satori* in Mahayana Buddhism resembled the attainment of *Tai-toku* in sport and art, discussed in Section 1.3.1. Just as exercise is necessary for attaining *Tai-toku*, actual action for conducting the common people to *Satori* in the real world is necessary for attaining complete *Satori*. In addition, just as a person who has attained *Tai-toku* gains new ability and produces a new self of him or her, so a Bodhisattva (or a Buddha) who has attained complete *Satori* produces a new land, called a Buddha's land, in which the common people who has been conducted to *Satori* by the Bodhisattva (or Buddha) dwelled.

New Buddhists in Mahayana Buddhism later designed another guide for Buddhist practice, called *Shi-shoho* (the proprieties consisting of four items)[3].

(1) Fuse, (2) Aigo (speaking kind words), (3) Rigyo (benefiting other persons by worthy deeds), (4) Doji (placing myself in the position of another person, having the same mind as another person)

Shi-shoho also teaches new Buddhists (or Bodhisattva's) how they can improve wisdom.

However, it is important to note that both *Roku-haramitsu* and *Shi-shoho* did not express the correct way to attain *Satori*. They were only what new Buddhists in Mahayana Buddhism invented artificially based on their experiences. Namely, it merely offered a superficial phenomenological way to attain *Satori*. The correct way to attain *Satori* had still not been clarified in those years.

Hosshin and Ohjin: Two aspects of a Buddha

The generalization of the principle of *Engi* also led to a great revolution in the concept of a Buddha. In Initial and Branch Buddhism, only Shaka-muni was a Buddha. On the other hand, in Mahayana Buddhism, a large number of Buddha's of various characters and abilities with and without names were created, together with a great number of

Bodhisattva's of various characters and abilities with and without names. Thus, Mahayana Buddhism constructed a profound Buddhist view of the world in which a huge variety of Buddha's and Bodhisattva's showed characteristic activity [1-4,7].

Why was such a drastic change brought about in the meaning of a Buddha? As mentioned earlier, new Buddhists in Mahayana Buddhism (or Bodhisattva's) aimed at approaching the ultimate goal step by step. Then, they first took a vow on a low level and accomplished the ability to realize the vow and then took a new vow on a higher level. In Mahayana Buddhism, anybody was able to become a Bodhisattva if he or she had sincere *Bodai-shin* (a mind to seek for the very truth), as mentioned earlier. When a Bodhisattva had accomplished the ability to realize his or her vow, the Bodhisattva was able to become a Buddha. Thus, the character and ability of a Buddha depended on what vow the Buddha took and accomplished. As a result, a large number of Buddha's with various characters and abilities emerged [*2].

[*2]This thought in Mahayana Buddhism is in harmony with the argument in Section 1.4.1 that the very truth we reach has various aspects and quality levels.

There was another important reason for the emergence of a huge diversity of Buddha's. Mahayana Buddhism invented an interesting idea of a Buddha. It says that a Buddha has two aspects: an aspect of essence called Dharma-kaya (*Hosshin* in Japanese) and an aspect of manifestation called Nirmana-kaya (*Ohjin* in Japanese) [1-4,7].

Hosshin is the ultimate essence of a Buddha. It is often simply called Buddha with no article. As mentioned earlier, a Buddha has the ability to conduct the common people to *Satori*. Thus, *Hosshin* is the internal ability to enable a Buddha to conduct the common people to *Satori*. In other words, *Hosshin* is the internal ability to make all people and things in nature and the world and their constituent elements realize a fully dynamic free independent harmonious state. *Hosshin* has no shape and color and is entirely invisible. However, *Hosshin*'s ability or *Hosshin*'s light arrives everywhere in nature and the world and penetrates into all people's mind. *Hosshin* also exists eternally, irrespective of whether people believe it or not.

Ohjin is a manifestation of *Hosshin* in this world. In contrast to *Hosshin*, *Ohjin* has a shape and color and displays birth and death. *Ohjin* works in this world for conducting all people to *Satori*. *Ohjin* can take a variety of shapes. For example, *Shaka-muni* can be regarded as *Ohjin*, who appeared in this world for letting people know a way to attain *Satori*. Similarly, new Buddhists in Mahayana Buddhism or Bodhisattva's can be regarded as *Ohjin*, who appeared in this world for conducting the common people to *Satori*. Even an evil person who did evil deeds in the past can be regarded as *Ohjin*, who appeared in this world for demonstrating to people how an evil person was able to attain *Satori*.

It is certain that the concept of *Hosshin* originated from the principle of *Engi* generalized in Mahayana Buddhism. As mentioned earlier, new Buddhists in Mahayana Buddhism revealed that the principle of *Engi* applied to all people and things in nature and the world and their constituent elements. Then, here probably arose a problem of why the principle of *Engi* applied to all people and things in nature and the world. For explaining this problem, they first created the concept of *In-en*. In fact, Mahayana Buddhism says as follows [1-4].

In the very truth, everything (including a living thing and a human idea) emerges from In-en

where *In-en* refers to complex connections among all people and things in nature and the world and their constituent elements, which acts as causes and conditions for their realizing a fully dynamic free independent harmonious state. It is very probable that the concept of *In-en* was further developed to the concept of *Hosshin*. Thus, *Hosshin* is the internal ability to make all people and things in nature and the world and their constituent elements realize a fully dynamic free independent harmonious state, as mentioned above.

We can say in the following way as well. Let us assume that an excellent (imaginary) Buddha has accomplished the internal wisdom to clearly look at the principle of *Engi* throughout nature and the world. Such a Buddha can look at a fully dynamic free independent harmonious state with no distinction and separation throughout nature and the world and hence has the ability to overcome all troubles and worries in nature and the world, as mentioned earlier. *Hosshin* is the internal ability that such a Buddha has acquired.

There are some important points to be noted about the above argument. Firstly, the concepts of *In-en* and *Hosshin* indicate that new Buddhists in Mahayana Buddhism clearly recognized that the internal world lying beyond the bounds of human consciousness really existed and acted as the basis for a person realizing a free independent harmonious state.

In this connection, it is interesting to remember the problem pointed out in Section A1.2.2. The problem is "If a person jumps to another ideal world upon the attainment of *Satori*, where is the mind of such a person?" *Shaka-muni*'s or Initial Buddhism's answer was that such a problem was beyond human understanding and was impossible to discuss. It seems that new Buddhists in Mahayana Buddhism dealt with this problem by the concepts of *Hosshin* (the essence of a Buddha) and *Ohjin* (its bodily form). Namely, they recognized that nature has a dual structure composed of the internal world and the external appearance of it, as discussed in Section 1.4.2. The concepts of *Hosshin* and *Ohjin* were an expression of such recognition.

Note, however, that Mahayana Buddhism emphasizes that *Hosshin* and *Ohjin* never exist separately. It says that *Hosshin* and *Ohjin* are not different but unity. Namely, they are different in appearance and action but not different in existence. *Hosshin* cannot be *Hosshin* without working as *Ohjin* in this world. In other words, *Hosshin* can be *Hosshin* only by working as *Ohjin*. There is no *Hosshin* who is separated from *Ohjin*. In the same way, there is no *Ohjin* who is separated from *Hosshin*. Here is unique logic of "different but not different" peculiar to Buddhism. Note that such a unique logic is necessary to correctly represent the very truth (the way that truly real things exist) [*3] because there is a non-removable gap between word-based understanding and the very truth (see Section 1.2.1).

[*3] In traditional word-based understanding, the above statement is usually understood as follows. *Hosshin* and *Ohjin* are first imagined separately and then they are regarded as unity. However, such understanding does not correctly represent what Mahayana Buddhism says. *Hosshin* and *Ohjin* originally have no distinction and separation. Therefore, they are unity.

Another important point to be noted is that the concepts of *In-en* and *Hosshin* indicate that *nature and the world are in essence of a peaceful harmonious character*. Thus, Buddhism asserts that in the very truth all people and things in nature and the world and their constituent elements realize a fully dynamic free independent peaceful harmonious stable state. Such a favorable view of nature and the world in Buddhism does not come from theoretical consideration. It comes from *the fact* that we achieve a wholly clear state of mind when we have arrived at the very truth about nature and the world. The same argument was also given in Section 1.4.1.

Hohjin: Another aspect of a Buddha

Mahayana Buddhism says that a Buddha has another aspect, called Sambhoya-kaya (*Hohjin* in Japanese) [1,2]. As mentioned earlier, when a Bodhisattva has conducted the common people to *Satori* and attained complete *Satori*, the Bodhisattva (or Buddha) produces a new land, called a Buddha's land, in which the common people who has been conducted to *Satori* by the Bodhisattva (or Buddha) dwell. Here, *Hohjin* is the internal ability the Bodhisattva (or Buddha) has acquired when he or she has attained complete *Satori*. In other words, *Hohjin* is the internal ability to control the common people (or in general all people and things) in a Buddha's land so that they realize a fully dynamic free independent harmonious state.

Interestingly, Mahayana Buddhism says that *Hohjin* continues to live long and work to conduct the common people to *Satori* even after a Buddha dies. This means that we human beings can live as *Hohjin* for a long time even after we die if we have become a Buddha during our life. Such an idea may look unbelievable in the present days when science and technology have made great progress. However, we have to note that word-based scientific understanding is only of an approximate character. Therefore, the above idea deserves careful consideration. It seems to me that the concept of *Hohjin* is not unreasonable. We should rather say that Mahayana Buddhism disclosed a profound view of human life by introducing the concept of *Hohjin*.

Naturally, the concept of *Hohjin never* means that an eternal soul is present in a person, apart from a body. As mentioned above, a Bodhisattva (or a Buddha) produces a Buddha's land when he or she has attained complete *Satori*. *Hohjin* is the internal ability to control the common people (or in general all people and things) in a Buddha's land so that they realize a fully dynamic free independent harmonious state. Thus, *Hohjin* can continue to work after a Buddha dies. According to this understanding, this world consists of a set of Buddha's lands which an enormous number of Buddha's in the past produced. This means that a huge diversity of *Hohjin* works in this world.

All people can attain Satori by the ability of Buddha

The establishment of the concepts of *Hosshin*, *Hohjin* and *Ohjin* made the philosophy of Mahayana Buddhism quite plentiful, diverse and familiar to people. The most important point of the establishment of these concepts was that it led to the following conclusion.

> *Hosshin, Hohjin and Ohjin always work to conduct all people to Satori and therefore all people can attain Satori by the ability of Buddha.*

This was the key conclusion of Mahayana Buddhism because new Buddhists in Mahayana Buddhism had long been making effort to establish *Bosatsu-jo* or *Butsu-jo*, i.e. the way to conduct the common people to *Satori*.

In fact, based on such a conclusion, Mahayana Buddhism has created a large diversity of *imaginary* Buddha's and Bodhisattva's, who completed *Chi-e* (the grand wisdom) long ago and now works in this world to conduct all people and things to *Satori*. *Amida* Buddha, *Kannon* Bodhisattva, *Monju* Bodhisattva, and so on, which enter in sutras of Mahayana Buddhism, are typical examples of such Buddha's and Bodhisattva's. The creation of *imaginary* Buddha's and Bodhisattva's was really important for making the above key conclusion widely familiar to people.

As mentioned in Section A1.2.1, Shaka-muni revealed that everybody was able to realize a free independent harmonious life if he or she had transcended individual things and attained *Satori*. This revelation gave people a bright hope but at this stage people were only able to attain *Satori* by their own ability. In this sense, people were left in a solitary situation. On the other hand, Mahayana Buddhism disclosed that Buddha always works for conducting all people to *Satori*. Thus, everybody can attain *Satori* by the ability of Buddha. It is certain that such teachings gave people great relief.

Hosshin (the grand internal ability) works in multiple ways

Finally, let us make some comments on *Hosshin*. Firstly, *Hosshin* can be regarded as being equivalent to the grand internal ability introduced in Section 1.4.1 and 2.1.3. In fact, the grand internal ability controls all things in nature and the world, including living things and human ideas, so that they as a whole realize a free independent harmonious stable state, in the same way as *Hosshin*.

Secondly, *Hosshin* works in multiple ways. An important difference between Initial Buddhism and Mahayana Buddhism is that the former deals with a one-centered world while the latter deals with a multi-centered world. Namely, Initial Buddhism deals with the problem of how an individual person can realize a free independent harmonious state. On the other hand, Mahayana Buddhism deals with how all people and things in nature and the world and their constituent elements can realize a free independent harmonious state. Therefore, *Hosshin* works in multiple ways. Naturally, the grand internal ability also works in multiple ways, as discussed in Section 1.4.1 and 2.2.

A1.3.2 Mahayana Buddhism in the Primary Stage

This section explains the teachings of Mahayana Buddhism according to individual sutras, though a considerable part of explanation is still given based on explanatory books. Readers will see that main characteristics of Mahayana Buddhism briefly surveyed in the preceding section are really embodied in sutras of it.

(A) The Concept of Emptiness (Kuh)

Hannya-kyo [1,2,7]

Hannya-kyo is the first-issued famous sutra of Mahayana Buddhism, where *Hannya* means *Chi-e* (the grand wisdom, the internal ability to look at the very truth about nature and the world clearly) and *kyo* means a sutra. The initial version of this sutra was compiled in around the late 1st century B.C. or the early 1st century A.D. This sutra is famous for the following words.

In the very truth, everything equals emptiness.

The concept of "emptiness" is called *Kuh* in Japanese and regarded as one of the most fundamental ideas of Mahayana Buddhism.

Note that the concept of "emptiness" does not mean "nothing". It means "the lack of an individual unchanging quality or entity". Therefore, the above words say that everything in the very truth has no individual unchanging quality. Accordingly, the above words are just a simple expression of the principle of *Engi* generalized in Mahayana Buddhism. The above words are also a simple expression of the aforementioned assertion that everything emerges from *In-en*. In fact, a famous theoretician in ancient India in the 3rd century, Nagarjuna (*Ryuju* in Japanese name), asserted that "emptiness" refers to *In-en*[2].

The concept of "emptiness" was primarily proposed as a firm criticism on a wrong thought that the most influential branch of Branch Buddhism declared in its theoretical system at the beginning of the 1st century B.C. This branch categorized constituent elements of human life and concluded that both physical and mental constituent elements had immutable entities on the essential level, in contrast to Shaka-muni's conclusion, as mentioned in Section A1.3.1. If this was the case, a Buddha, a Bodhisattva and the common people were clearly distinguished from one another and the argument that a Bodhisattva or a Buddha was able to conduct the common people to *Satori* lost its theoretical basis. Probably, new Buddhists in Mahayana Buddhism were surprised to hear such a thought. However, they were able to reconfirm the correctness of the principle of *Engi* Shaka-muni revealed in a generalized form.

Hannya-kyo thus emphasizes that Bodhisattva's (new Buddhists in Mahayana Buddhism) [*1] should look at "emptiness" in all people and things in nature and the world. The accomplishment of the internal wisdom to look at "emptiness" in all people and things, which is called *Hannya Haramita* (the completion of *Chi-e*), is nothing else but the attainment of complete *Satori* in *Hannya-kyo*.

> [1]Sutras in Mahayana Buddhism were in general compiled for the purpose of giving teachings to new Buddhists in Mahayana Buddhism or Bodhisattva's, not to the common people.

Now, what is meant by looking at "emptiness" in all people and things in nature and the world? The concept of "emptiness" was also proposed as a firm criticism on a traditional view of nature in our daily life. Namely, it was a proposal of the antitheses of our traditional view of nature. Therefore, the above problem becomes easy to understand if the thought of *Hannya-kyo* is compared with our traditional view of nature. As argued in Section 1.2.1, in our traditional view of nature, all things are regarded as having individual unchanging properties. Accordingly, we first look at individual things and then think of their connections. This is certainly a way of understanding familiar to us. On the other hand, in *Hannya-kyo*, "emptiness" or *In-en* is the basic governing reality and everything in the external visible world emerges from it. Therefore, *Hannya-kyo* says that we should first look at "emptiness" (or *In-en*, invisible complex connections in nature and the world) and then look at individual things in the external visible world as emerging from it. This way of understanding refers to looking at "emptiness" in all people and things in nature and the world.

In general, it is impossible for us to first look at invisible connections (*In-en*, "emptiness") and then to look at individual things as emerging from them. However, such a way of understanding becomes possible when we have arrived at the very truth because in this case we have formed a continuous internal connection of constituent elements (see Section 1.3.1 and 1.3.2). It is interesting to note here that quantum mechanics explained in Section 2.1.1 also request us to first look at an invisible connection. In fact, in quantum mechanics, a non-observable wave function is first obtained and various individual physical quantities are calculated from it.

In *Hannya-kyo*, a large number of Bodhisattva's display a diversity of activities. They all are typical Bodhisattva's in Mahayana Buddhism [7]. Namely, they are not Buddhist monks confined to monasteries but active, able, and rich heroes living in towns and villages. They have high ideals and strong leadership and busy themselves about solving problems upon emergency. They make up their mind not to become a Buddha until they conduct all people to *Satori* even if they have accomplished *Chi-e*. Bodhisattva's in *Hannya-kyo* well recognize that complete *Satori* does not exist separately from all people's *Satori*.

Such an attitude of Bodhisattva's in *Hannya-kyo* was in harmony with the concept of "emptiness", because the principle of *Engi* generalized in Mahayana Buddhism, from which the concept of "emptiness" arises, says that in the very truth Buddha's, Bodhisattva's and the common people are not different but unity even though they look greatly different in the external visible world. On the other hand, there were clear distinction and separation among a Buddha, an Arahant and the common people in Initial or Branch Buddhism. Accordingly, in Initial or Branch Buddhism, people had to attain *Satori* by their own ability and were left in a solitary situation, as mentioned earlier.

Hannya-kyo also emphasizes that all things exist with no distinction and separation and thus nothing should be understood separately. Thus, this sutra says that all practice items in *Roku-haramitsu* should be carried out with no distinction and separation. For example, *Fuse* (offering wealth, truth and peacefulness to other persons) should be carried out with no distinction and separation from *Zenjo* (deep meditation with a concentrated mind). This implies that *Fuse* should be carried out with no intention of offering. Also, *Zenjo* should be carried out with no distinction and separation from *Shojin* (making sincere effort) in the common people. The moment a Bodhisattva is attached to some item, even if it is a worthy deed such as *Fuse* and *Zenjo*, it is converted to the opposite, an evil deed.

Kongo-hannya-kyo [1,2,8)]

A sutra, *Kongo-hannya-kyo*, is presumed to have been issued a little later than *Hannya-kyo*. This sutra explains the concept of "emptiness" in a different way from *Hannya-kyo*. For example, this sutra says,

> Mr. Subhuti! They say Buddha's teachings, Buddha's teachings! However, they are not Buddha's teachings. Therefore, they are called Buddha's teachings.

These words contain strange logic, which is difficult to understand straightforwardly. Similar strange words are written repeatedly in the sutra.

> Bodhisattva is not Bodhisattva. Therefore, it is called Bodhisattva.
>
> A body is not a body. Therefore, it is called a body.

In my opinion, these words can be explained as follows.

In the very truth, all people and things in nature and the world emerge from *In-en* and have no individual unchanging quality. Therefore, they cannot be correctly expressed by words. Accordingly, what is expressed by words is not a truly real thing. For example, what are called "Buddha's teachings", which are expressed by words, are not truly real "Buddha's teachings". The first part of the above words thus indicates that anything expressed by words is not a truly real thing. However, what is expressed by words, for instance, what are called "Buddha's teachings" actually emerge from *In-en* in the real world according to the principle of *Engi*. Thus, what are called "Buddha's teachings" really exist in the real world and therefore they are called "Buddha's teachings". After all, the above words indicate that what are called "Buddha's teachings" exist in the real world in such a way that what are called Buddha's teachings are not truly real "Buddha's teachings".

Sutras of Mahayana Buddhism often teach strange logic such as "existing but not existing" and "different but not different". The concepts of *Hosshin* and *Ohjin*, discussed in Section A1.3.1, also represent strange logic of "different but not different". Presumably, new Buddhists in Mahayana Buddhism made every possible effort to explain "the very truth" by words though it is originally impossible to express by words. Strange logic is results of such efforts.

Yuima-kyo [1,2,7)]

A sutra, *Yuima-kyo*, was issued in the late 1st century A.D. or the early 2nd century A.D. This sutra explains the concept of "emptiness" in a further different way. In this sutra, a rich citizen and Bodhisattva, named Yuima-kitsu, discusses various key concepts of Buddhism such as wisdom, Dharma, unity, an expedient way, and so on with famous *Monju* Bodhisattva, who has accomplished *Chi-e* long ago. The discussion takes place in front of a large number of people including Buddhist monks of Branch Buddhism. Yuima then criticizes wrong ideas included in the traditional thought of Branch Buddhism one after another.

The main aim of Yuima's criticism is to completely deny a clear distinction in dualism, based on the concept of "emptiness". Yuima's idea is thus called the thought of *Funi* (which means "what looks different is not different"). He declares that mutually conflicting concepts such as beauty and ugliness, wisdom and ignorance, largeness and smallness, agony and pleasure, and so on are all not two things but unity (see a footnote #3 of Section A1.3.1 about how to understand *Funi*). This is because in the very truth all people and things in nature and the world and their constituent elements exist with no distinction and separation.

Let us look at some examples of Yuima's words, which exerted strong influences on Buddhism in later ages. Yuima says,

Bon-noh equal Bodai

where *Bon-noh* refers to a person's desires and attachments for individual things such as existence, living, wealth, fame, and so on, from which severe worries and agonies as well as unwise behavior such as avarice, rage, and idle complaints arise (see Section A1.2.2 and A1.3.1). On the other hand, *Bodai* has the same meaning as *Hannya* and *Chi-e* and stands for the internal wisdom to look at the very truth about nature and the world clearly (see Section A1.3.1). Thus, *Bon-noh* and *Bodai* are the complete antithesis of each other. Nevertheless, Yuima says that they are unity. He says that new Buddhists in Mahayana Buddhism (Bodhisattva's) cannot attain *Bodai* without entering into the wide sea of *Bon-noh* of the common people. Another important example is

"Life and death" equal Nehan

where "life and death" refers to actual human life in this world, which is full of illusions and *Bon-noh* and hence full of worries and agonies. On the other hand, *Nehan* (nirvana) has the same meaning as complete *Satori* and expresses an ideal peaceful harmonious state of human life. Thus, these two words are also the complete antithesis of each other. However, Yuima says that they are unity. He says that *Nehan* neither exists in deep contemplation in monasteries nor in hard ascetic practice in mountains but exists in daily life of the common people full of illusions and *Bon-noh*.

As mentioned earlier, Buddhism teaches strange logic which is quite difficult to grasp under word-based understanding. This is because truly real things in the very truth originally have a strange way of existing, as mentioned in Section 1.4.1. Buddhism catches such a strange way of existing as it is, in contrast to other philosophies, as mentioned at the beginning of this chapter, and for this reason Buddhist teachings look strange.

Hannya-shin-gyo [1,2,8]

The initial version of *Hannya-kyo* was issued in the late 1st century B.C. or the early 1st century A.D., as mentioned earlier. However, a large number of words were later added to this sutra repeatedly over a long period of time of more than 1,000 years. Therefore, *Hannya-kyo* now has quite a huge volume.

In about the 4th to 7th century, a famous shortened version of *Hannya-kyo*, entitled *Hannya-shin-gyo*, was compiled. Well known words in this sutra are

Shape equals emptiness, emptiness equals shape

where "shape" refers to physical substances or bodies with shapes and color, i.e. things we can look at or we can be conscious of. Thus, the above words can be understood as follows: Things in the external visible world, which we can look at or we can be conscious of, emerge from *In-en* and have no individual unchanging quality in the very truth. Therefore, shape equals emptiness. However, various things in the external visible world really emerge from *In-en* according to the principle of *Engi*. Therefore, emptiness equals shape.

The above words say that "emptiness" or *In-en* has a strange way of existing. It appears that *Hannya-shin-gyo* emphasizes this point. For example, this sutra says as follows.

"Emptiness" or In-en neither appears nor disappears, neither pure nor impure, and neither increases nor decreases.

Furthermore, *Hannya-shin-gyo* says as follows.

There is no unclearness, but unclearness is inexhaustible.

There is no "aging and death", but "aging and death" is inexhaustible.

Note that "unclearness" and "aging and death" are constituent elements of human life included in *Juni-in-en* (twelve causes and conditions) of Initial Buddhism (see Section A1.2.2). According to my understanding, the above words can be explained as follows. Constituent elements of human life such as "unclearness" and "aging and death" emerge from *In-en* and have no individual unchanging quality. Therefore, such constituent elements do not exist. Namely, there is no "unclearness" or no "aging and death". However, such constituent elements emerge from *In-en* in the real world according to the principle of *Engi*. Accordingly, they are not exhausted. After all, the above words say the same thing as the words of *"shape equals emptiness, emptiness equals shape"*.

(B) The Concepts of Buddha and Buddha's Land

The principle of *Engi* generalized in Mahayana Buddhism was first expressed by the concept of "emptiness" (*Kuh*) in sutras of Mahayana Buddhism, as explained thus far. However, it was later expressed also by the concept of Buddha. In fact, the concept of Buddha became the main subject in sutras issued in the late 1st century A.D. to the early 2nd century A.D. such as *Kegon-kyo*, *Hoke-kyo*, *Muryoju-kyo* and *Amida-kyo*. In these sutras, Buddha with infinite life and infinite ability enters as the central Buddha. Namely, these sutras are compiled under the thought that nature and the world are governed by Buddha.

The concept of "emptiness" explains how things exist in the very truth, while the concept of Buddha deals with how things are controlled or organized in the very truth. Thus, the above-mentioned change in the main subject in sutras from "emptiness" to "Buddha" indicates that the object of consideration in Mahayana Buddhism advanced from the recognition of "a state" to the recognition of "dynamics".

Kegon-kyo [1,2,7]

The complete form of *Kegon-kyo* was compiled as a set of a number of sutras with a huge volume in the 4th to 5th century in an area of the Tarim Basin of Central Asia. However, the earliest version of the sutra was issued in around the late 1st century A.D. or the early 2nd century A.D. in ancient India.

In *Kegon-kyo*, *Hosshin*, which is called Vairocana Buddha in this sutra, enters as the central Buddha. As explained in Section A1.3.1, *Hosshin* is the ultimate essence of all Buddha's in the whole world and is the internal ability to make all people and things in the whole world and their constituent elements realize a fully dynamic free independent peaceful harmonious state. Therefore, in the land of Vairocana Buddha, which is nothing else but this world, a great number of *Hohjin* Buddha's and Bodhisattva's live in their respective lands [*1] and display their own activities. Vairocana Buddha unifies all activities of these Buddha's and Bodhisattva's. Then, *Kegon-kyo* tells various stories about activities of Buddha's and Bodhisattva's on a grand scale.

[*1] It was mentioned in Section A1.3.1 that this world consists of a set of Buddha's lands which a large number of *Hohjin* Buddha's constructed in the past. This means that a large number of *Hohjin* Buddha's works in this world.

An important point of *Kegon-kyo* is that such stories are told under the following unique philosophy peculiar to this sutra:

Many equal one, one equals many.

The words mean that in the very truth many things are embodied in one thing and one thing is embodied in many things. In other words, they mean that in the very truth all people and things in nature and the world and their constituent elements penetrate into one another without any obstacle. It was mentioned in Section 1.4.1 that in the very truth nature and the world consist of a complexly-entangled, multi-fold and multi-dimensional dynamic network of a huge number of infinitely-spreading internal continuous connections. The above words in *Kegon-kyo* may be interpreted as an expression of such a situation in nature and the world. Note also that the above words in *Kegon-kyo* can be regarded as a simple expression of the principle of *Engi* generalized in Mahayana Buddhism, in the same way as the word in *Hannya-kyo*[*2].

> [*2] The above words in *Kegon-kyo* explain how things in the very truth exist in more detail than the words in *Hannya-kyo* or *Hannya-shin-gyo* such as "Shape equals emptiness, emptiness equals shape". This suggests that the philosophy of Buddhism advanced between these sutras.

In this way, *Kegon-kyo* teaches Bodhisattva's (new Buddhists in Mahayana Buddhism) to look at "many equal one, one equals many" in the whole world. The accomplishment of the internal wisdom to look at "many equal one, one equals many" is nothing else but to attain complete *Satori* in *Kegon-kyo*.

Kegon-kyo has another important feature. It stresses that any path toward complete *Satori* is within the light of Vairocana Buddha, i.e. any path toward complete *Satori* is controlled by Vairocana Buddha and thus belief in Vairocana Buddha is of key importance.

> *Belief is the basis for Satori and the mother of worthy beneficial deeds. · · · Belief is filled with delight. · · · Belief improves wisdom and leads to complete Satori. · · · Belief is not bound by difficulty and let people accomplish peacefulness. · · · Belief is the most worthy thing and the most difficult thing to attain.*

Such an emphasis of "belief in Buddha" in *Kegon-kyo* deserves careful attention because *Roku-haramitsu* in *Hannya-kyo* did not include any item about belief in Buddha (see Section A1.3.1). Thus, such an emphasis means that belief in Buddha was regarded as a new important item of Buddhist practice, apart from *Roku-haramitsu*.

Hannya-kyo divided the path toward complete *Satori* into four stages, while *Kegon-kyo* divided it into ten stages. Interestingly, *Kegon-kyo* says that the ten stages are not separated from one another. For example, the sutra says,

> *The moment you have reached the first stage, called the delight*

stage, you realize complete Satori of Buddha.

The words mean that when a person has arrived at the first stage, the person already reaches the final stage. Such a rather contradictory assertion comes from the aforementioned characteristic philosophy of Mahayana Buddhism, which says that in the very truth all people and things exist with no distinction and separation.

Kegon-kyo also emphasizes that Bodhisattva's should carry out Buddhist practice among people and with people, in the same way as *Hannya-kyo* and *Yuima-kyo*. Furthermore, *Kegon-kyo* says

This world is a result of activities of the mind.

The words are reasonable if "activities of the mind" are interpreted as referring to "activities of *Hosshin* (Vairocana Buddha)".

Hoke-kyo [1,2,7)]

In *Hoke-kyo*, Shaka-muni enters as the central Buddha. The sutra is composed of three parts. The first part proposes the concept of *Ichijo-myoho*, based on *Bosatsu-jo* or *Butsu-jo*. Namely, the first part teaches Bodhisattva's to clearly look at complex multi-fold connections in nature and the world, called *Ichijo-myoho*.

The second part teaches Bodhisattva's how they should carry out actual practice. In this sutra, Bodhisattva's are regarded as apostles of Buddha sent to this world to conduct the common people to *Satori* and are highly praised. Simultaneously, the sutra says that Bodhisattva's should tell the truth to people with a mind of kindness and compassion with no grudge and attachment.

The final part makes it open to the public that Shaka-muni is actually eternal Buddha. Shaka-muni accomplished complete *Satori* long, long ago and appears in this world for the purpose of displaying a path toward *Satori* to people.

Muryoju-kyo [1,2,7,9)]

In *Muryoju-kyo*, *Amida* Buddha enters as the central Buddha. This sutra played the key role in great advancement of the philosophy of Buddhism in North East Asia in later ages, as will be explained in the next chapter.

In general, sutras in Mahayana Buddhism teach Bodhisattva's (new Buddhists in Mahayana Buddhism) to accomplish the internal wisdom to clearly look at the principle of *Engi* generalized in Mahayana Buddhism (the very truth about nature and the world which Mahayana Buddhism reached). For example, *Hannya-kyo* teaches Bodhisattva's to achieve the internal wisdom to look at "emptiness". *Kegon-kyo* emphasizes the importance of accomplishing the internal wisdom to look at "many equal one, one equals many". *Hoke-kyo* stresses the importance of completing the internal wisdom to look at *Ichijo-myoho*. In this context, *Muryoju-kyo* emphasizes the importance of accomplishing the internal wisdom to clearly look at *Amida* Buddha, who always works in this world to conduct all people to *Satori*.

It is important to note that the teachings of *Muryoju-kyo* are largely different from those of the other sutras. Namely, the other sutras teach Bodhisattva's to accomplish the internal wisdom to look at the very truth about nature and the world by their own ability and effort, as mentioned above, while *Muryoju-kyo* teaches Bodhisattva's to attain *Satori* with the aid of *Amida* Buddha. It was *Muryoju-kyo* that for the first time directly taught the key conclusion of Mahayana Buddhism that all people can attain *Satori* by the ability of Buddha. Certainly, here was why *Muryoju-kyo* played the central role in the great advancement of the philosophy of Buddhism in North East Asia in later ages. For convenience in later discussions, let us here briefly summarize the teachings of *Muryoju-kyo*.

> *Shaka-muni (Gautama Buddha) has appeared in this world to tell you the following important fact. Long, long ago, a great Bodhisattva, named Hozo, was given a prediction and guarantee of becoming a Buddha in the future by past Buddha, named Sejizaio. Hozo Bodhisattva then took a vow consisting of forty eight items with the hearty aim of building up Buddha's land at which all people are necessarily conducted to Satori. Quite long ago, he had completed the ability to realize the vow through hard practice. He now works as Amida Buddha with infinite life and infinite ability in his land called "Jodo" for conducting all people to Satori. Accordingly, if you wish to attain Satori, simply want to come to Amida Buddha's land (Jodo) and think of Amida Buddha with a sincere mind and worthy deeds. Then, you necessarily attain Satori by the power of the accomplished vow of Amida Buddha.*

The Buddhist practice of "wanting to come to *Amida* Buddha's land and thinking of *Amida* Buddha with a sincere mind" is called *Nen-butsu*. Thus, *Muryoju-kyo* teaches people that they can attain *Satori* by *Nen-butsu*.

It is said that *Muryoju-kyo* was compiled in close relation to *Kegon-*

kyo [7)]. Thus, it is interesting to compare the teachings of *Hannya-kyo*, *Kegon-kyo* and *Muryoju-kyo*, which are representative sutras of Mahayana Buddhism, and investigate how the meaning and role of Buddha changed.

> *Hannya-kyo* teaches a Bodhisattva to carry out hard Buddhist practice such as *Roku-haramitsu* by a Bodhisattva's own ability, with the clear recognition that in the very truth a Buddha, a Bodhisattva and the common people exist with no distinction and separation.
>
> *Kegon-kyo* teaches a Bodhisattva to have the belief that all paths toward complete *Satori* are governed by *Hosshin* Buddha and carry out hard Buddhist practice similar to *Roku-haramitsu* by a Bodhisattva's own ability.
>
> *Muryoju-kyo* teaches a Bodhisattva to want to come to *Amida* Buddha's land and think of *Amida* Buddha with a sincere and enthusiastic mind.

It is clear that the weight of Buddha in the process of attaining *Satori* increases in the order of *Hannya-kyo*, *Kegon-kyo* and *Muryoju-kyo*.

As can be seen from the above summary of *Muryoju-kyo*'s teachings, *Amida* Buddha is not *Hosshin* but *Hohjin*. Probably, *Amida* Buddha as *Hohjin* of an imaginary character was created for the purpose of making the key conclusion of Mahayana Buddhism much clearer. Note that *Amida* Buddha is indeed *Hohjin* but has infinite life and infinite ability and has constructed Buddha's land at which all people necessarily attain *Satori*. Therefore, *Amida* Buddha's ability is nearly equal to that of *Hosshin*.

For making the teachings of *Muryoju-kyo* easier to understand, let us cite some items of *Amida* Buddha's vow written in this sutra. Of forty eight items of Amida Buddha's vow (in the Chinese version of *Muryoju-kyo*), the 1st to the 17th items and the 21st to the 48th items teach us how splendid *Amida* Buddha's ability and land are. On the other hand, the 18th to the 20th items teach us how people in other lands can come to *Amida* Buddha's land, namely, how people can attain *Satori*. Therefore, intense attention has been paid to these three items to date.

> *(18) Even if I can become a Buddha, I do not become a Buddha if persons in other lands want to come to my land with a sincere and enthusiastic mind and think of me, say, ten times and still fail to come to my land. However, Go-gyaku, Hibo and Sendai are excluded* [*3].
>
> *(19) Even if I can become a Buddha, I do not become a Buddha if persons in other lands have Bodai-shin and collect worthy*

deeds and take vows with a sincere mind and want to come to my land and still fail to see me when their lives come to end.

(20) Even if I can become a Buddha, I do not become a Buddha if persons in other lands hear my name and think of my land and besides have worthy deeds and send them to me and want to come to my land and still fail to accomplish their aim.

[3] *Go-gyaku* refers to a person who did one of five evils (killing father, killing mother, killing a saint, destroying the association of Buddhist monks, and wounding a Buddha's body). *Hibo* stands for a person who slanders Buddhism. *Sendai* refers to a person who has no mind to believe in Buddha.

The 18th item definitely teaches the attainment of *Satori* by *Nen-butsu* while the 19th and 20th items accept *Roku-haramitsu* as additional Buddhist practice for attaining *Satori*. Therefore, *Nen-butsu* and *Roku-haramitsu* co-exist in a mixed manner in *Muryoju-kyo* though the former is much more emphasized than the latter.

A1.3.3 Mahayana Buddhism in the Middle and the Last Stages

It became a central problem to clarify how all people were able to attain Satori

Mahayana Buddhism made big progress in the primary stage, as explained in the preceding sections. However, the Gupta dynasty, which was established in 320 A.D. and governed the whole India, supported Hinduism (descending from Brahmanism), not Buddhism. Accordingly, the number of Buddhists rapidly decreased in ancient India later.

Under such a severe situation, Mahayana Buddhism brought forth two new thoughts in the 4th to 6th centuries. One of them was based on the concept of *Nyorai-zo* or *Bussho* and the other was based on the concept of *Yui-shiki*. These new thoughts had a common feature in that they took notice of the bottom of the mind of the common people, in contrast to Mahayana Buddhism in the primary stage which paid attention to the mind of new Buddhists in Mahayana Buddhism or Bodhisattva's.

Historically, such a change in the object of consideration may be regarded as due to the advancement of the philosophy of Buddhism. New Buddhists in Mahayana Buddhism in the primary stage succeeded in generalizing the principle of *Engi* and creating the concepts of *Hosshin*, *Hohjin* and *Ohjin*, as discussed earlier. Then, they finally reached the key conclusion that all people were able to attain *Satori* by the ability of Buddha. However, they had not yet clarified *how* (i.e. via what processes) all people were able to attain *Satori* by the ability of Buddha. For this reason, the common people were still unable to attain *Satori*. Probably, the appearance of the above-mentioned severe social situation made this problem apparent. Then, Mahayana Buddhism in the middle stage came to pay attention to it.

The concepts of *Nyorai-zo* and *Bussho* [1,2,7]

The concepts of *Nyorai-zo* and *Bussho* had nearly the same meaning. *Nyorai-zo* refers to a receptacle for accepting *Nyorai* (where *Nyorai* means a person who has come from the very truth and has the same meaning as Buddha), while *Bussho* stands for the quality specific to a Buddha. Thus, both *Nyorai-zo* and *Bussho* represent a possibility of becoming a Buddha. The concept of *Nyorai-zo* was developed in sutras, *Nyoraizo-kyo* and *Shoman-kyo*, while that of *Bussho* was developed in a sutra, *Nehan-kyo*, all being compiled in the 4th century.

These sutras had a severe view that the mind of the common people was strongly covered with heavy illusions and intense *Bon-noh*. Nevertheless, they said, "All people hold *Nyorai-zo* or *Bussho* (a possibility of becoming a Buddha) at the bottom of the mind". It is certain that such an idea came from the thought of *Hosshin*, which says that the ability of *Hosshin* penetrates into the mind of all people. For example, *Nyoraizo-kyo* said as follows. *Nyorai-zo* exists deep in the mind of the common people and is covered with intense *Bon-noh*. Therefore, it may be difficult to understand that all people have *Nyorai-zo*. Accordingly, just believe that all people have *Nyorai-zo* even if this is not apparent. *Nyorai-zo* suddenly rises under a certain cause and condition and yields a driving force for having *Bodai-shin* (a mind to seek for the very truth). If *Bon-noh* have been removed, *Nyorai-zo* shines brilliantly.

The concept of *Yui-shiki* [1,2]

The concept of *Yui-shiki* was first proposed in sutras, *Ge-jinmitsu-kyo* and *Daijo-abidatsuma-kyo*, in the 4th century. Here, *Yui-shiki* means that only *Shiki* exists in the world, where *Shiki* refers to the act of knowing. In a broad sense, *Shiki* can be regarded as equal to the mind. It is said that the concept of *Yui-shiki* came from the aforementioned words of *Kegon-kyo*: "This world is a result of activities of the mind". Here, activities of the mind can be interpreted as activities of *Hosshin* and thus the concept of *Yui-shiki* also came from the thought of *Hosshin*.

The concept of *Yui-shiki* was later theoretically considered by

Asanga (Mujaku in Japanese) and his younger brother, Vasubandhu (Tenshin or Seshin in Japanese) and a theoretical system called *Yoga-gyo Yui-shiki* was constructed in the 5th century. Here, *Yoga-gyo* refers to the practice of *Yoga* (an Indian traditional ascetic practice for concentrating the mind). In the *Yui-shiki* theory, the mind of a person covered with heavy *Bon-noh* was analyzed with the aim of clarifying the origins of *Bon-noh* and discernment inherent in a human being and disclosing a way of attaining *Satori* by the control of a person's mind through ascetic practice of *Yoga* [4]. The theory of *Yuga-gyo Yui-shiki* constructed by Asanga and Vasubandhu is regarded as one of great theoretical systems of Indian Buddhism together with the theory of "emptiness" (*Kuh*) constructed by Nagarjuna (Ryuju in Japanese) in the 3rd century.

The *Yui-shiki* theory explains activities of the human mind by dividing *Shiki* into eight sub-*Shiki*'s: seeing, hearing, smelling, tasting, toughing, consciousness, *Mana-Shiki* and *Araya-Shiki*, where *Araya-shiki* lies in the utmost bottom of the mind and is regarded as the most fundamental sub-*Shiki*. However, I do not touch this theory in detail because I cannot understand how this theory is justified. In my opinion, the idea of attaining *Satori* by controlling a person's mind is incorrect. In fact, we cannot attain *Tai-toku* and theoretical creation, which are analogous events to *Satori* (see Section A1.2.2), by controlling our mind.

The rise of esoteric Buddhism (Mikkyo) [1,2]

In around the 6th century, a further new thought of Buddhism, called esoteric Buddhism (*Mikkyo* in Japanese), rose. The thought was proposed in sutras, *Dainichi-kyo* and *Kongocho-kyo*, in the 7th century.

Mikkyo means a secret teaching. It uses a symbol system to express the very truth, composed of a stamp-mark expressing Buddha (called *Inso* in Japanese), a mantra (spell, called *Shin-gon* in Japanese), an illustrated symbol of Buddha (called *Sanmayagyo* in Japanese), and Mandala (a drawing expressing the world, in which Buddha's, Bodhisattva's, and Heavens are arranged with a geometrical symmetry). With the aim of attaining *Satori*, ascetics continue special *Yuga* practice, having a stamp-mark in hand, chanting a mantra, and consolidating the mind so that they fuse with Buddha.

It is said that esoteric Buddhism was widely favored in India in the 8th to 13th centuries. However I do not touch it any further because it appears to contain mysterious teachings with no reasonable basis.

A1.4 Buddhism in North East Asia

Mahayana Buddhism showed marked developments in North East Asia such as China, Korea and Japan since around the 6th century A.D. Main events were the emergence of a variety of new original Buddhist teachings in China, including *Jodo* and *Zen* teachings, in the 6th and 7th centuries and their developments in China, Korea and Japan in later ages.

Buddhism in China[1-4,7]

The transmission of Buddhism to China started at around the 1st century A.D. A large number of sutras, theoretical books and commentary books were translated into Chinese. Buddhist monks Kumarajiva in the early 5th century and Genjo in the 7th century made great contributions to this hard task.

In the 3rd to 5th centuries, after the ruin of the Kan (Hàn, 202 B.C.-220 A.D.) dynasty, China was split into a number of small countries and battles among them broke out one after another. Such an unstable social situation promoted wide spread of Buddhism in China, and this led to the establishment of a number of new original Buddhist teachings and schools in the 6th and 7th centuries.

It seems that Chinese Buddhism has a unique feature in that Chinese people looked through the whole of Indian Buddhism and aimed at taking the essence of it. Presumably, such an attitude was necessary partly because Chinese people accepted Buddhism in comparison with their own traditional philosophies, Taoism and Confucianism, and partly because various teachings of Indian Buddhism were translated into Chinese and Chinese people needed to put them in order.

The first original Buddhist teachings in China were created by Buddhist monks, E-on, Chigi and Kichizo in the 6th century. E-on established new Buddhist teachings called *Jiron-shu*, where *shu* means a school or sect. Chigi constructed a profound Buddhist view of the world mainly based on *Hoke-kyo* and established the Buddhist teachings called *Tendai-shu*. A little later, Kichizo established the Buddhist teachings called *Sanron-shu*, based on the concept of *Kuh* (emptiness). In the late 7th century, a Buddhist monk, Hozo, thought *Kegon-kyo* to represent the most correct teachings of Shaka-muni and completed the Buddhist teachings called *Kegon-shu*. In nearly the same age, a Buddhist monk, Jion, completed the Buddhist teachings called *Hosso-shu* based on the *Yui-shiki* theory.

Buddhism in China showed another important development in two respects. One was the advancement of the *Jodo* teachings (*Jodo-kyo* in Japanese, where *Jodo* means a clean land and refers to *Amida* Buddha's land and *kyo* in this case means teachings) based on sutras such as *Muryoju-kyo*, *Amida-kyo* and *Kan-muryoju-kyo*. The other was the establishment of the *Zen* teachings and *Zen-shu* (where *Zen* means meditation) as the completely original Buddhist teachings in China.

The *Jodo* teachings (*Jodo-kyo*), which taught people the attainment of *Satori* by the ability of *Amida* Buddha, already had the origin in ancient India. For example, Nagarjuna (Ryuju in Japanese), theoretician in the 3rd century, stated in one of his books that Buddhism in ancient India had two ideas about the way to attain *Satori*: the idea of carrying out hard practice and the idea of performing easy practice. The former refers to the way to attain *Satori* by *Roku-haramitsu* (see Section A1.3.1) while the latter stands for the way to attain *Satori* by *Nen-butsu*, where *Nen-butsu* means "wanting to come to *Amida* Buddha's land named *Jodo* and deeply thinking of *Amida* Buddha with a sincere and enthusiastic mind" (see Section A1.3.2(B)). In the 5th century, another theoretician, Vasubandhu (Tenshin or Seshin in Japanese), wrote a book entitled "*Jodo-ron*" (where ron means theory) and argued that people can arrive at the wisdom of Buddha by *Nen-butsu*.

The *Jodo* teachings in China were started by a Buddhist monk, Don-ran, in the early 6th century. His excellent commentary on Vasubandhu's *Jodo-ron* exerted large influences on later *Jodo* teachings. Don-ran called a person's own ability *Ji-riki* and Buddha's ability *Ta-riki*. Then, Don-ran emphasized that the attainment of *Satori* by *Ta-riki* in the *Jodo* teachings provided the easiest way to attain *Satori* and was most suitable for the relief of the common people in the age of *Mappo*[*1].

[*1] A prediction about the future of Buddhism was formulated in the middle or the last stage of Mahayana Buddhism in India. According to the prediction, the history of Buddhism was divided into three stages: *Shoho* (correct teachings), *Zoho* (imaginary teachings), and *Mappo* (last teachings). In the stage of *Mappo*, only Buddhist teachings remain and the attainment of *Satori* becomes impossible. The stage of *Mappo* was predicted to start at 1,000 years after the death of Shaka-muni though there were other ideas about the time when *Mappo* started. Thus, the 6th century was regarded as the age of *Mappo* in China.

The thought of Don-ran was taken over by a Buddhist monk, Do-shaku, in the early 7th century and his follower, Zen-do, in the late 7th century. Do-shaku also stressed that the attainment of *Satori* by *Nen-butsu* best fitted the common people in the age of *Mappo*. Zen-do followed Do-shaku's thought and proposed a new idea of *Shomyo Nen-butsu* (where *Shomyo* means "calling *Amida* Buddha by name"). Then, he recommended people to call *Amida* Buddha by name repeatedly.

The *Zen* teachings and *Zen-shu* were originated by a Buddhist monk, Bodai-darma, in the early 6th century and established by Buddhist monks, Jin-shu and Enou, in the late 6th to the 7th century. In general, the *Zen-shu* teaches people the attainment of *Satori* by the method of deep meditation with a concentrated mind. The teachings were widely accepted in China and became prosperous in the ages of the Toh (Táng, 618-907) and Soh (Sòng, 960-1279) dynasties.

Buddhism in Korea [4]

Buddhism was transmitted to Korea in the late 4th or the early 5th century A.D. Buddhism was welcome in all countries of Korea. It is said that the understanding of Buddhist teachings in Korea in the 7th century, in which the Shiragi dynasty governed a main part of Korea, was promoted to a level exceeding China. Buddhism in Korea was also prosperous in the 10th to the 14th centuries in which the Korai dynasty governed the whole Korea. However, Buddhism in Korea declined after the Lee dynasty was established in 1392 because it adopted Confucianism as the national thought and excluded Buddhism.

Buddhism in Japan [1-4,7]

Buddhism was transmitted to Japan in the 6th century. Initially, Buddhism was accepted by persons of power as a symbol to protect the country from sufferings. Thus, the center of Buddhism moved with the transfer of the capital.

In the early 7th century, i.e. in the age of the Shotoku Prince, the capital was located in Asuka near Nara. The Shotoku Prince attached the highest importance to Buddhism. He established the first national constitution based on the Buddhist teachings. The words of "Harmony shall be the noblest", included in the constitution, are famous even in the present Japan. He also issued a commentary on Buddhist sutras and built many Buddhist temples such as the Horyu-ji temple and the Shitenno-ji temple.

In the 8th century, the capital was transferred to Nara and the center of Buddhism moved to the Todai-ji temple and the Kofuku-ji temple in Nara. In this age, a number of Buddhist teachings of China such as *Hosso-*, *Sanron-* and *Kegon-shu* were introduced to Japan.

In the 9th century, the capital was transferred to Kyoto and the Hei-an era started. At the beginning of this era, a Buddhist monk, Saicho, went to China and learned the teachings of *Tendai-shu*. After he came back to Japan, he built the Enryaku-ji temple in Mt Hi-ei near Kyoto and established *Tendai-shu* of a new form. This temple later played an important role in the advancement of Buddhism in Japan, as will be explained later. Another Buddhist monk, Kuh-kai, also went to China in nearly the same age as Saicho and learned esoteric Buddhism. He built the Kongobu-ji temple in Mt Koya-san located to the south of Osaka and established *Shingon-shu*.

The Hei-an era continued to the end of the 12th century. In the ages of the Asuka, the Nara and the Hei-an era, Japan had the political system of aristocracy. In the 10th century, Buddhism, in particular, the *Jodo* teachings (*Jodo-kyo*) gradually spread to general people. In 985, a Buddhist monk, Gen-shin, issued an explanatory book of the *Jodo* teachings. Around the 11th century, the political system of aristocracy began to show instability because of natural disasters and the rise of military families in districts. Then, the common people came to suffer from confusion and poverty. Also, the idea of *Mappo* spread. In such a social situation, i.e. in the late 12th century, a Buddhist monk, Hoh-nen, and his follower, Shin-ran, entered into the history of Buddhism. They started activities for relieving sufferings of the common people, based on the *Jodo* teachings, and the activities finally led to a big revolution in the philosophy of Buddhism, as will be explained in the next chapter.

On the other hand, *Zen-shu* was introduced from China by Japanese Buddhist monks, Eisai and Dogen, in the 12th to the 13th century. Since then, *Zen-shu* became prosperous in Japan, in particular, among military families (called *Bushi* or *Samurai*), which were the governing class in Japan from the end of the 12th century till the middle of the 19th century.

References

(1) H. Nakamura, M. Fukunaga, Y. Tamura, T. Konno, and F. Sueki (editors), *Encyclopedia of Buddhism, The Second Edition* (in Japanese), Iwanami-shoten, Tokyo, 2002.
(2) W. Hiromatsu, et al. (editors), *Encyclopedia of Philosophy and Thought* (in Japanese), Iwanami-shoten, Tokyo, 1998.
(3) S. Kamata, *What Buddha Looked At* (in Japanese), Kodansha-gakujutsu-bunko, Kodan-sha, Tokyo, 1977.
(4) M. Saigusa, *Introduction to Buddhism* (in Japanese), Iwanami-shinsho, Iwanami-shoten, Tokyo, 1990.
(5) Ryukoku Museum in Ryukoku University (edition), *Shaka-muni and Shin-ran, A Route from India to Japan* (in Japanese), Hozokan, Kyoto, 2011.
(6) H. Nakamura (translator), *Suttanipata (Buddha's Words)*,

Iwanami-bunko, Iwanami-shoten, Tokyo, 1984.
(7) Y. Takeuchi and T. Umehara (editors), *Buddhist Sutras in Japan* (in Japanese), Chuko-shinsho, Chuo-koron-sha, Tokyo, 1969.
(8) H. Nakamura and K. Kino (translators), *Hannya-shingyo and Kongo-hannya-kyo*, Iwanami-bunko, Iwanami-shoten, Tokyo, 1960.
(9) H. Nakamura, K. Hayashima, and K. Kino (translators), *Jodo Sanbu Kyo (Three Sutras of Jodo teachings) (I) Muryoju-kyo*, Iwanami-bunko, Iwanami-shoten, Tokyo, 1963.

A2. The Philosophy of Buddhism (II)
– The Disclosure of How the Internal Ability Works in This World –

In the late 12th century when a Japanese Buddhist monk, Hoh-nen, and his follower, Shin-ran, entered into the history of Buddhism, a diversity of sutras of big volumes together with a large variety of theoretical books and commentary books had been compiled in Mahayana Buddhism. Therefore, most of Buddhist monks in Japan in those years regarded the teachings of Mahayana Buddhism as being completed and devoted themselves to the learning and the practice of the teachings. Actually, however, Mahayana Buddhism had not yet been completed in those years. Hoh-nen and Shin-ran became aware of this fact while they learned the teachings of Mahayana Buddhism and Shin-ran succeeded in promoting them to a completed stage.

Shin-ran's work led to a great revolution in the philosophy of Buddhism. The traditional teachings of Mahayana Buddhism in his age had yet clarified no correct way to attain *Satori* or to arrive at the very truth. For this reason, they also had yet clarified no correct idea of the very truth or of the state of *Satori*. Thus, after all, all the basic issues in the philosophy of Buddhism had been left uncertain. Shin-ran became aware of such problems, overcame them, and promoted the teachings of Mahayana Buddhism to a completed stage, as mentioned above. It was Shin-ran that made the teachings of Mahayana Buddhism wholly effective in all areas of human activities including science.

The preceding chapter dealt with the teachings of Initial and Mahayana Buddhism before Hoh-nen and Shin-ran entered into the history of Buddhism. Therefore, in this chapter we deal with the advancement of the teachings of Mahayana Buddhism brought about by Hoh-nen and Shin-ran, in particular, the one by Shin-ran. It is to be mentioned here that the interpretation and the evaluation of Shin-ran's work described in this chapter are solely based on my original understanding. Fortunately, the pursuit of the origin of autonomous dynamic self-organizing ability or free independent harmonious spirit and wisdom of living things led me to the discovery of the great importance of Shin-ran's work.

A2.1 A Historical Background of Shin-ran's Work

No correct way to attain Satori had been clarified in Shin-ran's age

For correctly understanding Shin-ran's work, it is important to clarify what problems had been left unsolved in the philosophy of Buddhism in his age. The most important problem was that no correct way to attain *Satori* had been clarified and thus no correct idea of the very truth had also been revealed, as mentioned above. Naturally, the revelation of how to attain *Satori* or how to arrive at the very truth was a fundamentally important problem in Buddhism but it had long been left unsolved because it was difficult to overcome, as mentioned in Section 1.4.3, A1.1 and A1.2.2.

To make matters worse, nobody in Shin-ran's age had become aware that the Buddhist teachings were in such a serious situation. Most of Buddhist monks believed that the Buddhist teachings had been completed, as mentioned at the beginning of this chapter. In fact, various ways to attain *Satori* had been proposed in Shin-ran's age, such as *Hassho-do* (eight correct ways of living) in Initial Buddhism, *Roku-haramitsu* (the completion of six practice items) in Mahayana Buddhism, *Yoga-gyo* (*Yoga* practice for controlling the mind) in the *Yui-shiki* theory, *Zen* practice (deep meditation) in the *Zen* teachings, and *Nen-butsu* (wanting to come to *Amida* Buddha's land and deeply thinking of *Amida* Buddha with a sincere and enthusiastic mind) in the *Jodo* teachings, as explained in the preceding chapter.

In particular, Buddhist monks in Shin-ran's age were strongly recommended to carry out hard Buddhist practice such as *Roku-haramitsu* by their own ability (which is called *Ji-riki* in Japanese) and achieve a wholly clear state of mind completely released from illusions (incorrect views of nature arising from word-based understanding) and *Bon-noh* (a person's desires and attachments for individual things such as existence, living, wealth, fame, and so on) [1-4]. The ultimate goal was to transcend all individual ideas and things in the external visible world and accomplish the internal wisdom to look at the principle of *Engi* throughout nature and the world. Such a traditional way to attain *Satori* is in general believed to be the orthodoxy of the Buddhist teachings even in the present Japan.

It was Hoh-nen and Shin-ran that began to feel doubt about such a traditional way to attain *Satori*. A main reason was that it was very difficult or actually nearly impossible to attain *Satori* by such a traditional way. In fact, it is said that almost no Buddhist monk attained *Satori* in Japan in the age of Hoh-nen and Shin-ran [4]. In those years, the Japan society was thrown into severe political confusion and many people suffered from disorder and poverty. Thus, Buddhist monks were requested to work for relieving people's sufferings but they had no enough ability for it because he had not yet attained complete *Satori*. In the history of Mahayana Buddhism, the ineffectiveness of the traditional way to attain *Satori* already became apparent in the middle stage of it, as discussed in Section A1.3.3. However, no successful solution was discovered in that age. It seems that the teachings of Mahayana Buddhism had rather changed in a wrong direction since then. For example, the *Yui-shiki* theory emphasized the importance of controlling the mind through particular practice such as *Yoga-gyo*. Thus, Buddhist monks in later ages had a tendency to try to eliminate illusions and *Bon-noh* by particular practice instead of transcending them.

Indeed, Shaka-muni carried out hard ascetic practice by his own ability (*Ji-riki*) and achieved a peaceful harmonious state, as explained in Section A1.2.1. New Buddhists in Mahayana Buddhism in the primary stage also carried out hard Buddhist practice such as *Roku-haramitsu* by their own ability (*Ji-riki*) and achieved a peaceful harmonious state, as explained in Section A1.3.1 and A1.3.2. Accordingly, the traditional teachings of Mahayana Buddhism in Shin-ran's age followed these ways to attain *Satori*. However, in my opinion, here was a big mistake. It was necessary to take into account the historical progress of the Buddhist teachings. Not only Initial Buddhism but also Mahayana Buddhism in the primary stage was still in the initial stage of the history of Buddhism. In fact, new Buddhists in Mahayana Buddhism in the primary stage were in the midst of clarifying the very truth about nature and the world (i.e. the generalized principle of *Engi*). In such an initial stage, no sufficient knowledge about the way to attain *Satori* was obtained. Therefore, to carry out hard practice by *Ji-riki* was the *only* possible way to attain *Satori*. For example, for accomplishing the generalization of the principle of *Engi*, new Buddhists in Mahayana Buddhism had to conduct some of the common people in the neighborhood to *Satori* in the real world by their own ability, as mentioned in Section A1.3.1. (It will be explained in Section A2.3 why Shaka-muni and new Buddhists in Mahayana Buddhism in the primary stage were able to attain *Satori* before the correct way to attain *Satori* was revealed.)

However, the situation completely changed when the generalization of the principle of *Engi* had been completed and the concepts of *Hosshin*, *Hohjin* and *Ohjin* had been established. As mentioned in Section A1.3.1, new Buddhists in Mahayana Buddhism finally reached the key conclusion that people can attain *Satori* by the ability of Buddha because Buddha always works to conduct all people to *Satori*. When this conclusion was obtained, the traditional way to attain *Satori* by a person's own ability (*Ji-riki*) came to be in contradiction to the conclusion and lost its theoretical basis. Originally, such a traditional way to attain *Satori* was only what new Buddhists in Mahayana Buddhism artificially invented based on their experiences and had no definite theoretical basis. Namely, it only offered a superficial phenomenological way. In this way, the advancement of the Buddhist teachings came to demonstrate that the *Jodo* teachings (*Jodo-kyo*), which taught people the attainment of *Satori* by the ability of Buddha (*Ta-riki*), were the orthodoxy of the Buddhist teachings.

Nevertheless, amazingly, the traditional way to attain *Satori* by *Ji-riki* had long been regarded as the main way to attain *Satori* even after the above key conclusion was obtained. In fact, the traditional teachings of Mahayana Buddhism in the age of Hoh-nen and Shin-ran recommended Buddhist monks to carry out hard Buddhist practice such as *Roku-haramitsu* by *Ji-riki*, as mentioned earlier. Such a confused situation came from the fact that no correct way to attain *Satori* had been clarified. Indeed, the key conclusion of Mahayana Buddhism said that all people can attain *Satori* by the ability of Buddha, as mentioned above. However, Mahayana Buddhism had not yet revealed *how* (or via what processes) all people can attain *Satori* by the ability of Buddha. Namely, Mahayana Buddhism had not yet clarified how Buddha works in this world, how Buddha conducts people to *Satori*, or how Buddha's ability (*Ta-riki*) and a person's ability (*Ji-riki*) are related to each other. For this reason, Buddhist monks in those years were still unable to attain *Satori* according to the *Jodo* teachings. Thus, after all, they had no way but to rely on hard Buddhist practice such as *Roku-haramitsu* by *Ji-riki*. Even Buddhist monks following the *Jodo* teachings tried to improve the quality of *Nen-butsu* by adding results of hard Buddhist practice.

In the age of Hoh-nen and Shin-ran, nobody noticed that the traditional teachings of Mahayana Buddhism had such a severe problem, as mentioned earlier. We can see here how difficult it is to become aware of a limit of authorized traditional knowledge. Even Hoh-nen and Shin-ran initially did not notice a limit of the traditional teachings. Therefore, they first tried to attain *Satori* according to the traditional teachings. Then, they gradually came to feel doubt about the traditional teachings and Shin-ran finally succeeded in overcoming the limits of the teachings after strenuous effort.

These considerations indicate that a revolution in the philosophy of Buddhism by Hoh-nen and Shin-ran was an inevitable result of the historical progress of it, just in the same way as the case of the rise of

Mahayana Buddhism in ancient India. New Buddhists in Mahayana Buddhism in ancient India overcame the limits of Shaka-muni's teachings, based on Shaka-muni's teachings, as mentioned in Section A1.3.1. Similarly, Shin-ran overcame the limits of the traditional teachings of Mahayana Buddhism, based on the traditional teachings of Mahayana Buddhism.

No correct idea of the very truth had been clarified in Shin-ran's age

There was another important problem in the traditional teachings of Mahayana Buddhism in Shin-ran's age. As mentioned above, no correct way to attain *Satori* or no correct way to arrive at the very truth had been clarified yet. For this reason, no correct idea of the very truth or no correct idea of the state of *Satori* had also been clarified yet.

A fatal fault of the traditional teachings of Mahayana Buddhism about the idea of the very truth arose from the fact that they taught Buddhist monks to attain *Satori* by achieving a wholly clear state of mind completely released from illusions and *Bon-noh*, as mentioned earlier. Indeed, the achievement of a wholly clear state of mind guarantees arrival at the very truth, as discussed in Section 1.3.1 and 1.3.2. However, the very truth (i.e. the way that truly real things exist) embodied in a wholly clear state of mind completely released from illusions and *Bon-noh* only represents the very truth about a part of human life. Namely, this very truth does not include the very truth (the way that truly real things exist) about another part of human life in which a person has illusions and *Bon-noh* and suffers from mental pains. Accordingly, this very truth does not represent the very truth about the whole human life.

The above argument indicates that the state of *Satori* or the very truth which the traditional teachings of Mahayana Buddhism had taught was of an incomplete character. A similar conclusion is obtained by the following consideration as well. As mentioned in Section A1.3.1, it is actually impossible to accomplish the internal wisdom to clearly look at the principle of *Engi* throughout nature and the world, which are infinite in size and content. This means that a wholly clear state of mind (or the very truth embodied in it) which Buddhist monks actually achieved according to the traditional teachings of Mahayana Buddhism was of an approximate character.

Furthermore, it should be noted that the very truth about nature and the world is controlled by *Hosshin* (or the grand internal ability) and is of an incessantly changing dynamic character, as discussed in Section A1.2.2 and A1.3.1. On the other hand, a wholly clear state of mind (or the very truth embodied in it) which Buddhist monks actually achieved according to the traditional teachings of Mahayana Buddhism shows no such dynamic character. This is because a wholly clear state of mind (or the very truth embodied in it) in this case is only achieved in a particular quality level and aspect, as mentioned above.

Shin-ran overcame all problems that the traditional teachings of Mahayana Buddhism had faced

It was Shin-ran that overcame all problems that the traditional teachings of Mahayana Buddhism had faced. At first, he discovered the correct way to attain *Satori*. As will be explained later, Shin-ran considered how he was able to attain *Satori* and then attained *Satori*. This meant that he reached the very truth about how to attain *Satori*. Here is why we can say that Shin-ran discovered the correct way to attain *Satori*. Nobody had ever clarified how to attain *Satori* in such a way.

Furthermore, based on the discovery of the correct way to attain *Satori*, Shin-ran disclosed that a person achieves a wholly clear state of mind when he or she has acquired the internal ability of the same character as Buddha's ability and has been given belief in it. The achievement of a wholly clear state of mind guarantees arrival at the very truth, as mentioned earlier. Therefore, the above disclosure meant that *a person arrives at the very truth and attains Satori when he or she has acquired the internal ability of the same character as Buddha's ability and has been given belief in it*. This disclosure led to the clarification of the correct idea of the state of *Satori* and the correct idea of the very truth.

The difference in the state of Satori between the traditional teachings of Mahayana Buddhism and Shin-ran's thought

How did the above disclosure lead to the clarification of the correct idea of the state of *Satori* and the correct idea of the very truth? At first, let us consider the difference in the state of *Satori* between the traditional teachings of Mahayana Buddhism and Shin-ran's thought. The traditional teachings of Mahayana Buddhism aimed at attaining *Satori* by accomplishing the internal wisdom to clearly look at the principle of *Engi* throughout nature and the world, as mentioned earlier. Namely, they aimed at attaining *Satori* based on the principle of *Engi* generalized in Mahayana Buddhism or based on the very truth about nature and the world which Mahayana Buddhism reached.

On the other hand, Shin-ran aimed at attaining *Satori* based on the key conclusion of Mahayana Buddhism. Namely, he aimed at attaining *Satori* according to the *Jodo* teachings, which taught people the attainment of *Satori* by the ability of Buddha (*Ta-riki*). Therefore, he had long been asking where Buddha was and how Buddha conducted people to *Satori*. Then, he finally succeeded in clarifying the way that Buddha conducted people to *Satori* and catching it within his body. In this way, he acquired the internal ability of the same character as Buddha's ability and attained *Satori*, as mentioned above.

After all, there was a large difference in the underlying philosophy of Buddhism between the traditional teachings of Mahayana Buddhism and Shin-ran's thought. The principle of *Engi* generalized in Mahayana Buddhism represents "a result" of activities of Buddha, as can be seen from the arguments in Section A1.3.1. Therefore, the traditional teachings of Mahayana Buddhism taught Buddhist monks to attain *Satori* by catching "a result" of activities of Buddha. On the other hand, Shin-ran attained *Satori* by catching activities of Buddha themselves or Buddha itself.

Characteristics of the idea of the very truth Shin-ran revealed

The above-mentioned difference in the state of *Satori* between the traditional teachings of Mahayana Buddhism and Shin-ran's thought led to a great difference in the idea of the very truth between them. The idea of the very truth Shin-ran revealed had the following superior characteristics.

Firstly, Shin-ran discovered that a person achieves a wholly clear state of mind and attains *Satori* when he or she has acquired the internal ability of the same character as Buddha's ability and has been given belief in it. This means that a wholly clear state of mind a person achieves according to Shin-ran's thought and hence the very truth embodied in it are controlled by Buddha itself, in the same way as the very truth about nature and the world. Accordingly, the very truth embodied in such a wholly clear state of mind correctly reflects the very truth about nature and the world that lies in an incessantly changing dynamic state. In addition, for the same reason, it correctly reflects the very truth about nature and the world which are infinite in size and content. Furthermore, the very truth embodied in such a wholly clear state of mind correctly reflects the very truth about every stage of human life, including the very truth about a stage in which a person has illusions and *Bon-noh* and suffers from mental pains. In this way, the very truth in Shin-ran's thought best represents the very truth about nature and the world.

Secondly, to our surprise, a wholly clear state of mind a person achieves according to Shin-ran's thought (and the very truth embodied in it) still involve illusions (incorrect views arising from word-based understanding) and *Bon-noh* (a person's desires and attachments for individual things). Namely, the acquisition of the internal ability of the same character as Buddha's ability and belief in it allow a person to achieve a wholly clear state of mind without transcending or eliminating illusions and *Bon-noh*. This is because Buddha itself has achieved a wholly clear state of mind without transcending or eliminating illusions and *Bon-noh*. Buddha has achieved an undisturbed wholly clear state of mind in the presence of illusions and *Bon-noh* by accomplishing the high-level internal wisdom to freely control illusions and *Bon-noh*. For this reason, Buddha can govern all people and things in nature and the world and their constituent elements, which are full of illusions and *Bon-noh*, so that they realize a free independent harmonious stable state, as discussed in Section A1.3.1.

The attainment of a wholly clear state of mind (or the attainment of *Satori*) without transcending or eliminating illusions and *Bon-noh* was an unbelievable event in the traditional teachings of Mahayana Buddhism because to transcend or eliminate illusions and *Bon-noh* was nothing else but to attain *Satori* in the traditional teachings (see Section A1.2.2, A1.3.1, A1.3.2 and A1.3.3). Shin-ran revealed the complete antithesis of the idea of the very truth which the traditional teachings of Mahayana Buddhism had taught.

Thirdly, again surprisingly, a person who has acquired the internal ability of the same character as Buddha's ability and has been given belief in it lives a strange ideal life. Namely, such a person lives in accord with the very truth lying in the internal invisible world and simultaneously lives a practical lively life in this world full of illusions and *Bon-noh*. This is because such a person has arrived at the very truth without transcending or eliminating illusions and *Bon-noh*, as mentioned above. The above argument means that such a person lives the correct life with the dual structure of nature such as discussed in Section 1.4.2 as the base. On the contrary, the traditional teachings of Mahayana Buddhism in Shin-ran's age only allowed people to live based on the very truth lying in the internal invisible world. By the way, traditional word-based scientific understanding also only allows us to live based on the external appearance of the very truth (see Section 1.4.2). The traditional ideas both only allow us to live based on the *single* structure of nature.

Fourthly, a person who has acquired the internal ability of the same character as Buddha's ability and has been given belief in it continues to pursue the very truth about nature and the world even after attaining

Satori. This is because such a person has the internal ability of the same character as Buddha, who always works to make all people and things realize a free independent harmonious state. This is also because such a person still has illusions and *Bon-noh*, which act as the cause of arrival at the very truth, as will be explained in Section A2.3. Note that this argument explains why such a person can live in accord with the very truth about nature and the world which lies in an incessantly changing dynamic state.

Comparison of some representative ways of living proposed thus far
For making the meanings of Shin-ran's work clearer, let us compare some representative ways of living proposed thus far.

Descartes discovered the existence of a person's ego as a clear fact, together with the objective truth. Then, he proposed a new way of living in which a person lives a free independent exploratory and rational life with his or her ego and the objective truth as the base (see Section A3.1.1). This thought provides us with a fundamental way of living in the external visible world. A demerit of this thought is that it is based only on things we are conscious of, which are of an approximate mechanical character, the origins of a person's ego and the objective truth being left unclear.

Existentialism emphasizes the importance of the free will of a person's ego. However, this thought is also based on things we are conscious of, which are of an approximate mechanical character, and thus seems to fail to clarify how a person can achieve true freedom. In addition, it appears that this thought speaks of nothing about the truth and hence has a danger of going wrong.

Shaka-muni attained *Satori* and discovered that a person was able to transcend individual ideas and things and stand firmly without relying on anything in the external visible world. Namely, he discovered that the origin of human life is in the internal world we cannot be conscious of. Thus, he proposed a way of living in which a person transcends individual ideas and things in the external visible world and lives a free independent peaceful harmonious life with the very truth in the internal invisible world as the base.

New Buddhists in Mahayana Buddhism discovered that the principle of *Engi* Shaka-muni revealed applied to all people and things in nature and the world and their constituent elements. Thus, they proposed a way of living in which a person acquires the internal wisdom to clearly look at the principle of *Engi* throughout nature and the world and works to conduct all people as well as him- or herself to *Satori* (i.e. a free independent peaceful harmonious life).

Shin-ran discovered that a person was able to attain *Satori* by acquiring the internal ability of the same character as Buddha's ability and being given belief in it. The acquisition of such ability allows a person to live in accord with the very truth without transcending or eliminating illusions and *Bon-noh* and hence to live an entirely free natural life. Thus, Shin-ran proposed a way of living in which a person first acquires the internal ability of the same character as Buddha's ability and is given belief in it and then lives an entirely free natural life in accord with the very truth.9

A2.2 The Satori of Shin-ran

The philosophy of Shin-ran is constructed based on his experience, not on his thinking. Therefore, for correctly understanding his philosophy, it is indispensable to know what problems Shin-ran encountered, how he considered them, and how he solved them. In addition, how Shin-ran attained *Satori* is of much interest because it gives a good example of an experience of attaining *Satori*.

A brief survey of Shin-ran's life [1-4]

Shin-ran was born in 1173 and died in 1262. He was a child of an aristocrat in Kyoto but lost his parents when he was very young. Therefore, in conformity with the custom in that age, he was, at the age of nine, let enter into the Enryaku-ji temple in Mt Hi-ei near Kyoto as a Buddhist ascetic with a duty to labor. In those years, the Enryaku-ji temple was a center of Buddhism in Japan. A large number of Buddhist monks carried out Buddhist practice day after day.

Shin-ran will, in intervals of labor, have learned Buddhist sutras and theories and carried out Buddhist practice. In the Enryaku-ji temple, Buddhist monks were taught to carry out hard Buddhist practice such as *Roku-haramitsu* by their own ability (*Ji-riki*) and achieve a wholly clear state of mind completely released from illusions (incorrect views arising from word-based understanding) and *Bon-noh* (desires and attachments for individual things), as mentioned earlier. Shin-ran made sincere efforts for about twenty years. However, he was unable to obtain anything meaningful. He later honestly said that he was unable to remove *Bon-noh* by his own ability (*Ji-riki*).

While Shin-ran stayed in the Enryaku-ji temple, Japan was thrown into severe political confusion. In the 12th century, near the end of the

Hei-an era, military families (called *Bushi* or *Samurai* in Japanese) came to have independent power and frequently brought about battles in districts. The ruling power of aristocrats declined rapidly and they came to rely on the power of military families for putting down such battles. Finally, in 1192, a new political system by a military family was established instead of aristocracy and the Kamakura era started. Continuing battles and social confusion made the country full of severe mental pains and poverty.

In such a miserable situation, the *Jodo* teachings (*Jodo-kyo*), which taught people the attainment of *Satori* by *Nen-butsu* (wanting to come to Amida Buddha's land and deeply thinking of Amida Buddha with a sincere mind), gradually spread into the common people. In particular, Hoh-nen's simplified *Jodo* teachings became prosperous in Kyoto and its neighborhood. Not only the common people but also aristocrats followed his teachings. Hoh-nen previously learned the traditional teachings of Mahayana Buddhism in the Enryaku-ji temple and was called a Buddhist monk of the highest intelligence. However, in 1175, when Hoh-nen was forty three years old, he made up his mind to create new simplified *Jodo* teachings and came down from Mt Hi-ei to teach people the new *Jodo* teachings.

Hoh-nen's conversion was certainly a big event, from which a revolution in Buddhism in Japan really started. Until that time, all Buddhist monks, including Hoh-nen himself, followed the traditional teachings, which taught people to carry out hard Buddhist practice such as *Roku-haramitsu* by their own ability (*Ji-riki*) and transcend or eliminate illusions and *Bon-noh*, as mentioned earlier. Even Buddhist monks following the *Jodo* teachings had tried to improve the quality of *Nen-butsu* by adding results of hard Buddhist practice to it. However, Hoh-nen boldly chose new simplified *Jodo* teachings as the only way to conduct the common people to *Satori* in the age of *Mappo* (see a footnote #1 of Section A1.4 for the meanings of *Mappo*). Namely, he adopted the way to attain *Satori* by *Shomyo Nen-butsu*, which a Chinese Buddhist monk, Zen-do, first proposed (see Section A1.4). Here, *Shomyo* means calling Amida Buddha by name. Based on this idea, Hoh-nen set up *Jodo-shu* (*shu* means a school or sect). Hoh-nen's teachings are clearly described in his book entitled "*Senchaku Hongan Nen-butsu Shu*" [4-6], issued in 1198.

> *Nen-butsu is the essence of Ojo (Ojo means "coming to Amida Buddha's land" and has the same meaning as attaining Satori). If you want to attain Satori, put the way of carrying out hard*

> Buddhist practice by your own ability aside and choose the way of carrying out Nen-butsu in the Jodo teachings. If you choose the way of carrying out Nen-butsu in the Jodo teachings, put supplementary practice (such as thinking of Buddha's other than Amida Buddha and collecting worthy deeds) aside and adopt the correct practice. What is the correct practice? It is to call Amida Buddha by name. It necessarily leads you to Ojo because Amida Buddha has accomplished the vow of conducting all people to Satori.

Shin-ran in Mt Hi-ei suffered from difficulty in attaining *Satori*, as mentioned earlier. Therefore, the prosperity of Hoh-nen's teachings in down town will have been felt a remarkable situation. In 1201, three years after Hoh-nen's book was issued, Shin-ran made up his mind to come down from Mt Hi-ei and join Hoh-nen's school. At that time, Shin-ran was twenty nine years old.

Only four years after Shin-ran joined Hoh-nen's school, he met a serious trouble, which completely changed his life. In 1205, Buddhist monks of *Hosso-shu* in the Kofuku-ji Temple in Nara submitted a bill of complaint to the Emperor for the reason that some members of Hoh-nen's school spoke ill of the teachings of *Hosso-shu* and led people to ill behavior. By this bill of complaint, twelve members of Hoh-nen's school, including Hoh-nen and Shin-ran, were incriminated in 1207. Four members were killed and Hoh-nen was exiled to an undeveloped district, Tosa, in Shikoku. Shin-ran was also exiled to a savage district, Echigo, which was far from Kyoto. This was really an abnormal event in the long history of Buddhism.

The crime for Hoh-nen and Shin-ran was pardoned five years after they were incriminated. Hoh-nen came back to Kyoto and died soon. On the other hand, Shin-ran remained in Echigo and later moved to Hitachi, a district which was also far from Kyoto but relatively near Kamakura, the capital of the Kamakura era. Shin-ran was silent for about thirty years after his crime was pardoned [4]. Presumably, he had continued considering the criticisms of the *Jodo* teachings raised by the bill of complaint. It is said that the bill of complaint was compiled by an excellent Buddhist monk, Jokei, in *Hosso-shu* and full of theoretical criticisms on the highest level in those years. In this sense, the bill of complaint may have played an important role in driving Shin-ran to an innovative thought.

It is said that Shin-ran roughly completed a draft of his book, entitled "*Kyo Gyo Shin Sho*", in around 1231~1235 [4]. In this book, Shin-ran described his new thought as well as his answers to the criticisms of the *Jodo* teachings by the bill of complaint. At that time, he was about sixty years old.

The first step toward Satori – Arrival at limits of the traditional teachings of Mahayana Buddhism

Shin-ran's book, "*Kyo Gyo Shin Sho*"[5,6], consists of a collection of words cited from a large number of Buddhist sutras and theoretical books, his opinions and conclusions being inserted here and there in short sentences. Shin-ran will have examined the criticisms of Hoh-nen's *Jodo* teachings by the bill of complaint one by one. Then, he had to search for words representing his answers in authorized Buddhist sutras or theoretical books because otherwise his any rebuttal had no power to convince Buddhist monks in *Hosso-shu*[4]. Only words in the Buddhist literature were acknowledged as the correct opinions in those years.

On the other hand, it is also well known that Shin-ran's book, "*Kyo Gyo Shin Sho*", is filled with words full of firm confidence and great delight. Apart from the criticisms by the bill of complaint, Shin-ran must have possessed his own problem. Namely, he had not yet solved the problem of how he himself as well as the common people was able to attain *Satori* even after he joined Hoh-nen's school, as will be explained later. Thus, he must have continued asking and answering about this problem for about thirty years. Then, he finally reached a wholly clear state of mind full of delight and wrote his conclusion in his book, "*Kyo Gyo Shin Sho*", with firm confidence and great delight.

Now, let us consider how Shin-ran attained *Satori*, based on the literature [1-4], his personal history, and his conclusions described in his book, "*Kyo Gyo Shin Sho*"[5,6] and a book named *Tan-i-sho*[7]. It is said that *Tan-i-sho* was compiled by one of Shin-ran's followers, Yui-en, after Shin-ran's death. This book is famous as a collection of Shin-ran's words that clearly express the kernel of his thought.

The bill of complaint was composed of nine items[4]. One of them was an accusation of the lack of the genealogy of *Jodo-shu* which Hoh-nen set up. The others were all criticisms of Hoh-nen's teachings. Certainly, Hoh-nen's teachings deserved to get such criticisms because it had a clear feature of a revolution in the philosophy of Buddhism, as mentioned earlier. Hoh-nen said, "You need only rely on *Amida* Buddha" or "You need only call *Amida* Buddha by name". The teachings were very simple and willingly accepted by a large number of people. However, for this reason, Hoh-nen's teachings neglected many important concepts that were handed over in the long history of Buddhism.

Naturally, the bill of complaint severely criticized this point[4]. It says, "Hoh-nen's teachings involve serious misunderstandings of the way to attain *Satori*. People can go up to higher wisdom only by carrying out hard Buddhist practice. However, Hoh-nen's teachings neglect many of important items of Buddhist practice such as learning sutras, observing commandments, accumulating worthy deeds, and so on, which have long been handed down since Shaka-muni's *Satori*. As a result, Hoh-nen's teachings slander Buddhism, which is the largest evil in Buddhism."

The bill of complaint also says, "Hoh-nen's teachings misunderstand *Nen-butsu* itself. Why is it important only to call *Amida* Buddha by name? This is *Nen-butsu* on the lowest level. *Nen-butsu* which is not accompanied by the accumulation of worthy deeds is only *Nen-butsu* in name. By continuing to carry out hard Buddhist practice, people can have *Nen-butsu* on a higher level such as imaging *Amida* Buddha in mind and looking at *Amida* Buddha with a quiet state of mind."

The criticisms by the bill of complaint were certainly right if they were judged on the standard of the traditional teachings of Mahayana Buddhism. The opinions of the bill of complaint were thus supported by Buddhist monks in other Buddhist sects. For example, Myo-e in *Kegon-shu*, who was known as a Buddhist monk of high intelligence in those years, also severely criticized Hoh-nen's teachings. Myo-e initially highly respected Hoh-nen but was surprised to know that *Bodai-shin* (a mind to seek for the very truth) was neglected in Hoh-nen's book.

Hoh-nen himself in advance expected the appearance of such criticisms. However, he deliberately chose a way to attain *Satori* by *Shomyo Nen-butsu* as the only way to conduct the common people to *Satori* in the age of *Mappo*. It is said that almost no Buddhist monk in the Enryaku-ji temple had attained *Satori* in Hoh-nen's age [4], as mentioned earlier. Thus, it is highly probable that Hoh-nen came to think that a traditional way to attain *Satori* by *Ji-riki* was ineffective. In addition, more importantly, Hoh-nen must have become confident that the *Jodo* teachings (*Jodo-kyo*), which taught people the attainment of *Satori* by the ability of Buddha, best represented the key conclusion of Mahayana Buddhism and was the orthodoxy of the Buddhist teachings [3]. Quite sharp words in his book, "*Senchaku Hongan Nen-butsu Shu*", can be explained only by taking into account this point.

Originally, a confrontation between Hoh-nen' teachings and the teachings of *Hosso-shu* emerged from a contradiction included in the traditional teachings of Mahayana Buddhism, as mentioned in Section A2.1. Namely, the key conclusion of Mahayana Buddhism was in contradicttion to the traditional way to attain *Satori* by *Ji-riki*. Hoh-nen correctly caught the key conclusion of Mahayana Buddhism and expressed it in a simplified form. On the other hand, Jokei in *Hosso-shu* and Myo-e in *Kegon-shu* failed to correctly catch the key conclusion of Mahayana Buddhism though they may have had much knowledge about Mahayana Buddhism.

Now, how had Shin-ran considered the problems raised by the bill of complaint? He already ascertained through his own experience in Mt Hi-ei that the traditional way to attain *Satori* by *Ji-riki* was ineffective [4], in the same way as Hoh-nen. He made sincere efforts for about twenty years but was unable to gain anything meaningful, as mentioned earlier. Not only Shin-ran but also other Buddhist monks in the Enryaku-ji temple had not gained anything meaningful. Shin-ran saw that most of Buddhist monks in the Enryaku-ji temple were enthusiastic about promoting themselves to higher ranks in the organization of Buddhist monks [4]. Shin-ran thus clearly recognized that *Bon-noh* were too strong to be destroyed by *Ji-riki*.

Shin-ran will also have investigated theoretically why the traditional way to attain *Satori* by *Ji-riki* was ineffective. In the Enryaku-ji temple, Buddhist monks were taught to improve wisdom by carrying out hard Buddhist practice such as *Roku-haramitsu* by their own ability (*Ji-riki*), as mentioned earlier. Now, when Shin-ran watched other Buddhist monks making sincere effort to attain *Satori*, he will have noticed that here was no favor of Buddha [4]. Buddha only saw Buddhist monks to make sincere effort without offering any help. Such an attitude of Buddha was the opposite of the key conclusion of Mahayana Buddhism which says that Buddha always works to conduct all people to *Satori*. Thus, Shin-ran will have become confident that the traditional teachings made some essential mistakes. Probably for this reason Shin-ran made up his mind to join Hoh-nen's school.

Why were Hoh-nen and Shin-ran able to notice limits of the traditional teachings?

It was mentioned in Section A2.1 that it is not easy to notice a limit of authorized traditional knowledge. Thus, it is interesting to consider why Hoh-nen and Shin-ran were able to notice limits of the traditional teachings of Mahayana Buddhism. Most probably Hoh-nen looked at sufferings of the common people in those years and strongly desired to help them by the Buddhist teachings. Thus, he will have carefully investigated the Buddhist teachings and noticed the ineffectiveness of the traditional way to attain *Satori*. On the other hand, Shin-ran sincerely desired to attain *Satori* and made much effort according to the traditional way to attain *Satori*. However, he was unable to get anything meaningful. He later honestly said that he was unable to remove *Bon-noh* by his own ability (*Ji-riki*), as mentioned earlier. The words prove that he had really made sincere effort for a long time. Thus, both Hoh-nen and Shin-ran truly sincerely desired to attain *Satori* and carefully investigated the Buddhist teachings and then came to feel doubt about the effectiveness of the traditional way to attain *Satori*.

Other Buddhist monks in those years must also have suffered from difficulty in attaining *Satori* but did not come to feel doubt about the traditional teachings. In my opinion, this was because other Buddhist monks disregarded or hid their feeling of difficulty. For example, they may have thought that *Satori* was originally difficult to attain or that their learning and practice were still insufficient. In short, the desires of other Buddhist monks to attain *Satori* were not so sincere, or in other words, their dependence on the traditional teachings was so strong.

Interestingly, the position of Hoh-nen and Shin-ran against the traditional teachings of Mahayana Buddhism resembled that of new Buddhists in Mahayana Buddhism in ancient India against the traditional thought of Branch Buddhism. New Buddhists in Mahayana Buddhism livelily worked for the common people with high ideals and deep affecttion (see Section A1.3.1). For this reason, they were able to notice limits of the traditional teachings of Initial and Branch Buddhism and create new Buddhism.

The second step toward Satori – Arrival at limits of Hoh-nen's teachings

How had Shin-ran thought about *Satori* after he joined Hoh-nen's school? Hoh-nen simply said, "You need only rely on *Amida* Buddha", as mentioned earlier. Certainly, these words of Hoh-nen correctly expressed the key conclusion of Mahayana Buddhism that all people were able to attain *Satori* by the ability of Buddha. Therefore, Shin-ran must have been confident that he was not wrong to make up his mind to follow Hoh-nen's teachings. However, he still had a serious problem. He had not yet found a convincing answer to the problem of how the common people attained *Satori* by the ability of Buddha. First of all, he himself had not yet attained *Satori* even in Hoh-nen's school.

It seems that Shin-ran's true pursuit started from here. He will have reinvestigated the Buddhist teachings from the beginning. It is expected that his primary question was where Buddha was. Hoh-nen said, "You need only rely on *Amida* Buddha", as mentioned above. However, Shin-ran was unable to know where *Amida* Buddha was and how he was able to rely on *Amida* Buddha. In fact, here was a fundamental problem. As mentioned in Section A2.1, Mahayana Buddhism indeed disclosed that all people were able to attain *Satori* by the ability of Buddha but had not yet clarified *how* (or via what processes) people were able to attain *Satori* by the ability of Buddha. Shin-ran just faced this problem.

Hoh-nen also said, "*Nen-butsu* is the essence of *Ojo*", where *Ojo* refers to the attainment of *Satori*. Thus, Shin-ran will have considered what *Nen-butsu* was. The 18th item of the vow of *Amida* Buddha in *Muryoju-kyo* [8] taught people to want to come to *Amida* Buddha's land with a sincere and enthusiastic mind and think of *Amida* Buddha, say, ten times (see Section A1.3.2(B)). Then, Shin-ran will have actually tried to sincerely want to come to *Amida* Buddha's land and think of *Amida* Buddha, say, ten times. However, nothing notable happened. He was never able to attain *Satori*. In this way, Shin-ran will have asked, "What is wrong about my understanding? What is meant by a sincere and enthusiastic mind? What is *Nen-butsu*?"

The meaning of the 18th item of *Amida* Buddha's vow had been considered by a number of Buddhist monks in the *Jodo* teachings. In particular, the interpretation of a Chinese Buddhist monk, Zen-do, had a large influence on Buddhist monks in the *Jodo* teachings later. He interpreted the words of "to want to come to *Amida* Buddha's land ·····." in the 18th item as completely believing Buddha's ability and absolutely relying on it. Then, he recommended people to carry out *Shomyo Nen-butsu* (*Nen-butsu* by calling *Amida* Buddha by name repeatedly), as already mentioned earlier. Hoh-nen highly respected Zen-do and adopted his interpretation. Probably, the majority of Buddhist monks in Hoh-nen's school, including Shin-ran, adopted Zen-do's interpretation. In fact, one of Hoh-nen's grand-followers, Ippen, who set up new *Jodo* teachings later, recommended people to call *Amida* Buddha by name with an empty state of mind or with no intention of *Nen-butsu*. Ippen pushed forward the thought of *Shomyo Nen-butsu* to the extreme limit.

However, such an interpretation was not correct. Shin-ran later had an unexpected experience, which led him to the conclusion that he had no way to reach a state in which he absolutely relied on *Amida* Buddha. The experience is recorded in a letter of his wife, dated April 1231. It is roughly summarized as follows [4].

> *When Shin-ran caught a cold and was severely feverish in bed for four days, his wife saw he had a night-mare. After the fever was reduced, he said he had continued reading a sutra, Muryoju-kyo, by heart during the night-mare. He said, "Many years have*

passed after I made up my mind to follow Hoh-nen's teachings and solely rely on Amida Buddha. Nevertheless I continued reading sutras by heart in a dream under fever. I still have possessed a mind to rely on my own ability (Ji-riki). I have been deeply impressed with how strong a human mind to rely on Ji-riki is. I have to consider this issue again in much more detail."

This letter indicates that *Shin-ran* had not yet attained *Satori* completely at this stage, i.e. in April 1231.

After this experience, Shin-ran will have considered how he was able to completely remove the mind to rely on his own ability (*Ji-riki*). This is because he was unable to absolutely rely on *Amida* Buddha as far as he had such a mind. However, it is expected that this consideration led him to an absolutely difficult situation. To make effort to remove the mind to rely on *Ji-riki* was nothing else but to rely on *Ji-riki*. To return to the initial stage of consideration, *Shomyo Nen-butsu* itself was nothing else but to rely on *Ji-riki* because people had to continue calling *Amida* Buddha by name. Shin-ran will have noticed that such a way had the same limits as the traditional way to attain *Satori* through hard Buddhist practice such as *Roku-haramitsu* by *Ji-riki*. In fact, when Shin-ran watched Buddhist monks continuing *Shomyo Nen-butsu*, he must have clearly recognized that here was no favor of *Amida* Buddha. *Amida* Buddha only saw Buddhist monks to call *Amida* Buddha by name, without offering any help. This meant that Hoh-nen's teachings also did not correctly represent the key conclusion of Mahayana Buddhism, which says that Buddha always works to conduct all people to *Satori*. Hoh-nen's teachings also made some essential mistake.

In this way, Shin-ran was led to the conclusion that he had no way to reach a state in which he absolutely relied on *Amida* Buddha, as mentioned above. He was rather forced to conclude that a human being had no choice but to rely on *Ji-riki* in any situation. This conclusion, however, meant that Shin-ran was unable to attain *Satori* according to the *Jodo* teachings. After all, Shin-ran was neither able to transcend or eliminate illusions and *Bon-noh* by *Ji-riki* according to the traditional teachings of Mahayana Buddhism nor able to completely believe Buddha's ability and absolutely rely on it according to Hoh-nen's teachings. Thus, Shin-ran was driven into an absolutely difficult situation in which he was neither able to go forward nor able to go backward.

Shin-ran had another serious problem about the 18th item of *Amida* Buddha's vow in *Muryoju-kyo*. When he was exiled to Echigo and moved to Hitachi, he saw the majority of farmers in these districts to live a desperate life under extreme poverty [4]. Some of people even had done evils under unavoidable situations. Thus, Shin-ran sincerely desired to release these people from sufferings. However, the 18th item of *Amida Buddha*'s vow said that evil persons such as *Go-gyaku*, *Hibo* and *Sendai* were excluded from the light of Buddha (The meanings of *Go-gyaku*, *Hibo* and *Sendai* were explained in a footnote #3 of Section A1.3.2(B)). Then, Shin-ran may have asked [4], "If the ability of *Amida* Buddha is infinite, why are *Go-gyaku*, *Hibo* and *Sendai* excluded? If *Amida* Buddha could not give any help to such miserable people, *Amida* Buddha could no longer be called Buddha." Shin-ran's worries reached an extreme level.

The third step toward Satori – The attainment of Satori

It is unknown how long Shin-ran's worries continued. Probably, a solution suddenly appeared. Shin-ran had long been asking where Buddha was. Then, he abruptly reached a wholly clear state of mind and noticed that *Amida* Buddha always worked within him. He got a completely unexpected answer. How much his surprise was and also how much his delight was, at this moment.

> *All my worries and agonies have come from Amida Buddha. I have completely misunderstood the ability of Amida Buddha. Amida Buddha has given me plenty of illusions (knowledge) and Bon-noh and made me have serious doubts and problems. Through them, Amida Buddha has guided me to an absolutely difficult situation (the utmost limit of my ability) and finally conducted me to a wholly clear state of mind via the jump. I have been thoroughly guided by Amida Buddha from the beginning to the end.*

This conclusion of Shin-ran is clearly described in the volume of *Sho* of his book, "*Kyo Gyo Shin Sho*" [5,6].

> *When I consider the processes of attaining Satori according to Shin-shu, I clearly recognize that they are wholly due to activities of Amida Buddha. Therefore, all causes and results in the processes come from the ability of Amida Buddha.*

Here, Shin-shu refers to *Jodo-shin-shu* (the true *Jodo* teachings) which Shin-ran set up.

Shin-ran had long been asking where Buddha was, what *Nen-butsu* was, and what was meant by a sincere and enthusiastic mind. In other words, he had long been asking how Buddha worked in this world and how Buddha conducted people to *Satori*. Then, he was finally driven into an absolutely difficult situation. This was natural because Buddha, *Nen-butsu* and a sincere and enthusiastic mind were all beyond the bounds of human consciousness, as will be explained later. Shin-ran had long been tackling problems that were impossible to solve within the realm of word-based understanding, in the same way as Shaka-muni (see Section A1.2.1 and A1.2.2). Then, he achieved the jump in his understanding and attained *Satori*.

We can say in the following way as well. Before Shin-ran attained *Satori*, he understood *Ji-riki* (his own ability in the external visible world) and *Ta-riki* (Buddha's ability in the internal invisible world) individually and separately. For example, when he asked where Buddha was, he looked at Buddha in front of him. Here was clearly separation between *Ji-riki* and *Ta-riki*. Actually, not only Shin-ran but also all Buddhist monks or all people in those years understood *Ji-riki* and *Ta-riki* individually and separately. In reality, however, *Ji-riki* and *Ta-riki* were not separated from each other. Therefore, Shin-ran was finally led to an absolutely difficult situation and then led to the very truth:

Ji-riki equals Ta-riki

Indeed, Shin-ran attained *Satori* by their own ability (*Ji-riki*) when his behavior was looked at from the outside. However, in reality, he attained *Satori* under the control of Buddha's ability (*Ta-riki*).

A2.3 The Correct Way to Attain Satori

Shin-ran discovered the correct way to attain Satori

Now, let us consider characteristics and meanings of Shin-ran's *Satori*. The first important conclusion is that Shin-ran discovered the correct way to attain *Satori*. A way to attain *Satori*, which Shin-ran discovered, was already described at the end of the preceding section. It can be rewritten in the following systematic way. Shin-ran did not describe his conclusion in such a systematic way. However, he really reached such a conclusion, as will be explained later.

> 1. *Amida Buddha makes us possess heavy illusions (incorrect views arising from word-based understanding) and intense Bon-noh (desires and attachments for individual things) and lets us show lively activity in this world.*
> 2. *Amida Buddha then lets us become aware of limits of our*

knowledge and ability. Namely, Amida Buddha lets us feel interest, unclearness, unease, doubt, worry, agony, etc. and leads us to carry out thorough consideration and pursuit about the origins of such feelings.
3. Amida Buddha finally leads us to an absolutely difficult situation (the utmost limit of our ability, the boundary between word-based understanding and the very truth or the boundary between this world and Buddha's land) and furthermore makes us leap over the limit of our ability and come to Buddha's land, in which all things are continuously connected with one another and in a peaceful harmonious state. Thus, the moment we have come to Buddha's land, we achieve a wholly clear state of mind and attain Satori.

Note that the above working of *Amida* Buddha is essentially the same as that of the grand internal ability discussed in Section 1.4.1 and 2.1.3, indicating that Buddha's ability refers to the grand internal ability.

Shin-ran's way to attain *Satori* or to arrive at the very truth, described above, can be expressed in the following way if it is viewed from the side of human consciousness.

1. We human beings by nature have desires for various individual things such as existence, living, wealth, knowledge, fame, love, etc. and show lively activity in this world.
2. Through such activities, we human beings come to feel interest, unclearness, unease, doubt, worry, agony, etc. and come to carry out thorough consideration and pursuit about the origins of such feelings.
3. Such thorough consideration and pursuit finally lead us to an absolutely difficult situation (the utmost limit of our ability), at which we suddenly and unexpectedly attain Satori and arrive at the very truth, i.e. we create a new idea and achieve a wholly clear state of mind full of delight.

Shin-ran's way to attain *Satori*, mentioned above, consists of three steps. We can express them simply as follows.

Learning and activity, doubt-pursuit, and the jump

This way to attain *Satori* or arrive at the very truth is the same as the way to achieve creation, discussed in Section 6.2. This is natural because all discussions in this book are given based on Shin-ran's way to attain *Satori*, as already stated in Section 1.4.3.

Here, let us examine whether Shin-ran's way to attain *Satori* is really correct. As already mentioned in Section A2.1, Shin-ran investigated how he was able to attain *Satori* and attained *Satori*. This meant that he reached the very truth about how to attain *Satori*. Here is why we can say that Shin-ran's way to attain *Satori* is correct. Interestingly, Shin-ran's way to attain *Satori* (the three steps) is entirely different from *Hassho-do* in Initial Buddhism (Section A1.2.2) and *Roku-haramitsu* in Mahayana Buddhism (Section A1.3.1), both of which were artificially invented by Buddhist monks.

The correctness of Shin-ran's way to attain *Satori* is also given by the fact that Shin-ran himself attained *Satori* by the above three steps. He at first learned the traditional teachings of Mahayana Buddhism, then felt doubt about them, carried out thorough consideration and pursuit, reached an absolutely difficult situation, and achieved a wholly clear state of mind via the jump.

Moreover, not only Shin-ran but also many other Buddhist monks attained *Satori* by the above three steps. For example, Shaka-muni attained *Satori* by similar three steps, as discussed in Section A1.2.2. He initially lived a happy life but came to feel intense mental pain to see the death of a person. Then, he carried out thorough consideration and pursuit, fell in an absolutely difficult situation, and attained *Satori* via the jump. As another example, it is said [9] that a Chinese *Zen* monk, Rinzai, who established *Rinzai-shu*, attained *Satori* by similar three steps. When he learned Buddhism under the supervision of his master, Oh-baku, he was called and asked by the master, "What is the essential meaning of Buddhism?" Rinzai was not able to answer the question at once. He then carried out thorough considerations and finally reached a wholly clear state of mind and attained *Satori*. Moreover, a Japanese *Zen* monk, Dogen, who lived a little later than Shin-ran and established *Sodo-shu*, also attained *Satori* in a similar way [3]. While he learned Buddhism in Mt Hi-ei, he faced a question of "The Buddhist teachings say that everybody has *Bussho* (the nature of Buddha). If so, why does a Buddhist monk have to carry out hard Buddhist practice to become a Buddha?" He carried out thorough consideration and attained *Satori*.

Furthermore, the correctness of Shin-ran's way to attain *Satori* is given by theoretical consideration as well, as will be discussed later.

The discovery of a positive role of illusions and Bon-noh

A prominent feature of Shin-ran's way to attain *Satori* is that it revealed a positive role of illusions (incorrect views arising from word-based understanding) and *Bon-noh* (desires and attachments for individual things), both of which the traditional teachings of Mahayana Buddhism in Shin-ran's age taught people to transcend or eliminate intention- ally by their own ability (*Ji-riki*). Shin-ran reached the completely oppos- ite conclusion to the traditional teachings.

Shin-ran and the traditional teachings of Mahayana Buddhism had a common thought in that the mind of the common people was strongly covered with heavy illusions and intense *Bon-noh*, from which serious worries and agonies arose. Then, the traditional teachings taught people to transcend or eliminate such illusions and *Bon-noh* by their own ability (*Ji-riki*). On the other hand, Shin-ran thought in a different way. Indeed, illusions and *Bon-noh* cause worries and agonies. However, if a person feels worries and agonies, he or she will unawares make effort to overcome them and finally come to an absolutely difficult situation (the utmost limit of his or her own ability), at which he or she can attain *Satori*. In fact, Shin-ran himself attained *Satori* in this way. Thus, illusions and *Bon-noh* play an important positive role in attaining *Satori*. If they are eliminated, a person loses a clue to arrival at the utmost limit of his or her ability and hence loses a clue to the attainment of *Satori*.

The clarification of a positive role of illusions and *Bon-noh* had an indescribably important meaning. It enabled people to attain *Satori* or arrive at the very truth in daily life full of illusions and *Bon-noh* without carrying out any special Buddhist practice such as *Roku-haramitsu*. *In Shin-ran's thought, activities in daily life are nothing else but practice for attaining Satori.* By this thought, the philosophy of Buddhism was made wholly effective in all areas of human activity including science. The traditional teachings of Mahayana Buddhism were entirely incompatible with scientific understanding because they claimed that word-based understanding is of an imaginary character and should not be relied on.

The discovery of the way to attain Satori by Ta-riki

Why did such a great difference emerge between Shin-ran's thought and the traditional teachings of Mahayana Buddhism? This is because Shin-ran discovered a new way to attain *Satori*. Mahayana Buddhism reached the key conclusion that all people were able to attain *Satori* by the ability of Buddha (*Ta-riki*). However, it had not yet clarified *how* (or via what processes) all people were able to attain *Satori* by the ability of Buddha, as mentioned in Section A2.1. For this reason, people were actually unable to attain *Satori* by the ability of Buddha and had no way but to rely on hard Buddhist practice such as *Roku-haramitsu* by *Ji-riki*. When people were placed in such a situation, *Shin-ran* really succeeded in attaining *Satori* by the ability of *Amida* Buddha and proved that all people were able to attain *Satori* by the ability of Buddha. Shin-ran was the first person to attain *Satori* based on the key conclusion of Mahayana Buddhism or based on the *Jodo* teachings.

Let us look back again on how Shin-ran attained *Satori*. He initially made strenuous effort to attain *Satori* according to the traditional teachings of Mahayana Buddhism. However, he was unable to attain *Satori*. Then, he next made earnest effort to attain *Satori* according to *Hoh-nen's* teachings, but he was again unable to attain *Satori*. Thus, Shin-ran was finally driven into an absolutely difficult situation. However, strangely, when he was in an absolutely difficult situation, he suddenly and unexpectedly attained *Satori*. Shin-ran will have been surprised to look at such a result. By this experience, Shin-ran will have noticed that there was an essentially new way to attain *Satori*, which was quite different from the traditional way to attain *Satori* by *Ji-riki*. In addition, he will have noticed that this was just the way to attain *Satori* by *Ta-riki*, which Hoh-nen taught him. This is because he was unawares driven into an absolutely difficult situation and unawares attained *Satori* there, without carrying out any Buddhist practice.

Sangan Tennyu (Satori via three gates) [4-6)]

Shin-ran explains his career in his book, "*Kyo Gyo Shin Sho*" [5,6)], as follows. This explanation is well known as *Sangan Tennyu* (*Satori* via three gates).

> *I first entered into an incorrect gate in which I carried out hard Buddhist practice by my own ability. Then, I entered into another incorrect gate in which I continued Nen-butsu by my own ability. Finally, I have entered into the correct gate in which I am wholly conducted by Amida Buddha from the beginning to the end.*

The first incorrect gate refers to the traditional teachings of Mahayana Buddhism in the Enryaku-ji Temple, while the second incorrect gate refers to Hoh-nen's teachings. The final correct gate stands for *Jodo-shin-shu* (the true *Jodo* teachings) Shin-ran set up. In the first and the second gates, *Ji-riki* and *Ta-riki* are separated from each other. In the final gate, the very truth of "*Ji-riki equals Ta-riki*" is realized.

Interestingly, Shin-ran says that *Amida* Buddha has prepared two incorrect gates using *Ji-riki* as expedient gates in order to lead people to the correct gate. This is an ingenious way of *Amida* Buddha. The incorrect gates are necessary because people, who have illusions and *Bon-noh*, always want to solve problems by their own ability (*Ji-riki*). If people enter into incorrect gates, they will soon notice that their ability has limits. Then, they are necessarily guided to the correct gate. In Shin-ran's thought, *Hassho-do* and *Roku-haramitsu* are expedient gates to guide people to the correct gate.

Akunin Shouki (Just evil persons are on the correct path toward Satori)

Shin-ran's way to attain *Satori* has a prominent feature in that it has clarified a positive role of illusions and *Bon-noh*, as mentioned earlier. This feature is clearly expressed in Shin-ran's famous idea of "*Akunin Shouki*", which says that not good persons but evil persons are on the correct path toward *Satori*. The third paragraph of *Tan-i-sho* [7] says as follows.

> *Even a good person can attain Satori, much more an evil person. However, it is usually said that even an evil person can attain Satori, much more a good person. The latter opinion looks reasonable but it is in disagreement with the thought of "Hongan Ta-riki" (i.e. the thought that a person attains Satori by the ability of Amida Buddha). The reason is that a good person who accumulates worthy deeds by his or her own ability completely lacks the idea of relying on Amida Buddha and therefore is off the aim of Amida Buddha. Amida Buddha has felt deep compassion to persons who can by no means escape from illusions and Bon-noh and has taken the vow of helping such persons. Therefore, just evil persons who heartily rely on Amida Buddha are on the correct path toward Satori.*

The idea of *Akunin Shouki* expresses the essence of Shin-ran's thought. Unfortunately, it has been misunderstood in various ways. In my opinion, the idea of *Akunin Shouki* is understood as follows.

It is certain that the idea of *Akunin Shouki* came from Shin-ran's experience. The kernel of his experience is that he was driven into an absolutely difficult situation and suddenly and unexpectedly attained *Satori*. By this experience, Shin-ran clearly recognized that just an absolutely difficult situation was a place at which a person was able to attain *Satori*. Furthermore, he recognized that only an "evil" person, who had heavy illusions and intense *Bon-noh*, was able to come to such an absolute difficult situation. Thus, not "good" persons but "evil" persons are on the correct path toward *Satori*.

The above understanding can be explained as follows. At first, let us consider the statement that an absolutely difficult situation is a place at which a person can attains *Satori*. An important point to be noted here is that there is a non-removable gap between word-based understanding and the very truth (see Section 1.2.1). Therefore, a person cannot attain *Satori* or arrive at the very truth without transcending his or her word-based understanding. However, it is very difficult or nearly impossible for a person to transcend his or her word-based understanding by his or her own ability (*Ji-riki*). This is because a person's word-based understanding is nothing else but his or her own ability (*Ji-riki*). It is in principle impossible for a person to transcend his or her own ability by his or her own ability. Moreover, a person in general has a strong attachment to his or her word-based understanding because it is nothing else but his or her own ability, as mentioned above.

Now, how can a person attain *Satori*? There is one possibility. A person can transcend his or her word-based understanding when he or she has arrived at the utmost limit of it, i.e. when he or she has fallen into an absolutely difficult situation. In such a situation, a person no longer has anything to do by his or her own ability (*Ji-riki*). Thus, a person has no attachment to it.

An absolutely difficult situation has another important meaning. In such a situation, a person's ability reaches the utmost limit, as mentioned above. This means that a new internal possibility, which is superior to a person's original ability, is formed and becomes dominant, as discussed in Section 6.2 and 6.3. With no formation of a new internal possibility, everything is felt natural and reasonable and nobody goes to an absolutely difficult situation. Therefore, when a person has fallen into an absolutely difficult situation, he or she has a largely-growing new internal possibility, i.e. he or she has "something internal" which strongly attracts his or her mind. Accordingly, a person in an absolutely difficult situation can easily jump to such a new internal possibility. This means that a person can easily attain *Satori* in such a situation.

In harmony with the above considerations, interesting words are recorded at the top of the first paragraph of *Tan-i-sho*. The words represent a situation at which a person is about to attain *Satori*.

> *As soon as you are about to call Amida Buddha by name with the belief that you can necessarily attain Satori with the aid of unaccountable merciful Buddha's ability, you are immediately promised to attain Satori.*

The words of "with the belief that you can necessarily attain *Satori* with the aid of unaccountable merciful Buddha's ability" can be interpreted as "with something internal, which strongly attracts your mind". On the other hand, the words of "As soon as a person is about to call *Amida Buddha* by name" can be interpreted as "the moment a person is about to jump to something internal".

A next issue to be clarified is who can come to an absolutely difficult situation. Shin-ran's answer is that just an "evil" person who can by no means escape from illusions and *Bon-noh* can come to such a situation, as mentioned earlier. This answer can be explanied as follows. A "good" person is one who simply understands and accepts the traditional teachings and obediently follows them. Such a person believes that everything can be controlled within the realm of *Ji-riki* (or understanding by words). Namely, such a person has not yet become aware that there are many problems that are impossible to overcome within the realm of *Ji-riki*. Therefore, such a person does not come to feel serious doubt about the traditional teachings and hence does not go to an absolutely difficult situation.

The situation is quite different for an "evil" person who can by no means escape from illusions and *Bon-noh*. For example, Hoh-nen and Shin-ran sincerely desired to attain *Satori* and carefully investigated the Buddhist teachings and then came to feel doubt about the effectiveness of the traditional way to attain *Satori*, as discussed in Section A2.2. We can say that the continuation of sincere desires leads to the emergence of a feeling of doubt. Originally, the concept of desire appears to tacitly include the concept of doubt. It should be noted here that other Buddhist monks in Hoh-nen's and Shin-ran's age were not able to feel such doubt. This is not because they did not desire to attain *Satori*. This is because their desires were not so sincere that they exceeded the mind to rely on the traditional teachings. A strong mind is needed to notice a limit of authorized traditional teachings.

Once a person has felt doubt, he or she necessarily comes to an absolutely difficult situation as far as he or she does not hide or disregard it because such doubt comes from intrinsic limits of existing knowledge or things. In fact, Shin-ran was driven into an absolutely difficult situation.

Shin-ran's understanding exceeded the level of sutras of Mahayana Buddhism

Based on the above discussions, let us consider Shin-ran's idea of *Akunin Shouki* in more detail. In my opinion, this idea indicates that *Shin-ran*'s understanding exceeded the level of sutras of Mahayana Buddhism. In fact, no such idea as *Akunin Shouki* is described in *Muryoju-kyo* and other sutras of Mahayana Buddhism. This idea represents Shin-ran's original understanding of Buddha's wisdom and mind. As mentioned in Section A2.1, Shin-ran overcame all problems that the traditional teachings of Mahayana Buddhism had faced. Therefore, it is natural that *Shin-ran*'s understanding exceeded the level of the traditional teachings of Mahayana Buddhism.

The kernel of Shin-ran's *Satori* is that he was driven into an absolutely difficult situation and suddenly and unexpectedly attained *Satori*, as mentioned earlier. Before he attained *Satori*, he was led to the severe conclusion that he had no way to reach a state in which he absolutely relied on *Amida* Buddha, i.e. he was unable to attain *Satori* according to the *Jodo* teachings (see Section A2.2). Nevertheless, he suddenly and unexpectedly attained *Satori* when he was in an absolutely difficult situation. Thus, by this experience, Shin-ran must have clearly recognized that *Nen-butsu* (i.e. the situation in which a person absolutely relies on *Amida* Buddha) is originally impossible to achieve intentionally by *Ji-riki*. *Nen-butsu* is given by *Ta-riki* when a person has arrived at an absolutely difficult situation. In this way, Shin-ran gained the truly correct understanding of *Nen-butsu* and was wholly convinced of the words in the 18th item of *Amida* Buddha's vow.

Now, it was "evil" persons who were by no means able to escape from illusions and *Bon-noh* that came to an absolutely difficult situation. Shin-ran watched this fact clearly and then concluded that *Amida* Buddha's mind lay in helping such "evil" persons. The idea of *Akunin Shouki* was an expression of this conclusion. Here, if some words are added, an "evil" person is one who displays lively activity in the real world for overcoming various severe problems with high ideals. In fact, such a person can by no means escape from illusions and *Bon-noh*. In addition, such a person is also one who can heartily rely on Buddha because he or she is conducted so as to become aware that the world is full of problems that are impossible to overcome within the realm of *Ji-riki* (word-based understanding).

Satori is attained by Ta-riki, not by Ji-riki

The considerations of the idea of *Akunin Shouki* given above made the meaning of Shin-ran's way to attain *Satori* much clearer. Then, let us consider it again in further detail.

Firstly, a person attains *Satori* by Buddha's ability (*Ta-riki*), not by his or her own ability (*Ji-riki*). Note that the attainment of *Satori* by *Ta-riki* never means that a person can attain *Satori* only by *Ta-riki* without using *Ji-riki*. It only means that all processes of attaining Satori, even if they look as if they are driven by *Ji-riki*, are completely controlled by

Buddha's ability (*Ta-riki*) on an unconscious level.

The traditional teachings of Mahayana Buddhism have taught people the attainment of *Satori* by *Ji-riki*. They have overlooked the underlying activity of Buddha, only looking at human activities in the external visible world. Shin-ran first revealed the important role of the underlying activity of Buddha. Thus he stresses that the attainment of *Satori* is of an entirely strange character. He states in his book, "*Kyo, Gyo, Shin, Sho*", and in *Tan-i-sho* as follows.

> *No planning is planning in Nen-butsu. Nen-butsu is a non-callable, unexplainable and unaccountable thing.*

The words indicate that the accomplishment of *Nen-butsu* or the attainment of *Satori* by *Nen-butsu* is far beyond a person's own ability (*Ji-riki*), as mentioned earlier. Shin-ran also says in his book, "*Kyo Gyo Shin Sho*", as follows [4].

> *Buddha is everywhere in this world but it is difficult to meet Him.*

Meeting Buddha means attaining *Satori* [4]. *Amida* Buddha works in this world but is shapeless and invisible and therefore it is difficult to meet Him.

In relation to the above argument, there is one thing to be noted. Certainly, a person attains *Satori* by *Ta-riki*, as mentioned above, but this does not mean that a person who follows the traditional way to attain *Satori* by *Ji-riki* cannot attain *Satori*. This is because even such a person unawares moves to the way to attain *Satori* by *Ta-riki* on a path toward *Satori*. In fact, Shin-ran himself initially followed the traditional way to attain *Satori* by *Ji-riki* and at the last stage moved to the way to attain *Satori* by *Ta-riki*. In Shin-ran's thought, the traditional way to attain *Satori* by *Ji-riki* is an expedient gate to guide people to the correct way to attain *Satori*, as mentioned earlier.

A person who can feel interest or doubt about existing knowledge or things can attain Satori

Secondly, a person who can feel interest, unclearness, unease, doubt, etc. about existing knowledge or things can attain *Satori*, irrespective of whether or not he or she knows the correct way to attain *Satori*. This is because a person who has felt such a feeling necessarily comes to an absolutely difficult situation as far as he or she does not hide or disregard it, as mentioned earlier.

In fact, Shaka-muni attained *Satori* by feeling unendurable unease about human life. New Buddhists in Mahayana Buddhism in the primary stage were in the midst of clarifying the very truth about nature and the world (such as the generalized principle of *Engi*) and were unawares able to feel unclearness or doubt about existing teachings. Therefore, they were able to attain *Satori*.

It should be noted here that the above argument implies that the attainment of *Satori* becomes difficult when existing teachings or knowledge has come to have a completed form. This is because in such a situation it is difficult to feel unclearness, unease, doubt, etc. about existing teachings or knowledge. In fact, it was mentioned in Section A2.2 that almost no Buddhist monk in the Enryaku-ji temple had attained *Satori* in the age of Hoh-nen and Shin-ran. This was most probably because the traditional teachings of Mahayana Buddhism in that age had a completed form on their own level. A similar situation is seen in Buddhist monks in Branch Buddhism because the teachings of Initial Buddhism had a completed form in the age of Branch Buddhism. Furthermore, a similar situation may also be seen in Mahayana Buddhism in the middle and last stages.

Illusions and Bon-noh are intrinsic to human beings

Thirdly, illusions (incorrect views arising from word-based understanding) and *Bon-noh* (desires and attachments for individual things) are given by *Amida* Buddha and are inherent in human beings. This means that *all* people can attain *Satori*, in harmony with the key conclusion of Mahayana Buddhism, because just a person who can by no means escape from illusions and *Bon-noh* can attain *Satori* in Shin-ran's thought, as mentioned earlier.

Shin-ran clearly recognized by his experience in the Enryaku-ji temple that he was never able to remove illusions and *Bon-noh* by his own ability (*Ji-riki*). He also saw in the Enryaku-ji temple that nobody was able to escape from illusions and *Bon-noh* by *Ji-riki*. In fact, most of Buddhist monks there were enthusiastic about promoting themselves to higher ranks in the organization of Buddhism monks. Furthermore, when he had achieved a wholly clear state of mind and attained *Satori*, he noticed that all worries and agonies of him came from *Amida* Buddha. This clearly indicated that illusions and *Bon-noh* were given by *Amida* Buddha.

If illusions and *Bon-noh* are given by *Amida* Buddha and are inherent in human beings, it is natural that a person holds them even after he or she attained *Satori*, as argued in Section A2.1. In fact, the following words are recorded in the ninth paragraph of *Tan-i-sho*[7].

Yui-en (one of Shin-ran's followers) asked Shin-ran, "I am already given belief in Amida Buddha and promised to come to Buddha's splendid land when I leave this world. However, I still desire to live in this world and do not want to come to Buddha's land. Why?" Shin-ran answered, "I have the same question. This is certainly because we have Bon-noh. I feel easy to know that I have Bon-noh, because this indicates that Amida Buddha, who has felt deep compassion to persons who can by no means escape from Bon-noh, necessarily helps me. If I had no Bon-noh, I would wonder whether or not Buddha helps me and feel uneasy."

The words clearly indicate that *Shin-ran* recognized that a person has illusions and *Bon-noh* even after attaining *Satori*.

A2.4 The Way to Live in This World in Accord with the Very Truth

Shin-ran's discovery of the correct way to attain *Satori* led to a great revolution in the philosophy of Buddhism. As mentioned in Section A2.1, it led to the disclosure of the correct idea of the state of *Satori* and the correct idea of the very truth (the way that truly real things exist or natural things themselves exist). In this section we consider what correct ideas Shin-ran disclosed.

Oh-so and Kan-so

At first, let us consider what basic view of nature and the world Shin-ran had. A prominent feature of Shin-ran's way to attain *Satori* is that illusions and *Bon-noh* play an important positive role, as mentioned in the preceding section. Therefore, our life (or the world of *Ji-riki*) full of illusions and *Bon-noh* is located at the *equivalent* position to Buddha's land filled with the grand wisdom (*Chi-e*) and the grand compassion (*Jihi*). Thus, Shin-ran's philosophy is constructed based on the dual structure of nature (see Section 1.4.2) or based on the concepts of *Hosshin* and *Ohjin* (see Section A1.3.1).

Another important feature is that we human beings and *Amida* Buddha live in entirely different worlds. We human beings always live in this world because we cannot escape from illusions and *Bon-noh*. On the other hand, *Amida* Buddha always lives in Buddha's land because it is the internal ability.

Therefore, the first issue to be considered is how we human beings are connected with *Amida* Buddha. Shin-ran says at the top of his book, "*Kyo Gyo Shin Sho*", as follows.

> In Jodo-shin-shu (the true Jodo teachings), there are two kinds of Buddha's Eko (Buddha's acts). One is Oh-so (the movement of a person from this world to Buddha's land) and the other is Kan-so (the return of a person from Buddha's land to this world).

The terminology of *Oh-so* and *Kan-so* was first adopted by a Chinese Buddhist monk, Don-ran (see Section A1.4). An outstanding feature of Shin-ran's thought is that both *Oh-so* and *Kan-so* are conducted by *Amida* Buddha. On the other hand, in the traditional *Jodo* teachings, *Oh-so* was understood to be achieved by a person's own ability (*Ji-riki*). As argued earlier, Shin-ran clearly recognized that *Nen-butsu* was impossible to achieve intentionally by *Ji-riki* and was given by *Amida* Buddha at an absolutely difficult situation. Therefore, all paths toward *Satori* are governed by *Amida* Buddha.

According to Noma's book [4], *Oh-so* and *Kan-so* are explained as follows. A person at first goes from this world to Buddha's land. Then, a person meets Buddha at Buddha's land and attains *Satori*. This is *Oh-so*. If a person has attained *Satori* and become a Buddha, the person immediately returns to this world and works as *Ohjin* (i.e. a Buddha who works in this world to conduc all people to *Satori*, see Section A1.3.1). This is *Kan-so*.

The above processes can be explained based on the dual structure of nature discussed in Section 1.4.2 as follows. When a person has come to Buddha's land via the jump, the person immediately takes a continuous internal connection and achieves a wholly clear state of mind. This is because Buddha's land is filled with continuous internal connections. If a person has taken a continuous internal connection, i.e. has formed a continuous internal connection, the person immediately returns to this world. This is because a person who has formed a continuous internal connection produces a thing that realizes a free independent harmonious state, i.e. produces a thing with individual unchanging properties in an averaged or common form, which cannot exist in Buddha's land.

In this way, *Oh-so* and *Kan-so* occur in an instant. In addition, *Oh-so* and *Kan-so* proceed on an unconscious level. Thus, when we have experienced *Oh-so* and *Kan-so*, we cannot be conscious of the processes. We only experience a sudden change from a state of mind full of worries and agonies to a wholly clear state of mind, together with the creation of a new idea or thing.

Kyo, Gyo, Shin, and Sho

Shin-ran next proposes the concepts of *Kyo*, *Gyo*, *Shin* and *Sho* to explain *Oh-so*. He simply says as follows.

> *In Buddha's Eko of Oh-so, there are Kyo (teaching), Gyo (practice), Shin (belief), and Sho (proof) of the true meaning.*

The words mean that *Oh-so* (the attainment of *Satori* by the aforementioned three steps of *learning and activity, doubt-pursuit, and the jump*) is accomplished by the concepts of *Kyo*, *Gyo*, *Shin* and *Sho*. Here, "the true meaning" implies that all these concepts are given by *Amida* Buddha. In Shin-ran's thought, all paths about *Satori* are governed by *Amida* Buddha, as mentioned earlier.

Kyo refers to the teachings of *Muryoju-kyo*, which say that *Amida* Buddha with infinite ability and infinite life always works to conduct all people to *Satori*. Shin-ran attained *Satori* according to the *Jodo* teachings, as mentioned earlier. Namely, he attained *Satori* by knowing the teachings of *Muryoju-kyo*. If it were not for the teachings, he would not have been able to ask where Buddha was and thus would not have been able to attain *Satori*. This explains why *Kyo* is necessary in *Oh-so*. Naturally, *Kyo* is given by *Amida* Buddha.

Gyo refers to the practice of "calling *Amida* Buddha by name". In Shin-ran's thought, this practice is not carried out for asking *Amida* Buddha for His favor. This practice is carried out to express praises and thanks to *Amida* Buddha. In Shin-ran's thought, people are always conducted by *Amida* Buddha so that they attain *Satori*. Therefore, people express their praises and thanks to *Amida* Buddha. In fact, when Shin-ran achieved a wholly clear state of mind, he recognized that he had been entirely conducted by *Amida* Buddha from the beginning to the end. Therefore, he sincerely expressed his praises and thanks to *Amida* Buddha. Thus, the practice of *Gyo* is given by *Amida* Buddha.

Shin refers to belief in *Amida* Buddha. In Shin-ran's thought, *Shin* is also given by *Amida* Buddha, in the same way as *Gyo*. *Shin* (belief) in Shin-ran's thought never means usual intentional belief for asking *Amida* Buddha for His favor. When Shin-ran achieved a wholly clear state of mind, he clearly recognized that he had been entirely conducted by *Amida* Buddha from the beginning to the end, as mentioned above. Then, he was given belief in (the existence and the splendid abilities of) *Amida* Buddha.

Sho refers to the proof of the attainment of complete *Satori*. The proof is given by the achievement of a wholly clear state of mind, as argued thus far. Shin-ran achieved a wholly clear state of mind when he was in an absolutely difficult situation. This meant that a wholly clear state of mind, i.e. *Sho* was given by *Amida* Buddha.

From these explanations, we can see that the kernel of the concepts of *Kyo, Gyo, Shin* and *Sho* is *Kyo* (the teachings of *Muryoju-kyo* that *Amida* Buddha always works to conduct all people to *Satori*). In fact, if a person acquires the truly correct understanding of *Kyo*, the person is immediately given *Gyo, Shin* and *Sho*. Accordingly, Shin-ran emphasizes the importance of gaining the truly correct understanding of *Kyo* or the truly correct understanding of *Amida* Buddha, which is the key entity of *Kyo*. In Shin-ran's thought, acquiring the truly correct understanding of *Amida* Buddha, i.e. being given belief in *Amida* Buddha is equal to attaining *Satori*.

The concepts of *Kyo, Gyo, Shin* and *Sho* in Shin-ran's thought play the same role as *Roku-haramitsu* in the traditional teachings of Mahayana Buddhism (see Section A1.3.1). Just as the traditional teachings assert that *Satori* is attained by the completion of the six items of *Roku-haramitsu*, so Shin-ran affirms that *Satori* is attained by the accomplishment of *Kyo, Gyo, Shin* and *Sho*. In contrast to *Roku-haramitsu*, the concepts of *Kyo, Gyo, Shin* and *Sho* involve no special Buddhist practice item. Nevertheless, Shin-ran says that the concepts of *Kyo, Gyo, Shin* and *Sho* are much superior to *Roku-haramitsu*. This is because these concepts are given by *Amida* Buddha[4], namely, they are given by Shin-ran's real experience of attaining *Satori*. On the other hand, *Roku-haramitsu* was only what new Buddhists in Mahayana Buddhism in ancient India artificially invented and had no firm theoretical basis.

The concepts of *Kyo, Gyo, Shin* and *Sho* lead to the correct idea of the state of *Satori*

Shin-ran's assertion that the concepts of *Kyo, Gyo, Shin* and *Sho* are much superior to *Roku-haramitsu* has another important meaning. The discovery of the correct way to attain *Satori* led to the disclosure of the correct idea of the state of *Satori* and the correct idea of the very truth, as mentioned earlier.

Shin-ran affirms that the attainment of *Satori* by the concepts of *Kyo, Gyo, Shin* and *Sho* leads a person to the center of Buddha's land because it is guided by *Amida* Buddha Himself[4]. Thus, such a person can realize a completely free peaceful harmonious state with no restriction. Shin-ran says that a person who has attained *Satori* by the concepts of *Kyo, Gyo, Shin* and *Sho* is absolutely free like a ship freely sailing in the great ocean [4]. On the other hand, the attainment of *Satori* through hard Buddhist practice such as *Roku-haramitsu* by *Ji-riki* only leads a person to a margin of Buddha's land [4]. Shin-ran says that a person who has attained *Satori* by *Roku-haramitsu* has to sit still endlessly at a state he or she attained [4]. There is a great difference in the state of *Satori* between *Satori* attained by *Kyo*, *Gyo*, *Shin* and *Sho* and that attained by *Roku-haramitsu*.

Why does such a great difference appear? It is not due to a difference in the way to attain *Satori* because a person always attains *Satori* by *Ta-riki*, as discussed in the preceding section. Such a great difference arises from a difference in the underlying philosophy of Buddhism, as already mentioned in Section A2.1. The traditional teachings of Mahayana Buddhism taught Buddhist monks to attain *Satori* by accomplishing the internal wisdom to clearly look at the principle of *Engi* throughout nature and the world and achieving a wholly clear state of mind completely released from illusions and *Bon-noh*. For this reason, a person who had attained *Satori* according to the traditional teachings had to stay still at a state he or she attained endlessly. On the other hand, Shin-ran revealed that a person who had attained *Satori* by the concepts of *Kyo*, *Gyo*, *Shin* and *Sho* was able to achieve a wholly clear state of mind without transcending or eliminating illusions and *Bon-noh*. Therefore, such a person was able to realize a completely free peaceful harmonious state with no restriction.

The state of Satori in the Zen teachings

For getting better understanding of the above argument, let us consider what state of *Satori* the *Zen* teachings teach, as an example. The *Zen* teachings are regarded as one of Buddhist schools following the traditional teachings of Mahayana Buddhism. In fact, the *Zen* teachings aim at directly catching the very truth about nature and the world (such as expressed by the words of *"shape equals emptiness, emptiness equals shape"* or *"all equal one, one equals all"* (see Section A1.3.2)) through meditation or mental control under a concentrated mind. Contrary to the true *Jodo* teachings (*Jodo-shin-shu*) of Shin-ran, the *Zen* teachings assert *Zettai Ji-riki* (absolutely relying on *Ji-riki*). The *Zen* teachings also emphasize *Furyu-monji* (never relying on words).

There are famous words of a Chinese *Zen* monk who lived in a quiet mountain, which are regarded as a good example of words expressing the state of *Satori* in the *Zen* teachings [10].

> *When I get hungry, I simply eat. When I become sleepy, I simply sleep.*

The words are regarded as a clear expression of the state in which *Kuh* (emptiness) is realized, i.e. the state in which a person has completely transcended desires and attachments for individual things and is absolutely free from them.

There is also a famous story about a Chinese *Zen* monk [9,10], which is also regarded as a good example of stories expressing the state of *Satori* in the *Zen* teachings. In the 13th century, Mongolian forces invaded into China. When a Chinese *Zen* monk, named *Mugaku Sogen*, continued *Zen* practice (meditation) in a temple, a Mongolian soldier entered into the temple and was about to kill him. At this moment, the *Zen* monk did not show any color of surprise in face and told the following poem with a calm and self-possessed attitude.

> *There is no opening in the heaven and on the earth, in which even a stick can be placed*
> *Pleasantly, a human being equals emptiness and everything equals emptiness*
> *I prize a three-feet sword of a Mongolian soldier*
> *Like a flash of lightening, it only cuts a spring wind.*

The poem represents "a completely free peaceful state of mind with no attachment to individual things", to which the *Zen* teachings attach the highest importance. It is said that a Mongolian soldier saw a highly self-possessed monk in the front of him and was unable to do anything. Then, he went away without doing anything.

There is a similar story in Japan, too. In the late 16th century, *Oda* forces invaded into the country of *Takeda* and fired a temple in which *Zen* monks continued meditation. Even when the fire became big, *Zen* monks did not leave the temple and continued meditation. Then, one of the monks said

> *If body and mind are in a completely extricated state, a fire is still cool.*

The monk really transcended all things in this world.

The above examples of the state of *Satori* in the *Zen* teachings indeed indicate that *Zen* monks had transcended individual things in their lives and achieved a completely free peaceful state of mind. However, the examples also indicate that *Zen* monks were not free from the restriction that they had to maintain "a completely free peaceful state of mind" in any situation. Namely, they were not free from the *Zen* teachings themselves. For this reason, they were unable to take any effective action when they may have lost their life. Certainly, a person who attained *Satori* in the *Zen* teachings had to sit still endlessly at the state of *Satori* he or she attained, as Shin-ran said. In other words, here was clearly an "attachment" to the idea of maintaining a completely free peaceful state of mind. This was evidently a big fault of the *Zen* teachings, which arose from the thought of *Zettai Ji-riki* (absolutely relying on *Ji-riki*).

The state of Satori in the true Jodo teachings (Jodo-shin-shu) of Shin-ran

Next, let us look at what state of *Satori* a person achieves when he or she has attained *Satori* by the concepts of *Kyo*, *Gyo*, *Shin* and *Sho*. Shin-ran says that such a person realizes a completely free harmonious state with no restriction, as mentioned earlier. How can we understand this assertion?

Shin-ran attained *Satori* according to the *Jodo* teachings and clearly recognized that the attainment of *Satori* was given by *Amida* Buddha. Therefore, he emphasized the importance of gaining the truly correct understanding of *Amida* Buddha, as mentioned earlier. Now, *Amida* Buddha has accomplished the high-level internal wisdom to freely control heavy illusions and intense *Bon-noh* and has achieved an *undisturbed* wholly clear state of mind in the presence of illusions and *Bon-noh*. In fact, *Amida* Buddha makes people possess heavy illusions (word-based understanding) and intense *Bon-noh*, lets them feel unclearness, unease or doubt, leads them to the utmost limit of their ability, and conducts them to *Satori* via the jump, as argued in the preceding section. Thus, *Amida* Buddha has achieved a completely free harmonious state with no restriction. Accordingly, if a person has acquired the internal ability of the same character as *Amida* Buddha and has been given belief in it, he or she can accomplish a completely free harmonious state with no restriction.

Then, how can a person acquire the internal ability of the same character as *Amida* Buddha and be given belief in it? Let us look at Shin-ran's *Satori* again. He had long been asking where Buddha was, what *Nen-butsu* was, how Buddha worked in this world and how Buddha conducted people to *Satori*. Thus, when he attained *Satori*, he clearly recognized that he had been entirely guided by *Amida* Buddha from the beginning to the end. In addition, he also clearly recognized that *Amida* Buddha conducted him to *Satori* by the aforementioned three steps of "*learning and activity, doubt-pursuit, and the jump*". Furthermore, when he attained *Satori*, he achieved a wholly clear state of mind full of delight. This meant that Shin-ran's mind and Buddha's mind completely agreed with each other. Therefore, we can clearly say that when Shin-ran attained *Satori*, he acquired the internal ability of the same character as

Buddha's ability [*1] and was given belief in Buddha.

> [*1] It should be noted that the very truth itself *Shin-ran* reached or the internal ability itself he acquired is of an invisible character and we cannot express it by words. We can correctly understand it when we have had essentially the same experience as Shin-ran.

In this way, Shin-ran demonstrated by his own experience that the attainment of *Satori* by the concepts of *Kyo*, *Gyo*, *Shin* and *Sho* leads to the acquisition of the internal ability of the same character as Buddha's ability and belief in it and allows a person to realize a completely free harmonious state with no restriction. In fact, in Shin-ran's thought, being given belief in Buddha is equal to gaining the truly correct understanding of *Amida* Buddha and is also equal to having the same mind as Buddha. Shin-ran says in his book entitled *Matto-sho* as follows [4].

> *A person, who is given belief in Amida Buddha, has the mind equal to Nyo-rai even though he or she has a disgraceful unclean body which may do evil.*

where *Nyo-rai* means a person who has come from the very truth and has the same meaning as Buddha.

Shin-ran's Satori is on the highest level in the history of Buddhism

There remains an important question. How can we explain theoretically that a person who has attained *Satori* by the concepts of *Kyo*, *Gyo*, *Shin* and *Sho* acquires the internal ability of the same character as Buddha's ability? As argued in Section 1.3.2 and 1.4.1, when a person has overcome a problem and achieved a wholly clear state of mind, he or she forms a continuous internal connection of various things (constituent elements) and simultaneously acquires the internal ability to control them so that they as a whole realize a free independent harmonious state. In other words, a thing existing as a continuous internal connection, which is produced by the grand internal ability, catches the grand internal ability within it and thus has the internal ability to organize itself so that it realizes a free independent harmonious state (see Section 1.4.1, 2.1.3, and so on). Therefore, when a person has overcome a diversity of problems, he or she acquires the internal ability in a diversity of quality levels and aspects.

Here, if a person has been asking how people and things in this world are controlled, in the same way as Shin-ran, he or she will form "the unified set of the internal ability formed in a diversity of quality levels and aspects" [*2]. This unified set is nothing else but the internal ability of the same character as Buddha's ability. In fact, Shin-ran had been asking where Buddha was or how Buddha conducted people to *Satori*, as mentioned earlier. For this reason, he was able to acquire the internal ability of the same character as Buddha's ability. It should be noted that such a unified set is formed only in a person who has been asking where Buddha is or how people and things in this world are controlled[*2]. In addition, a person who has been asking such a question cannot achieve a wholly clear state of mind until he or she has formed such a unified set even if he or she has formed a continuous internal connection in a diversity of quality levels and aspects.

> [*2]This statement is supported by the following argument. As argued in Section 4.4 (see also a footnote #1 of Section 6.2), a human being or in general a multi-cellular living organism has the internal ability to organize it itself so that it as a whole realizes a well-organized free independent harmonious state. Therefore, it is highly probable that a person forms "the unified set of the internal ability formed in a diversity of quality levels and aspects" when he or she asks how people and things in this world are controlled. Note also that a continuous internal connection and the internal ability have a relation similar to one between hen and egg.

The above consideration indicates that it is of key importance to ask how people and things in this world are controlled. On the other hand, Shaka-muni and new Buddhists in Mahayana Buddhism met the problem of how worries and agonies in human life were removed, i.e. how people and things in nature and the world existed. Similarly, *Tai-toku* (see Section 1.3.1) and theoretical creation in science (see Section 6.2) are achieved when we ask how people and things in nature and the world exist. In such a case, a clear state of mind is achieved by forming a continuous internal connection of various things (constituent elements), as discussed in Section 1.3.2, 1.4.1, A1.2.2 and so on. In other words, in such a case, a clear state of mind is achieved by catching "a result" of activities of Buddha's ability, as discussed in Section A2.1. In general, we human beings first ask how people and things exist and then ask how they are controlled.

Thus, we can safely say that Shin-ran's *Satori* is on the highest level in the history of Buddhism. It should be noted also that it was Shin-ran's *Satori* that proved the key conclusion of Mahayana Buddhism that Buddha's ability arrives everywhere and penetrates into all people. In addition, it was Shin-ran's *Satori* that proved the existence of Buddha itself.

Jinen Hohni (Living a natural life with belief in Buddha as the base)

The establishment of the concept of "belief in Buddha" led to the disclosure of an excellent way of living, called "*Jinen Hohni*"[4]. The latter half of the first paragraph of *Tan-i-sho* says as follows.

> *Hongan of Amida Buddha (the vow of Amida Buddha of conducting all people to Satori) neither distinguishes old persons and young ones nor distinguishes good persons and evil ones. Only Shin (belief in Amida Buddha, belief in Hongan) is the point. This is because the aim of Hongan is the relief of people full of intense Bon-noh and evil deeds. Accordingly, if you believe Hongan, other worthy deeds are not the point because there is no worthy deed which is superior to Shin. Also, you need not be fearful of any evil deed because there is no such heavy evil as hinders Hongan of Amida Buddha.*

Shin-ran's assertion is clear. Buddha's aim is to help persons with heavy illusions and intense *Bon-noh*. Therefore, for us human beings full of illusions and *Bon-noh*, just belief in Buddha is the point.

The superior features of a novel way of living which Shin-ran discovered are summarized in Section A2.1. Buddha is the internal ability to make all people and things in nature and the world and their constituent elements realize a fully dynamic free independent harmonious stable state (see Section A1.3.1). Therefore, Shin-ran asserts that belief in Buddha is the kernel of human life. If we have such belief, we can live a completely free *natural* life with no worry. In my opinion, Shin-ran's idea of *Jinen Hohni* refers to such a way of living.

Shin-ran's thought of *Hongan Ta-riki* (the thought that a person can attain *Satori* by the ability of *Amida* Buddha or the thought that human life is entirely governed by *Amida* Buddha) has often been misunderstood to date. In particular, this thought has often been incorrectly interpreted as recommending a passive way of living in which a person needs only rely on *Amida* Buddha or a person needs only pray to *Amida* Buddha. Such a misunderstanding arose from the fact that *Ji-riki* and *Ta-riki* were understood individually and separately. *Hongan Ta-riki* never denies *Ji-riki*, as emphasized in Section A2.3.

Hongan Ta-riki provides us with the firm theoretical basis for our life by *Ji-riki*. Just belief in Buddha allows us to live an undisturbed peaceful life. An important thing is not to pray to *Amida* Buddha but to acquire the internal ability of the same character as the ability of Buddha and embody it on our words and behavior.

To live as Ohjin

The establishment of the concept of "belief in Buddha" led to the revelation of not only an excellent way of living (*Jinen Hohni*) but also a splendid aim of human life (the aim of living as *Ohjin*). A person who has been given belief in *Amida* Buddha has the internal ability of the same character as Buddha's ability and hence lives as *Ohjin* (i.e. a person who has Buddha's wisdom and compassion and works in this world to conduct all people to *Satori*, see Section A1.3.1). The fourth paragraph of "*Tan-i-sho*" says as follows.

> *There is a difference in the meaning of Jihi (the grand compasssion) between the traditional teachings of Mahayana Buddhism and the true Jodo teachings. Jihi in the traditional teachings is that a Buddhist monk offers his "help and compassion" to the common people. However, it is difficult to offer sufficient "help and compassion" to satisfy the common people. Jihi in the true Jodo teachings is that a Buddhist monk, first of all, becomes a Buddha by Nen-butsu as soon as possible and favors the common people so nicely with Buddha's wisdom and compassion.*
> *Even if people feel how piteous and miserable the common people are, if they cannot give them sufficient help, this Jihi is meaningless. If so, to become a Buddha by Nen-butsu is just the greatest Jihi with no limit.*

Shin-ran recommends people to attain *Satori* as soon as possible and live as *Ohjin* with Buddha's wisdom and compassion.

It is to be noted also that the above words of Shin-ran say that *Jihi* in the traditional teachings has a limit. As mentioned in Section A2.1, the traditional teachings taught Buddhist monks to accomplish the internal wisdom to clearly look at the principle of *Engi* throughout nature and the world and then work to conduct the common people to *Satori*. This meant that the traditional teachings taught Buddhist monks to conduct the common people to *Satori* based on "a result" of activeities of Buddha. Therefore, the teachings had a limit. On the other hand, Shin-ran recommends Buddhist monks to catch Buddha's ability itself and then conduct people to *Satori* by it.

The disclosure of the new correct dynamic Buddhist view of nature and the world

Furthermore, the establishment of the concept of "belief in Buddha" led to the construction of the new correct dynamic Buddhist view of nature and the world. Originally, Mahayana Buddhism in the primary stage taught people the following strange logic (see Section A1.3.2), which indicates that nature and the world are in dynamic harmony.

Shape equals emptiness, emptiness equals shape

Hosshin equals Ohjin, Ohjin equals Hosshin

Bon-noh equal Bodai

"Life and death" equal Nehan

Many equal one, one equals many

Such strange logic was deduced from the principle of *Engi* generalized in Mahayana Buddhism (i.e. the very truth about nature and the world Mahayana Buddhism reached).

Unfortunately, Mahayana Buddhism in the primary stage failed to disclose the correct way to attain *Satori* and hence overlooked the very truth of

Ji-riki equals Ta-riki.

which Shin-ran disclosed. For this reason, the traditional teachings of Mahayana Buddhism made a mistake of teaching Buddhist monks to attain *Satori* by transcending or eliminating illusions and *Bon-noh*, as discussed in Section A2.1. Namely, they actually taught Buddhist monks "only emptiness", "only *Bodai*" and "only *Nehan*", instead of "shape equals emptiness", "*Bon-noh* equal *Bodai*" and "life and death equal *Nehan*".

Shin-ran revealed the correct way to attain *Satori* and constructed the philosophy of Buddhism which correctly expressed "shape equals emptiness", "*Bon-noh* equal *Bodai*" and "life and death equal *Nehan*". In fact, in Shin-ran's thought, illusions and *Bon-noh* play an important positive role in attaining *Satori*. Also, a person who has attained *Satori* still has illusions and *Bon-noh*. In this way, Shin-ran succeeded in constructing the new correct dynamic Buddhist view of nature and the world.

The same conclusion as the above is obtained from the following consideration as well. As mentioned earlier, Shin-ran's way to attain *Satori* consists of *Oh-so* and *Kan-so*. A person who has attained *Satori* does not remain in Buddha's land but immediately returns to this world and works as *Ohjin*. Thus, a person repeats *Oh-so* and *Kan-so* many times in his or her life. On the contrary, in the traditional teachings of Mahayana Buddhism, a person who has attained *Satori* remains in Buddha's land forever. An important point is that the repetition of *Oh-so* and *Kan-so* leads to improvement in wisdom (*Chi-e*) and compassion

(*Jihi*). Thus, the concepts of *Oh-so* and *Kan-so* give us an excellent explanation to how a human being can improve its ability and how natural things can improve their qualities. In Shin-ran's philosophy, the history of nature and the world can be regarded as the self-movement of Buddha's ability. This conclusion is in harmony with the discussions given in Chapter 1 to 6 if we remember that Buddha's ability is the same as the grand internal ability introduced in Section 1.4.1 and 2.1.3.

References

(1) H. Nakamura, M. Fukunaga, Y. Tamura, T. Konno, and F. Sueki (editors), *Encyclopedia of Buddhism, The Second Edition* (in Japanese), Iwanami-shoten, Tokyo, 2002.
(2) W. Hiromatsu, et al. (editors), *Encyclopedia of Philosophy and Thought* (in Japanese), Iwanami-shoten, Tokyo, 1998.
(3) Y. Takeuchi and T. Umehara (editors), *Buddhist Sutras in Japan* (in Japanese), Chuko-shinsho, Chuo-koron-sha, Tokyo, 1969.
(4) H. Noma, *Shin-ran* (in Japanese), Iwanami-shinsho, Iwanami-shoten, Tokyo, 1973.
(5) Shin-ran, *Kyo Gyo Shin Sho* (in Japanese), with comments by D. Kaneko, Iwanami-bunko, Iwanami-shoten, Tokyo, 1957.
(6) Shin-ran, *Kyo Gyo Shin Sho* (in Japanese and Chinese), annotated by G. Hoshino, M. Ishida, and S, Ienaga, "The Original Work The Philosophy of Japanese Buddhism 6", Iwanami-shoten, Tokyo, 1990.
(7) D. Kaneko, *Tan-i-sho* (in Japanese), Iwanami-bunko, Iwanami-shoten, Tokyo, 1981.
(8) H. Nakamura, K. Hayashima, and K. Kino (translators), *Jodo Sanbu Kyo (Three Sutras of Jodo teaching) (I) Muryoju-kyo*, Iwanami-bunko, Iwanami-shoten, Tokyo, 1963.
(9) D. Suzuki, *Zen Buddhism and its Influence on Japanese Cultures*, The Eastern Buddhist Society, Otani Buddhist College, Kyoto, 1938. Translated into Japanese by M. Kitagawa, Iwanami-shinsho, Iwanami-shoten, Tokyo, 1940.
(10) S. Kamata, *What Buddha Looked At* (in Japanese), Kodansha-gakujutsu-bunko, Kodan-sha, Tokyo, 1977.

A3. Human Understanding
– The Past, the Present and the Future –

In this chapter we look back on the history of human understanding and consider how it advances, how it is controlled by the internal ability, and where the final goal of it is. A main purpose is to clarify how we gain the truly correct understanding of nature and the world and what role word-based scientific understanding plays in it.

A3.1 Original Philosophies in Europe and East Asia

It is well known that quite different philosophies and cultures have been developed in Western and East Asian countries. Interestingly, investigations of original philosophies in these regions have revealed that such a large difference emerged in close relation to the features and the limits of word-based scientific understanding discussed in Chapter 1. In Western countries, the feature of word-based understanding of usefulness became important and thus science has made great progress. On the other hand, in East Asian countries, the intrinsic limit of word-based understanding of an approximate mechanical character became important and thus unique philosophies such as Confucianism, Taoism and Buddhism have made large progress. Human history clearly indicates that word-based understanding has features and limits.

A3.1.1 Original Philosophies in Europe

Philosophies and cultures in ancient Greece

Philosophies and cultures in ancient Greece are regarded as the origin of those in Europe. They had outstanding characteristics in that they demonstrated large advancements of an objective (scientific) view of nature and a rational way of thinking together with the exaltation of the spirit of praising intellectual and bodily power of human beings. Why did such characteristic philosophies and cultures emerge in ancient Greece?

A prominent feature of ancient Greece was that people in this region gained wealth by marine trade via the Mediterranean Sea. This region in the ancient age was very much blessed with geographical conditions for marine trade because the region was located close to areas with advanced civilizations such as Mesopotamia and Egypt and had the quiet Mediterranean Sea favorable for safe voyage. Another important feature was that people in ancient Greece were free from strong governmental control in contrast to people in Mesopotamia and Egypt. Thus, a large number of cities or city states were built near shores and people were able to display free and lively activity in marine trade with such cities as a base.

For gaining much wealth in marine trade, it must have been important to extend trade routes to unknown areas and find new producers and guests. Therefore, people in ancient Greece are expected to have lived a highly exploratory and adventurous life with a wide geometrical view of the world. In addition, people in this region must have paid keen attention to products and goods, methods for transporting them, and countries and their geography. Namely, they directed their eyes to "materials" such as goods, handcarts, ships, oceans, islands, and so on. Even cows and sheep were "materials" in commerce. They will also have paid attention to periodic motion of heavenly bodies for knowing accurate time and positions for safe voyage. In this way, for people in ancient Greece, nature was a "material world" and human beings were the subject that played active parts with the "material world" as a stage. Here was clearly an objective (scientific) view of nature. People in ancient Greece unawares obtained such a view of nature, based on an exploratory and adventurous life such as marine trade. This was a characteristic event in the history of human beings.

An exploratory and adventurous life such as marine trade had another important effect on philosophies and cultures in ancient Greece. It is very likely that successes in marine trade strongly depended on people's free creative plans and strategy for business, in sharp contrast to farming in which people's activities strongly depended on season and weather conditions. Besides, people who were engaged in marine trade will have often encountered various dangerous or hard situations. Such experiences must have strongly impressed on people the importance of strong rational ego, independent spirit, reliable objective knowledge and strong bodily power. Thus, people in this region came to attach the highest importance to these items.

In fact, such an attitude is clearly embodied in the way of understanding nature in ancient Greece. It is said that people in the Ionia district at the initial stage of ancient Greece understood natural phenomena in a rational way, i.e. to be due to properties of natural things themselves, in sharp contrast to people in Mesopotamia and Egypt who understood them to be due to the will or plan of gods. In addition, Pythagoras' school in south Italy made theoretical study of numbers and geometry.

Leucippus in Miletus and Democritus in Abdera Thrace proposed atomic theory. Probably, the development of elementary geometry happened because people paid attention to geometrical relations of islands and shores or those of heavenly bodies for knowing accurate time and position for safe voyage, as already mentioned earlier. Atomic theory will have come from daily acts of carefully inspecting various individual materials. According to literature [1], atomic theory came from the fact that people in ancient Greece paid attention to "units" of materials by analogy with exchange of goods and money in commerce or trade.

The attitude of making much of strong rational ego, independent spirit, reliable objective knowledge and strong bodily power is also clearly reflected in philosophies and cultures of ancient Greece in later ages. For example, Aristotle writes the following words at the top of his book, "Metaphysics".

All people desire intellect by nature.

On the other hand, Homeros says in his epic, "Ilias" ("Iliad"), as follows.

Always be the bravest and excel other people.

The spirit of praising strong bodily power of human beings is also clearly embodied in a large number of beautiful sculptures of men and women in ancient Greece as well as the rise of marathon races and Olympic Games.

Philosophies and cultures in West Europe in the Renaissance age

The Roman empire was a farmers' country [1]. Science, art, and philosophy in it did not exceed the level of ancient Greece though its military systems, law systems, and civil engineering such as the construction of roads and water supply were excellent [1]. The center of scientific study then moved to Arabian regions in the medieval time of Europe and came back to West Europe near the end of the medieval time. Interestingly, people's lives in West Europe near the end of the medieval time came to resemble those in ancient Greece. Because of the development of farming technologies in Europe in the medieval time and the accumulation of wealth, people who lived near shores of Europe came to gain wealth by marine trade [1]. Then, a large number of cities or city states were built in West South and West North Europe in a similar way to ancient Greece.

The development of marine trade in West Europe finally led to the emergence of an epoch-making era, called "the Age of the Grand Voyage", in the 15th to the 17th century. In addition, the accumulation of much wealth by the advancement of marine trade brought about a great cultural movement, called the Renaissance movement. As is well known, it was people who gained wealth by marine trade that propelled and supported the Renaissance movement. They loudly praised intellectual and bodily power of human beings, in a similar way to people in ancient Greece.

There was, however, an important difference between ancient Greece and West Europe because old social organizations in the medieval age as well as the authority of Christianity governed societies in West Europe in those years. The Renaissance movement was thus developed in the form of resistance to the old organizations, with the cultures, sciences and philosophies in ancient Greece taken as an ideal model. It appears that struggles of new thoughts in the Renaissance movement against old philosophies were effective for brushing up the new thoughts. In particular, the struggles seem to have impressed on people the importance of clarifying a way to acquire the "correct" understanding. Thus, a large number of philosophers such as F. Bacon, R. Descartes, B. Pascal, B. Spinoza, G. Leibniz, and J. Locke proposed their own original philosophies about this problem in the 17th century [1,2].

Naturally, such an atmosphere must have affected scientific study as well. It is very likely that the spirit of clarifying new truths stimulated the mind of scientists such as Copernicus, Kepler, Galilei, Descartes and Newton. In fact, fundamental ways of scientific research such as idealization and analysis methods were established in the Renaissance age, as discussed in Section 1.2.1.

The philosophy of Descartes

R. Descartes is regarded as the originator of modern philosophy in Europe. It is said that the most innovative point in his thought is that he started from the statement of "Any philosophy should be constructed based on a clear fact that anybody can no further doubt." He paid attention to clear logic in Euclidian geometry and aimed at constructing a philosophy having the same clearness as it [3].

Another important point in Descartes' thought is that he invented the analytic method as a way to discover new facts. It is said [3] that he invented it by investigating a way to solve problems in Euclidian geometry. The analytic method of Descartes has played a leading role in scientific study in later ages together with Galilei's method of idealization, as mentioned above.

Now, Descartes' philosophical study started from a question of where a clear fact was, as stated above. He refused to acknowledge what he perceived by his senses to be a clear fact for a reason that a human being had hallucinations [3]. He said that he was unable to prove that a sight of a building in front of him was not an illusion. Thus, he regarded sensationalism (the belief that only sensory images are reliable) as a prejudice. Descartes also did not acknowledge even logic or mathematical principles as a clear fact even though they themselves were completely correct because their correctness depended on human recognition and memories. Thus, Descartes was unable to find any clear fact. Probably, he fell into an entirely difficult situation. Then, he finally hit on an idea, "I continue doubting every day. So I cannot doubt myself that continues doubting in this way". Namely, he reached a famous statement, "I doubt and therefore I am". In this way, he succeeded in discovering the existence of human ego as a clear fact.

In Descartes' thought, a person's ego is completely free from anything, only depending on pure thinking [3]. It is the subject that looks at the outer world including his or her ego itself, obtains the correct objective knowledge, and rationally judges based on it. Thus, a person who has established his or her ego can live a free independent and rational life. It is said that Descartes' ultimate aim was to achieve the highest good for people through the control of emotions by his strong rational ego and the objective truth [3].

Certainly, Descartes' philosophy clearly embodied the spirit and wisdom of people who were engaged in marine trade. For Descartes and people of marine trade, nature was the "material world" and thus properties of nature were what they had to clarify and utilize. Human beings stood face to face to vast unknown nature and showed free lively activity with unknown nature as a stage. Descartes established the concept of a person's rational ego and the objective view of nature and proposed a new positive way to live a free independent exploratory and rational life with them as the base (see Section A2.1).

Features of original philosophies in West Europe

It will be evident that main features of philosophies and cultures in West Europe were established in the Renaissance age based on a free independent and exploratory life such as marine trade, which people in this region were successfully able to live. A person's strong ego, reliable objective knowledge, and a rational way of thinking were all indispensable for maintaining a free independent and exploratory life such as marine trade. It has often been claimed that philosophies and cultures in West Europe have been created based on pasturage in contrast to farming in East Asia. However, it seems that prominent characteristics of European philosophies are impossible to explain by this idea because there is no essential difference between pasturage and farming.

The progress of scientific knowledge and a scientific way of understanding in West Europe can thus be explained as a natural result of people's free independent and exploratory life in this region. Namely, science has made progress as an effective way to acquire reliable objective knowledge which is useful and indispensable for a free independent and exploratory life, as discussed in Section 1.2.1.

European philosophies in the Renaissance age are also clearly characterized by a mechanistic view of nature, in which all things are regarded as moving by physical forces. The advancement of marine trade brought about a large development of machines such as handcarts, ships, clocks and water mills, which moved simply by physical forces. Thus, people in the Renaissance age came to understand natural phenomena by analogy with machines, as discussed in Section 6.3. For example, it is famous that Descartes asserted that animals were machines. Not only Descartes but also other able persons such as Galilei and Newton had a clear mechanistic view of nature. It is said that Galilei started study on the motion of bodies for confirming an opinion that all motion in nature was caused by physical forces.

Note that the development of a mechanistic view of nature in the Renaissance age represented a marked advancement of the understanding of nature, though a mechanistic view of nature itself is of an approximate character, as discussed in Section 1.1.1 and others. In ancient Greece, natural phenomena were in general understood to be controlled by aims, wills or plans of natural things themselves by analogy with human society [1]. Such a way of understanding is called vitalism, animism or teleology (see Section 4.1.1 for vitalism and teleology). On the other hand, in a mechanistic view of nature, only a physical force was accepted as the cause of motion, as mentioned above, with complete exclusion of aims, wills or plans of natural things. Thus, for example, Descartes stressed, "Every natural phenomenon should be explained by existing laws and conditions alone, without adding any mental reason and evaluation" [3]. Reductionism in modern science strictly follows this opinion (see Section 1.1.2 and 4.1.1 for reductionism).

Limits of original philosophies in West Europe

The clear establishment of Descartes' ego caused a serious problem, called the dualism of body and mind (see Section 1.1.2, 1.2.3 and 1.4.2), together with separation between the subject and the object (see Section 1.2.3). In Descartes' thought, a person's ego was completely free from anything, only depending on pure thinking, as mentioned earlier. On the other hand, Descartes acknowledged the existence of the objective truth that obeys inevitable laws. Thus, here arose the problem of how the free will of a human being comes from a human body that follows the objective truth of a mechanical character. Descartes himself clearly recognized the existence of such a problem. It is said that the problem was already pointed out by Princess Elizabeth in Fürth, who highly respected Descartes [3]. Nevertheless, Descartes retained the dualism of body and mind most probably because both a person's ego and the objective truth were important conclusions of his considerations. In fact, reliable objective knowledge and free will (human ability to freely create plans and strategy) were both indispensable for an exploratory adventurous life such as marine trade even though they were theoretically incompatible with each other.

An important fault of Descartes' philosophy is that he caught a person's ego and the objective world individually and separately. Namely, he regarded a person's ego and the objective world as given a priori, without clarifying the origin of them. Such a way of understanding is characteristic of word-based scientific understanding (see Section 1.2.3). Descartes was unable to find a way to overcome the dualism of body and mind. It appears that later European philosophers such as Hegel and Sartre were also unable to find a way to overcome the dualism of body and mind.

According to my understanding, the dualism of body and mind is impossible to overcome without transcending word-based understanding because it comes from mutually conflicting concepts of freedom and inevitability (see Section 1.2.3). However, to transcend word-based understanding was extremely difficult for people in west Europe who clearly recognized the usefulness of such understanding in their free independent and exploratory life. In addition, originally, to transcend word-based understanding was very difficult for human beings because it is the inherent ability in human beings (see Section 1.2.1). Success in transcending word-based understanding in ancient India was achieved mainly because people in this region were forced to transcend word-based understanding under a severe social situation, as explained in the next section.

A3.1.2 Original Philosophies in East Asia

An objective view of nature had not advanced in East Asia

An objective view of nature as well as a scientific way of understanding had not advanced in East Asia such as India, China, Korea and

Japan, though technology developed largely in these regions. Instead, unique philosophies such as Buddhism, Confucianism and Taoism had advanced. Why had an objective view of nature not advanced in East Asia?

People in ancient Greece and west Europe, in particular, those who were engaged in marine trade had a wide unknown world before their eyes, into which they were able to make their way. They were successfully able to live a free independent and exploratory life, as discussed in the preceding section. On the other hand, people in East Asia were engaged in farming, in contrast to marine trade. In addition, in East Asia, a large number of people dwelled in a limited area of a large continent in a relatively closed manner. For this reason, people in East Asia were unable to live a free independent and exploratory life because such a way of living necessarily led to battles among people or countries. The difference in philosophies and cultures between ancient Greece and East Asia arose mainly from this difference in social conditions.

Roughly speaking, the development of farming technologies by the invention of bronze and iron implements in an ancient age led to the accumulation of wealth, which in turn brought about the advancement of commerce and trade on the one hand and continuing battles on the other hand, as already mentioned in Section 6.3. The advancement of commerce and trade really became important in ancient Greece, as discussed in the preceding section, while continuing battles became a severe problem in East Asia. In fact, the 6th to 3rd century B.C. in ancient China was called "the age of civil wars". Ancient India was in a similar situation.

Why did battles break out one after another? This is not because people liked to do battles. This is because people desired to live a free independent and exploratory life. If a large number of people dwelling in a closed area wanted to live such a life, clashes necessarily emerged about their opinions or interests, finally leading to battles. Thus, in ancient China and India, it became a key problem how battles were avoided or how peace was kept and hence how people were able to control their free independent and exploratory spirit. Such an atmosphere naturally led to the suppression of the free spirit of a person's ego and an objective view of nature, finally resulting in the suppression of a scientific way of understanding.

There was another important reason for why an objective view of nature had not advanced in East Asia. People in this region paid attention to living things. Namely, they looked at the behavior of human beings to keep peace on the one hand and watched plants and animals to gain wealth in farming on the other hand. Thus, for people in East Asia, nature consisted of living things, quite contrary to people in ancient Greece, for whom nature was the "material world", as mentioned earlier. Living things were not objective existents (see Section 1.1.2) and thus an objective (scientific) view of nature was useless for people in East Asia.

The rise of unique philosophies in East Asia

A matter of main concern for people in ancient India and China was how to avoid battles, how to keep peace, and how to control the free independent and exploratory spirit of individual persons, as mentioned above. Thus, the people of wisdom in these regions concentrated their considerations on these problems.

For example, Confucius, originator of Confucianism, who lived in the 6th to 5th century B.C. i.e. in "the age of civil wars" in China, taught people the importance of *Jin* (deep affection for other people, which everybody commonly has)[2]. Presumably, Confucius felt intense mental pain to see continuing battles and considered why battles broke out and how peace was kept. Thus, he deeply considered the mind of individual persons and finally found the importance of *Jin*. Confucianism is constructed with this concept as the base. It seems that Confucius found that an ideal human life was in a peaceful harmonious society of the very ancient age.

Another Chinese philosopher, Lao-tzu, who lived in nearly the same age as Confucius, awakened to the importance of Tao (the principles of the universe, which we human beings can neither look at nor hear and touch). Then, he taught people to separate from the common thoughts and cultures offered by rulers and to live with Tao as the base[2]. Namely, he taught people to live in harmony with nature rather than artificial civilization. Probably, he also felt pain to see continuing battles and considered how to keep peace. It seems that Lao-tzu found that an ideal human life was in natural scenery. His thought was later developed to a philosophy called Taoism.

On the other hand, Shaka-muni in ancient India, originator of Buddhism, who also lived in nearly the same age as Confucius and Lao-tzu, felt intense mental pain to see the death of a person or to notice the temporariness of human life, as explained in Section A1.2.1. He considered the origin of his mental pain and finally discovered that it arose from intrinsic limits of word-based understanding. Thus, he taught people to transcend word-based understanding. Shaka-muni found that an ideal human life was in the internal invisible world.

The above brief survey indicates that Confucius, Lao-tzu and

Shaka-muni taught people similar thoughts to one another. A common feature in these philosophers is that they lived in the age of a confused social situation and became aware that human civilization had a serious limit in that it led to endless disputes, battles and agonies, in contrast to natural things. Thus, these philosophers felt doubt about the correctness of human knowledge and carried out thorough consideration and pursuit about the origin of such a feeling.

Interestingly, such considerations of philosophers in East Asia were completely in the opposite direction to activities of people in ancient Greece and West Europe. People in ancient Greece and West Europe wanted to acquire new knowledge as much as possible and utilize it effectively for their life, while philosophers in East Asia felt doubt about the correctness of knowledge and wanted to transcend the limits of it.

Features of original philosophies in East Asia

The philosophies of Confucius, Lao-tzu, and Shaka-muni were later developed largely by their followers or successors. This was probably because it was always the central problem in East Asia how battles were avoided or how peace was kept.

In East Asia, individual things such as the freedom of individual persons were regarded as having a limit because they led to endless disputes or battles, as already mentioned. Thus, emphasis will have been placed on finding a fundamental principle that governs individual things so that they realize a peaceful harmonious state. An outstanding feature of philosophies in East Asia is that they really succeeded in revealing such a fundamental principle. In particular, it seems that this feature is clearly seen in Buddhism, as explained in Chapter A1 and A2.

It should be noted also that philosophies in East Asia have made progress by considering living things including human beings, in contrast to science which has made progress mainly by investigating non-living things. Therefore, philosophies in East Asia provide important concepts for correctly understanding the basic qualities of living things.

Limits of original philosophies in East Asia

It is evident that a main limit of original philosophies in East Asia is that they were unable to contribute to the advancement of scientific understanding. Because of frequently happening social confusions, people in this region attached the highest importance to the stability of society rather than its progress. For this reason, they had made enormous effort to control their free independent and exploratory spirit and such effort resulted in the suppression of scientific understanding. In fact, there was a continuing atmosphere in which professions dealing with the human mind were regarded as the noblest, much nobler than farming, manual industries and commerce.

To make matters worse, the lack of a reasonable advancement of scientific understanding seems to have led to the suppression of sound development of original philosophies in East Asia. Philosophers in East Asia succeeded in disclosing the very truth lying in the internal invisible world but their successors were unable to develop such novel ideas sufficiently.

Summary

The foregoing considerations indicate that a large difference in the original philosophies and cultures between Europe and East Asia came from a difference in social conditions.

People in ancient Greece and West Europe after the Renaissance age were successfully able to live a free independent and exploratory life such as marine trade. People who lived such a life needed much knowledge. Thus, they clearly recognized the usefulness of objective (scientific) knowledge and have made science advance to a great extent. However, for this reason, they failed to correctly understand inherent limits of objective knowledge and failed to disclose the very truth lying in the internal invisible world.

On the other hand, people in East Asia were engaged in farming in a large continent and fell into a severe social situation in which battle broke out one after another. Thus, people in this region were forced to consider why battle broke out and how peace was maintained. As a result, they came to become aware that objective knowledge had inherent limits. Accordingly, they made effort to overcome them and finally succeeded in disclosing the very truth lying in the internal invisible world. However, for this reason, they failed to correctly understand the usefulness of objective knowledge and thus failed to develop scientific understanding.

A3.2 How Does Human Understanding Advance?

Three stages of human understanding

We have considered how human understanding advances in the foregoing sections and chapters. Based on these considerations, we can say that there are three stages in human understanding.

   1. The stage at which a person has fundamental understanding

based on senses and memory.
2. The stage at which a person has word-based understanding, i.e. logical understanding by the use of ideas (consciousness) and words (knowledge).
   2.1. The stage at which a person only has simple word-based understanding without taking into account the internal invisible world.
   2.2. The stage at which a person has advanced word-based understanding by taking into account the internal invisible world.
3. The stage at which a person has arrived at the very truth and acquired the internal wisdom that works spontaneously and unconsciously when it is necessary as if it were instinctive ability.
   3.1. The stage at which a person has formed a continuous internal connection of various ideas in the brain, which agrees with an internal continuous connection of things in nature and the world.
   3.2. The stage at which a person has acquired the internal ability of the same character as Buddha's ability (or the grand internal ability) and has been given belief in it.

The grand internal ability in Stage 3.2 was explained in Section 1.4.1, 2.1.3, 2.2, and so on. It is essentially the same as Buddha's ability discussed in Chapter A1 and A2.

Where is the final goal of human understanding?

Traditional scientific understanding has remained at the stage of simple word-based understanding (Stage 2.1). Originally, it appears that the existence of the internal invisible world has not been clearly recognized to date. On the other hand, a new interpretation of modern science by taking into account the internal invisible world such as developed in Chapter 2 to 6 of this book is at Stage 2.2. A prominent feature of the understanding at Stage 2.2 compared with that at Stage 2.1 is that it provides convincing solutions to many important problems that have been left unsolved to date, as mentioned earlier. The Buddhist teachings, which explain what the very truth is and how people can arrive at the very truth, also belong to Stage 2.2.

However, the understanding at Stage 2.2 is still not the final goal of human understanding. Indeed, considerations by taking into account the internal invisible world lead to convincing solutions to many important problems, as mentioned above, but such solutions only give us "explanations". For example, it was stated at the last part of Section 1.1.2 that quantum mechanics and quantum thermodynamics only give us an explanation of the basic qualities of a living thing and do not give us the basic qualities themselves of a living thing. In fact, knowledge of a living thing, which quantum mechanics and quantum thermodynamics give us, exhibits no free independent harmonious spirit and wisdom, in sharp contrast to a real living thing. This means that we cannot correctly understand the basic qualities of a living thing by such knowledge. The very truth (the way that truly real things exist) expressed by words is not the very truth itself.

Thus, we have to transcend word-based understanding and arrive at the very truth, i.e. we have to proceed to Stage 3. Just arrival at the very truth allows us to grasp the true meanings of individual words and provides us with the truly correct understanding of nature and the world, as mentioned at the beginning of Chapter A1. Scientific knowledge displays its real worth only when we have arrived at the very truth.

Now, a problem is how we can transcend word-based understanding. There is a great difference between the very truth that is understood by words and the very truth that is grasped by real arrival at it, as argued in Section 1.4.3. Therefore, for example, even if we have completely understood arguments given in Chapter 2 to 6 and have been fully convinced of them, if we have only understood them by words, i.e. if truly real ideas (nervous patterns) in our brain are not continuously connected with one another, we shall unawares rely on individual words and stand on them and throw doubt on opinions coming from arguments in Chapter 2 to 6. It is impossible for us to escape from such a situation without forming continuous internal connections of truly real ideas in our brain.

An important point is that such a feeling of doubt gives us a clue to arrival at the very truth. Namely, we can start thorough consideration and pursuit based on such a feeling. When we have achieved a wholly clear state of mind full of delight, we arrive at the very truth. If we sincerely desire to gain the truly correct understanding, we shall necessarily come to have a feeling of doubt about knowledge understood by words (see Section A2.2 and A2.3).

The importance of arriving at the very truth was summarized in Section A1.1. The philosophy of Buddhism has revealed that everything in nature and the world, including a human being and a human idea, is governed by Buddha's ability (or the grand internal ability) so that it as a whole realizes a free independent harmonious stable state. Thus, if we have arrived at the very truth, we can realize a free independent harmonious stable state (Stage 3.1). Furthermore, we can catch Buddha's ability within ourselves and gain the internal wisdom to organize everything in nature and the world so that it as a whole realizes a free independent harmonious stable state (Stage 3.2).

The most important aspect of arrival at the very truth will be the achievement of the fusion of the subject and the object. When we remain within the realm of word-based understanding, we stand face to fact to nature and the world. Here is clearly separation between the subject and the object. Therefore, we only gain objective knowledge about nature and the world and utilize it for our life. Such a way of living may look as if we human beings act as the subject in nature and the world and positively control them. However, actually we human beings only follow the way that nature and the world exist at the present time. In addition, such a way of living tends to lead to a human-centered idea. On the other hand, when we have arrived at the very truth, we achieve the fusion of the subject and the object. Thus, we human beings can become the true subject of nature and the world and play an absolutely subjective leading role in their evolution and development. Science and technology will also take the truly correct form when they are developed under the unification of the subject and the object.

Word-based scientific understanding is always important

Note, however, that the above argument never means that word-based scientific understanding has a less important meaning. We have to note that human understanding advances via circulation. Namely, we proceed to high-level understanding by having an experience of arriving at the very truth repeatedly. If we take into account this point, word-based scientific understanding is always important in human life. In particular, the following points should be kept in mind.

Firstly, word-based understanding is intrinsic ability to human beings. We have no choice but to rely on word-based understanding in any situation on a conscious level, as discussed in Section 1.2.1 and A2.2.

Secondly, the acquisition of word-based understanding is a necessary condition for arriving at the very truth. A person at first gains sensory images (Stage 1), obtains ideas or words (Stage 2), and then arrives at the very truth (Stage 3). The very truth is never separated from word-based understanding.

Thirdly, word-based understanding is indispensable for explaining the very truth we have reached, which we cannot be conscious of. In addition, such an explanation is useful or rather inevitably necessary for us to arrive at the very truth in a new quality level or aspect or on a higher level.

A3.3 The Internal Ability in Human Life and Society

Finally, we in this section consider in what situation our life in the present world is, how human life and understanding are controlled by Buddha's ability (or the grand internal ability), and what life we can live in the present world.

Large possibilities and severe difficulties coexist in the present world

At first, let us consider in what situation our life in the present world is. Recent great progress of science and technology has brought us plenty of wealth and strong power for living. People's activities in the present days really spread throughout the world and even into the universe. In addition, we are now completely released from mysterious or irrational ideas about natural phenomena and can interpret them on a reasonable ground [*1]. The progress of science has enabled us to live a free independent creative life based on our own intelligence and wisdom. Such a situation really promises a brilliant future ahead of human beings.

> [*1] A mysterious opinion about the origin of life is proposed even in a well-developed country in the present days under an assumption of the existence of supernatural power probably because science has failed to clarify this issue to date (see, for example, Section 1.1.2). However, it is now unnecessary to rely on such a mysterious opinion because this book has revealed how the first living organism emerged on the primitive earth and how it evolved later, based on modern science on a reasonable ground (see Section 4.2 to 4.4).

On the other hand, great progress of science and technology has also brought us a diversity of severe difficulties on a global scale. The development of nuclear, chemical and bacteriological weapons has brought us a possibility of leading to the extinction of human beings. Strong utilitarianism, commercialism and materialism combined with the development of various high technologies have brought severe damage to the earth environment. Many of animals and plants are now in danger of extinction. In addition, strong utilitarianism, commercialism and materialism in the present days have also led to large distances between the rich and the poor in the world and to severe confusion and instability in society.

Furthermore, we cannot overlook the fact that great progress of science and technology has caused serious problems in the fields of ethics and morality. The clarification of the genome of living things has opened a possibility of producing a variety of new animals and plants. Studies of genome will finally lead to the control of human beings themselves. In addition, recent advancement of brain science has shown a possibility of controlling the human mind and ability.

In this way, great progress of science and technology in the present days has brought us large possibilities and severe difficulties together. There has been no age like now in the scale of possibilities and difficulties.

The emergence of large possibilities and severe difficulties in the present world proves that nature and the world are governed by Buddha's ability

How have such large possibilities and severe difficulties emerged in the present world? As argued in the preceding chapters, all people and things in nature and the world and their constituent elements are wholly governed by Buddha's ability (or the grand internal ability). Therefore, the emergence of large possibilities and severe difficulties in the present world is solely due to acts of Buddha's ability.

In fact, Buddha's ability makes various things interact with one another according to their inherent properties and motion, forms as widely-spreading fully-continuous internal connections of them as possible under given conditions in diverse quality levels and aspects, and completes the formation of such continuous internal connections (see Section 1.4.1, 2.1.3, 2.2, etc. and Chapter A1 and A2). Therefore, if nature and the world are governed by Buddha's ability, such processes occur in every area of nature and the world repeatedly. Accordingly, it is natural that all kinds of possibilities and difficulties emerge in nature and the world, irrespective of whether they are good or evil or they are constructive or destructive. The emergence of large possibilities and severe difficulties in the present world proves that nature and the world are wholly governed by Buddha's ability.

Somebody may ask why evil or destructive things or severe difficulties emerge in this world if nature and the world are governed by Buddha's ability that produces things which each realize a free independent harmonious stable state. This is because Buddha's ability works in multiple ways (see Section 1.4.1, A1.3.1, and so on). Certainly, everything in nature and the world is in a free independent harmonious stable state if it is viewed from a quality level and aspect in which it has been produced. However, even such a thing becomes an evil or destructive thing when it is viewed from other quality levels and aspects. For example, a bicycle realizes a free independent harmonious state in a person who rides a bicycle. However, such a bicycle is nothing else but an obstacle to pedestrians in the street. In addition, even a thing realizing a free independent harmonious stable state at the present time becomes an evil or destructive thing in the future as a result of historical progress of nature or society. As argued in Section 6.2, everything comes to have a limit if a new internal possibility is formed.

These considerations indicate that the emergence of evil or destructive things or severe difficulties in this world is an unavoidable matter[*2]. Thus, we have to make efforts to remove evil or destructive things one after another.

[*2]It is interesting to note that the emergence of all kinds of possibilities and difficulties, irrespective of whether they are good or evil or constructive or destructive, appears to be a necessary condition for sound evolution of nature and the world. It was argued in Section 2.2 that simultaneous realization of all possible microscopic states is a necessary condition for the formation of stable atoms, molecules and crystals or a stable equilibrium or stationary state. This argument will hold for nature and the world themselves.

Large possibilities and severe difficulties in the present world involve profound qualities beyond human personal abilities

Another important point to be noted about the emergence of large possibilities and severe difficulties in the present world is that they involve profound qualities beyond human personal abilities such as intellect, reason and creativity. This is also an inevitable result of the argument that all people and things in nature and the world and their constituent elements are wholly governed by Buddha's ability.

It was argued in Section 4.4 that the formation of an information-based connection of a large number of continuous internal connections leads to the formation of a continuous internal connection on a higher level (see also Figure 4-6). Let us recall some examples discussed there. A multi-cellular living organism, which is produced by forming an information-based connection of a large number of individual living cells, has a diversity of high-level functions such as adaptive behavior and intentional motion with an aim, which is far beyond the ability of individual living cells. The human brain, produced by forming an information-based connection of a large number of individual neurons, has a diversity of high-level functions such as thinking, creating and judging, which is far beyond the ability of individual neurons. The ecological system, produced by forming an information-based connection of a large number of individual living things, has a diversity of high-level functions such as co-evolution between water melons and animals or between blooming plants and insects, which is far beyond the ability of individual living things.

By the same principle, human society, produced by forming an information-based connection of a large number of individual persons, has a diversity of high-level functions that lie far beyond abilities of individual persons. Therefore, it is highly probable that large possibilities and severe difficulties in the present world involve profound qualities that are far beyond human personal abilities.

We human beings are unconsciously controlled by Buddha's ability when we deal with large possibilities and severe difficulties

Now, how can we deal with large possibilities and severe difficulties in the present world? As an example, let us consider how we can overcome an evil or destructive thing in the present world. The discussions given below may be too simplified but will express some essence of a relevant issue.

As mentioned earlier, the emergence of an evil or destructive thing in this world is an unavoidable matter. Therefore, such a thing emerges in this world one after another. Now, in general, it is not easy to overcome or deal with such a thing. First of all, difficulty arises from the fact that we cannot clearly recognize what is evil and what is good or what is destructive and what is constructive. For example, when a new thing has emerged in the world, we cannot know whether it is good or evil at the initial stage. Such judgment becomes possible later when a new thing has come into various relations in society. In addition, all things in nature and the world exist in infinite connection with others, i.e. all things have the reason for being at least in some aspects. Accordingly, we usually cannot reach agreement with one another about whether a certain thing is good or evil. Therefore, a controversy usually happens.

How is such a controversy settled? In general, a variety of opinions are proposed by many people and a controversy goes on. The continuation of a controversy and the proposal of a variety of opinions gradually make the essence of a problem in a controversy clear. Then, they finally lead to the creation of a good idea that a large number of people can agree with.

An important point is that such social processes of creating a good idea can be explained reasonably as due to acts of Buddha's ability to make various things interact with one another, form a new continuous internal connection (or a new internal possibility), and produce a thing that realizes a free independent harmonious stable state. The continuation of a controversy means that people carry out thorough consideration and pursuit, i.e. various opinions are made interact with one another.

Therefore, it is reasonable that the continuation of a controversy leads to the creation of a good idea, as argued in Section 6.2. Amazingly, we are unconsciously controlled by Buddha's ability when we deal with evil or destructive things in society. We human beings really play a part as the subject or commandant in this world but we play such a part based on acts of Buddha's ability.

The above example also indicates that human personal abilities such as intellect, reason and creativity have no sufficient ability to solve a problem in society. For this reason, a controversy happens. However, the above example further indicates that human personal abilities are effective enough to solve a problem in society through controversy. As mentioned earlier, large possibilities and severe difficulties in the present world arise from interactions (an information-based connection) between activities of a large number of individual persons and involve profound qualities beyond human personal abilities. Therefore, a controversy (thorough consideration and pursuit by many people, interactions between various ideas or opinions) is necessary to overcome them.

We can say in the following way as well. Certainly, human personal abilities such as intellect, reason and creativity come from Buddha's ability, as discussed in Chapter 5. However, Buddha's ability works in multiple ways, in various quality levels and aspects, as also discussed in Chapter 5. Therefore, human personal abilities can only act as one moment or element in society. They can display enough power when human personal abilities of many people work together.

It is important to acquire the internal ability of the same character as Buddha's ability and be given belief in it

We have thus far considered in what situation our life in the present world is and how human society is controlled by the internal ability. Through these considerations, we are strongly impressed with the fact that nature and the world are quite ingeniously organized by Buddha's ability. Indeed, we can be conscious of only individual ideas and things of an approximate mechanical character. However, such individual ideas and things are useful and effective for human life and thus we can show lively activity in this world, based on them. Such activities lead to the formation of a new continuous internal connection or a new internal possibility on an unconscious level. In this way, various possibilities and difficulties emerge in this world and they are reflected in our mind as feelings of interest, unclearness, unease, doubt, and so on. Then, we are unawares led to tackle such possibilities and difficulties. These activities lead to the realization of possibilities or the resolution of difficulties and lead to the advancement of society. In addition, the activities also lead us to the very truth about nature and the world and provide us with the spleendid internal wisdom and compassionate mind.

It was argued in the preceding chapters that all phenomena and events in nature including the emergence and the evolution of life and activities of our ego and consciousness happen under the control of Buddha's ability (or the grand internal ability). Similarly, all phenomena and events in society happen under the control of Buddha's ability, as mentioned above. Therefore, we can correctly treat them when we have acquired the internal ability of the same character as Buddha's ability and have been given belief in it. Indeed, phenomena and events in nature and the world include serious ones. Some of them look too irrational or cruel. However, the emergence of such phenomena or events in this world is probably necessary for sound evolution of nature and the world (see a footnote #2 of this section). In other words, adequate resolutions of such phenomena or events will lead to sound evolution of nature and the world. Therefore, we are requested to successfully deal with them.

Arrival at the very truth never means that we gain some eternal invaluable principles. It only means that our understanding jumps over the bounds of word-based understanding and extends deep into the internal world we cannot be conscious of. There is nothing visible in the internal world. However, everything including life, spirit and wisdom emerges from it. The internal world is the fountain of everything.

References

(1) For example, S. Mason, *A History of the Sciences – Main Currents of Scientific Thought*, Lawrence & Wishart Ltd., London, 1953: The Japanese edition translated by S. Yajima, Iwanami-shoten, Tokyo, 1955.
(2) W. Hiromatsu, et al. (editors), *Encyclopedia of Philosophy and Thought* (in Japanese), Iwanami-shoten, Tokyo, 1998.
(3) M. Noda, *Descartes* (in Japanese), Iwanami-shinsho, Iwanami-shoten, Tokyo, 1966.

A4. Explanations of Words

Arrival at the very truth: The very truth is in the internal invisible world and hence we cannot arrive at the very truth by the use of words. In general, we arrive at the very truth when we met a problem and have overcome it via the jump after carrying out thorough consideration and pursuit or thorough exercise. Arrival at the very truth is guaranteed by the achievement of true freedom, a wholly clear state of mind full of delight, free independent spirit and wisdom, etc. because these events demonstrate that human ideas and the real world are fully continuously connected with each other and fused into unity. When we have arrived at the very truth, we gain the internal wisdom that works spontaneously and unconsciously when it is necessary as if it were instinctive ability. We can arrive at the very truth only in a particular quality level and aspect. See Section 1.3, 1.4 and 6.2 and Chapter A1, A2 and A3 for further details.

Bodai (Japanese): The same as *Chi-e*.

Bodai-shin (Japanese): A mind to seek for the very truth.

Bodhisattva: A person who carries out hard Buddhist practice with the aim of completing the ability to conduct the common people as well as him- or herself to *Satori*. In general, a Bodhisattva takes a vow of conducting the common people as well as him- or herself to *Satori* and carries out hard Buddhist practice for accomplishing the ability to realize the vow.

Bon-noh (Japanese): A person's desires and attachments for individual things such as existence, living, wealth, love, social position, social power, fame, and so on.

Bosatsu-jo (Japanese): The way that a Bodhisattva conducts the common people to *Satori* or the way that the common people attain *Satori* by the power of a Bodhisattva.

Buddha: (1) A Buddha with an article: A person who has arrived at the very truth about human life or about nature and the world; a person who has attained *Satori*. (2) Buddha with no article: The essence or the ultimate essence of a Buddha; the same as Buddha's ability.

Buddha's ability: The same as the grand internal ability.

Bussho (Japanese): The quality specific to a Buddha.

Butsu-jo (Japanese): The same as *Bosatsu-jo*.

Chi-e (Japanese): The internal wisdom (the mind eye) to look at the very truth about nature and the world clearly. In the traditional teachings of Mahayana Buddhism, the completion of *Chi-e* leads to the achievement of complete *Satori*.

Complete *Satori* (Japanese): The accomplishment of the internal ability to conduct all people as well as myself to *Satori*.

Consciousness: Consciousness in this book mainly refers to the ability to recognize things by using ideas and words.

A **continuous connection**: A state in which things with a freely changing dynamic quality (i.e. truly real things or natural things themselves) are continuously connected with one another and fused into unity. Continuously connected things work simultaneously and harmoniously so that they as a whole realize a free independent harmonious stable state. Therefore, a continuous connection has the internal ability to organize itself so that it as a whole realizes a free independent harmonious stable state.

A **continuous internal connection**: See "a continuous connection". The word of "internal" is added to show that a continuous connection is in the internal world we cannot be conscious of.

Creation: Creation refers to discovering or inventing something new, solving a problem that has not been solved, or disclosing a new field that nobody has ever imagined or expected.

The **dualism of body and mind**: The belief that both a human body and the human mind exist. The belief involves the serious problem of why a human body following the objective truth of a mechanical character can display free will. The problem was for the first time clearly recognized in the philosophy of R. Descartes and has been a central problem in traditional western philosophies since then.

Dynamic harmony: Dynamic harmony between mutually conflicting powers or concepts means that they are simultaneously and harmoniously realized in an actual situation.

Ego: The subject of conscious or intentional activities of an individual person. A person's ego is in contact with the outer and the inner worlds, recognizes things there, and freely evaluates, judges, and plans. It also unifies all things in a harmonious way.

"**Emergence**": The appearance of a complex system with new structures, properties and functions that cannot be explained by existing things and conditions; the same as self-organization.

Engi (Japanese): This word means "emerging through or existing in dependence on others".

The **grand internal ability**: The universal and eternal internal ability immanent in nature and the world. It is, for example, embodied in the basic qualities of microscopic particles such as electrons and atomic nuclei (see Section 2.1.3 for details). The grand internal ability governs all people and things in nature and the world and their constituent elements so that they as a whole realize a free independent harmonious stable state. In other words, the grand internal ability makes various things interact with one another based on their inherent properties and motion, forms as widely-spreading fully-continuous internal connec- tions of them as possible under given conditions in diverse quality levels and aspects, and completes the formation of such continuous internal connections.

Holism: The belief that a whole is more than the sum of its parts. See also a footnote #3 of Section 1.1.2.

Hosshin (Japanese): The ultimate essence of a Buddha; essentially the same as the grand internal ability or Buddha's ability.

Illusion: Illusions in Buddhism refer to incorrect views of nature and the world arising from the intrinsic limits of word-based understanding.

An **information-based connection**: An indirect connection between things by the use of signal transmission or analogs.

In-en (Japanese): Complex connections among all people and things in nature and the world and their constituent elements, acting as causes and conditions for their realizing a fully dynamic free independent harmonious stable state.

The **internal ability**: A continuous internal connection or a thing existing as a continuous internal connection has the internal ability to organize it itself so that it realizes a free independent harmonious stable state. The grand internal ability which is embodied in a particular individual thing is in general called the internal ability.

The **internal wisdom**: We acquire the internal wisdom when we have transcended word-based understanding and arrived at the very truth. See "arrival at the very truth." See also Section A1.1 and A3.2 for details.

The **internal (invisible) world**: Human consciousness only catches common or repeatedly observed properties of natural things. Therefore, natural things themselves or truly real things are in the internal (invisible) world we cannot be conscious of. The very truth is also in the internal (invisible) world.

Jihi (Japanese): The grand compassion; a hearty wish to conduct all people to *Satori*.

Ji-riki (Japanese): A person's own ability. See "*Ta-riki*".

The *Jodo* **teachings**: One of Buddhist teachings. It teaches people that they can attain *Satori* by the ability of Amida Buddha (*Ta-riki*).

Knowledge: A linguistic expression of things we human beings are conscious of and their logical connections.

A **linear-response region**: A region in which the rates of processes in a non-equilibrium system follow the law of linear response (equation (3-10)).

A **logical connection**: A state in which things with individual unchanging properties such as words and constituent elements of machines are connected with one another according to facts (or results of experiences or experiments). A logical connection, which leads to the production of a machine, is the antithesis of a continuous connection, which leads to the production of a living thing.

Natural things themselves: The same as truly real things.

Nehan (Japanese): Nirvana; the same as *Satori*.

Nen-butsu (Japanese): An item of Buddhist practice; it refers to wishing to come to *Amida* Buddha's land named *Jodo* and deeply thinking of *Amida* Buddha with a sincere and enthusiastic mind.

A **nonlinear-response region**: A region in which the rates of processes in a non-equilibrium system do not follow the law of linear response (equation (3-10)).

Nyo-rai (Japanese): A person who has come from the very truth; the same as Buddha.

The **objective truth**: The truth expressed by objective knowledge, which is independent of human intention. The scientific truth is a typical example of the objective truth.

The **objective world**: The world consisting of objective existents. It consists of what human consciousness has created, such as human ideas and words, scientific knowledge, technology, human civilization, etc.

Objectivism: The belief that properties of natural things are independent of the mind of an observer.

Ohjin (Japanese): A manifestation of *Hosshin* in this world. *Ohjin* is a person who has Buddha's mind and ability and works in this world for conducting all people to *Satori*.

Quantum thermodynamics: The terminology which is newly introduced in this book. It refers to thermodynamics directly derived from quantum mechanics without using Boltzmann's statistical interpretation.

Real things: Real things in traditional understanding and philosophies refer to common or repeatedly observed properties of natural things or imaginary things with such properties. We human beings can be conscious of real things. See "truly real things".

Reductionism: A traditional way of scientific understanding in which an object of research is first divided into parts, each part is investigated in detail, and the original object of research is understood by a combination of parts thus investigated.

Roku-haramitsu (Japanese): A system of practice items for attaining complete *Satori*, which new Buddhists in Mahayana Buddhism designed. See Section 1.3.1.

Satori (Japanese): The attainment of *Satori* is equal to arrival at the very truth about human life or about nature and the world. See "arrival at the very truth". Roughly speaking, there are three cases: (1) Shakamuni's *Satori* (arrival at the very truth about human life by forming a continuous internal connection of constituent elements of personal life), (2) a Bodhisattva's *Satori* (arrival at the very truth about nature and the world by forming a continuous internal connection of all people and things in nature and the world and their constituent elements), and (3) Shin-ran's *Satori* (arrival at the very truth about nature and the world by acquiring the internal ability of the same character as Buddha's ability and being given belief in it).

Self-organization: The same as "emergence".

Shomyo Nen-butsu (Japanese): *Shomyo* means "calling *Amida* Buddha by name". See "*Nen-butsu*".

Tai-toku (Japanese): The acquisition of the ability beyond word-based understanding by thorough exercise or training. This word is mainly used in the fields of sport, art and technical professions. The attainment of *Tai-toku* leads us to the very truth.

Ta-riki (Japanese): Buddha's ability. See "*Ji-riki*".

The **truly correct understanding**: Human consciousness only catches common or repeatedly observed properties of natural things and fails to catch natural things themselves or truly real things. The truly correct understanding refers to catching natural things themselves as they are, or to be accurate, catching properties and behavior of natural things themselves and the internal ability that governs them as they are. In general, the truly correct understanding is obtained by transcending word-based understanding and arriving at the very truth. See "arrival at the very truth". See also "word-based understanding".

Truly real things: Truly real things refer to natural things themselves, which human consciousness fails to catch. See "real things".

The **truth**: The truth in traditional understanding and philosophies refers to the way that real things exist. See "real things". See also "the very truth".

Uncompensated heat: Heat discharged by an irreversible process.

The **very truth**: Human consciousness only catches common or repeatedly observed properties of natural things and fails to catch natural things themselves or truly real things. The very truth refers to the way that such natural things themselves exist. See "the truth".

Word-based understanding: Understanding by the use of words; the same as understanding based on things we are conscious of or based on things with individual unchanging qualities. See "the truly correct understanding".

Words: A linguistic expression of things we are conscious of.

The ***Zen* teachings**: One of Buddhist teachings, which teaches people the attainment of *Satori* through deep meditation with a concentrated mind.

About the Author
(nakato@chem.es.osaka-u.ac.jp)

Yoshihiro Nakato was born in Matsusaka, Mie, Japan at November 1942. He graduated from Osaka University in 1965 and received his Dr from Osaka University in 1972. He got a permanent position for research and education in Osaka University in 1969 and was promoted to full professor in Osaka University in 1990. He was a guest researcher at Prof. H. Gerischer's laboratory, Fritz-Haber-Institut der Max-Planck-Gesellschaft, Berlin in 1976 to 1978 and a guest professor at Prof. H. Tributsch's laboratory, Hahn-Meitner-Institute, Berlin in 1990. He supervised scientific research for solar to chemical energy conversion in a research center of Osaka University as the director. After he retired from Osaka University in 2006, he was a specially appointed professor at Institute of Scientific and Industrial Research, Osaka University from 2006 to 2016. He was also a Guest Professor at Graduate School of Science and Technology, Kwansei Gakuin University, Japan from 2006 to 2008.

His research has been in physical chemistry, in particular, in physical photochemistry, semiconductor photoelectrochemistry, solar energy conversion, and nonlinear chemical dynamics.

He published a large number of research papers in international academic journals. He also wrote many books in collaboration. In addition, he wrote some books by himself: *"A Method of Improving Creativity (in Japanese)"*, Daigaku Kyo-iku Shuppan, Okayama (2001); *"Electrochemistry – A Basis for Light Energy Conversion (in Japanese)"*, Tokyo-Kagaku-Dojin, Tokyo (2016).

He has been awarded the Progress Award from the Chemical Society of Japan in 1977, the Photochemistry Association Award in 1991, and the Electrochemical-Society-of-Japan Award in 2004.

Author Index

Aristotle, Aristotelian 22, 79, 179, 242, 355
Arrhenius, S. 25, 141
Asanga (*Mujaku*) 304
Belousov, B. P. 143
Boltzmann, L. 113, 114, 131
Born, M. 88, 96
Calvin, M. 175, 189
Clausius, R 126
Columbus, C. 244
Compton, A. H. 86
Confucius 361
Darwin, C.R. 171, 242
de Broglie, L. 86
de Donder 125, 126
Democritus 355
Descartes, R. 8, 10, 24, 51, 170, 213, 318, 356
Dogen 331
Don-ran 306, 341
Dostoevskii, F. 36
Driesch, H. 170
Einstein, A. 22, 86, 225
Eldredge, N. 172
Engels, F. 32
Fick, A. E. 134
Fisher, R. A. 171
Fourier, J. 133
Freud, S. 220, 226
Galilei, G. 22
Gautama Siddhartha
  → Shaka-muni
Gibbs, W. 127
Gordon, W. J. J. 226
Gould, S. 172
Haldane, J. S B. 171, 174
Hardy, G. H. 171
Hegel, G. 82, 222, 359
Heisenberg, W. 100
Heitler, W. 98
Hertz, H. 93
Hoh-nen 310, 320, 325
Homeros 355
Infeld, L. 22
James, W. 25
Jung, C. 220
Kauffman, S. 13, 145, 160, 177
Kawakita, J. 226
Kimura, M. 172
Lamarck, J. B. 170
Lao-tzu 361
Leucippus 355
London, F. 98
Luisi, P. 11, 149, 175, 177
Lysenko, T. D. 172
Margulis, L. 197
Maturana, H. 149
Mendel, G. J. 171
Michurin, I. V. 172
Miller, S. L. 174
Mitchell, M. 12, 14, 161, 173
Monod, J. 8, 82, 176
Nagarjuna (Ryuju) 266, 304, 306
Onsager, L. 125, 128
Oparin, A. I. 174, 175
Osborne, A. F. 225
Pasteur, L. 171
Planck, M. 86
Polanyi, M. 223
Prigogine, I. 125, 126, 135, 143
Prince, G. M. 226
Pythagoras 354
Rinzai 331
Röntgen, W. 244
Rutherford, E. 25
Sakai, K. 215
Sartre, J. P. 36, 359
Schrödinger, E. 87, 131

Shaka-muni 62, 231, 238, 240, 251, 255, 318, 331, 361
Shannon, C. 131
Shin-ran 72, 238, 240, 253, 310, 319
Shirakawa, H. 244
Taketani, M. 24
Tanaka, H. 209
Thomson, J.J. 25
Thomson, W. 136
Turing, A. 145
Urey, H. 174
Varela, F 149
Vasubandhu (Tenshin or Seshin) 304, 306
Venter, J. C. 28
Weinberg, W. 171
Wright, S. 171
Yukawa, H. 62, 232, 238
Zen-do 306, 320, 326
Zhabotinsky, A.M. 144

Subject Index

Accidental discovery 244
Activation energy 141
Aerobic bacteria 196
Akunin shouki 334
Analytic method 10, 24
Archaebacteria (archaea) 181
Artificial bacterial cell 28
Artificial intelligence (AI) 3, 29, 164, 165
Artificial life 3, 29
Autocatalysis 144, 177, 206, 246, 247
Autopoiesis 149
Belief in Buddha 298, 315, 342, 349, 371
Bodhisattva 275, 278, 286
Bon-noh 270, 311, 317
Born's interpretation 88, 96
Bosatsu-jo 275, 281
Brainstorming 225
Buddha 251, 285, 296
Bussho 303
Butsu-jo 275, 281
BZ reaction 143
Cambrian explosion 199, 205
Catalyst 141, 182
Catalytic action 145, 189
   Collective ····· 13, 147
Causality 223
Cell membrane 193
Cellular automaton 13
Central Dogma 171, 173, 203
Chaos 145
   Edge of ····· 133
Chemical potential 110, 113
Chemical reaction
   Rate constant for ····· 138, 141
   Rate of ····· 138
Chi-e 282, 292

Chloroplast 197
Coacervate 175
Common descent 183
Compound
   High energy ····· 184, 186
   High function ····· 188
Complexity science 12, 15, 132, 224
Constitutional equation 128
Confucianism 361
Contingency 10, 177
Continuous connection 45, 48, 55, 69, 95, 106, 117, 154, 160, 190, 199, 201, 207, 214, 228, 272, 274, 330, 341, 347, 364
   ····· of a completed form 233
   ····· of a non-completed form 233
Cooperation 58, 67, 211
Coupling 129, 134, 144
Creation 32, 221, 227, 237
   Theoretical ····· 231
Creativity 42, 246
Cyanobacteria 196
Darwinism 171, 203
Depth psychology 220
Determinism 10, 177
Dialectic 32, 222
Dissipative structure 143, 164, 177
Divergent thinking 227
Dual nature 67, 85, 88
Dual structure 69, 317, 340
The dualism of body and mind 9, 35, 70, 358
Dynamic harmony 58
Ediacaran period 199, 205
Ego 213, 219, 270, 357
Eigen (proper) function 90, 91, 115
Electromagnetic field 93, 102

Emergence 12, 13, 32
Emptiness 291
Entropy 112, 126
   Law of increase of ····· 14, 111, 118
   Law of a minimum ····· production 135, 153
   Negative ····· 131
   ····· production rate 129, 139, 157, 186, 191, 192
Equilibrium 140, 155, 160
   ····· state 110, 116, 117
   Local ····· 128, 152, 157, 187
Existentialism 36, 318
Eubacteria 181
Eukaryotes 181
Evolution 10, 12, 32, 65, 184, 190, 201
Fermentation 186, 196
Flux 128
Free energy 111, 127
Funi 294
Generation 32, 222
Gene switching 173, 204
Growth 32, 222
Hamiltonian operator 87, 108
Hassho-do 271
Heredity 171, 203
Heterotrophic 174
Hierarchical structure 12, 56
Holism 12, 16, 102
Homo sapiens 206
Hohjin 288, 301
Hosshin 286, 297, 340
Humanoid robot 2, 164, 165
Hydrothermal vent 180
Idealization 22, 23
Illusion 270
Indistinguishableness 101
In-en 287, 291
Infinitesimal 32

Information 131
   ····· based connection 207
Interference pattern 86, 97
Internal ability 57, 107, 120, 155, 160, 184, 189, 214, 347, 364
   Grand internal ability 59, 108, 119, 154, 178, 201, 215, 217, 330, 347, 364, 368, 372
Internal power 106, 116, 119
Internal wisdom 79, 252, 282
Internal world 46, 52, 69, 95, 211, 218, 227
Irreversibility 118, 130
   Irreversible process 110, 126, 130
Jihi 283
Ji-riki 306, 311, 329
Jodo teachings (*Jodo-kyo*) 305, 320
Jump 44, 231, 234, 330
Juni-in-en 264
KJ method 226
Kuh → emptiness
Linear response
   ····· region 132, 142
   Law of ····· 129, 140
   Nonlinear response 143
Logical connection 6, 27, 47, 272
Logical thinking 235
Mitochondria 197
Molecular biology 8, 172
Molecular genetics 172
Mutation 171, 203
Mutually conflicting concepts 34, 59, 204
Natural selection 171, 203
Nen-butsu 300, 326, 337
   Shomyo ····· 306, 320
New internal possibility 234, 241, 335
Newtonian mechanics 22, 99, 102

Nonlinear dynamics 13, 46, 177, 230
Nyorai-zo 303
Objectivism 8, 10
Objective
..... knowledge 25
..... truth 25
..... world 25, 69
Observation 91, 92, 94, 103
Ohjin 286, 340, 349
Onsager's reciprocity theorem 129
Oscillation, chemical 13, 144
Pattern formation 144
Photosynthetic bacteria 195
Physical quantity
Observed value of 90, 103
Operator for 90
Polypeptide (protein) structure
Control of 192
Positive feedback → autocatalysis
Pragmatism 25
Principle of *Engi* 263
Generalized 279, 287, 291, 296
Probability
..... of observation 88, 89, 90
..... law 91, 118
Progenote 183
Prokaryotes 181
Psychoanalysis 220
The rate of flow → Flux
Reaction
..... affinity 138
..... kinetics 162, 163, 178
..... network 145, 156, 189
Re-creation 77
Reductionism 8, 10, 170, 358
Ribosome 9
RNA-world hypothesis 175
Roku-haramitsu 284
San-po-in 271

Satori 251
Shaka-muni's 257, 272
Bodhisattva's 282
Shin-ran's 315, 328
The correct state of 315, 343, 346
Complete 275
Schrödinger equation 87, 108, 178
Self-organization 12, 143, 177, 198
Self-reproduction 194, 198
Self-stabilization 192, 198
Separation
..... between the known and the unknown 49, 50
..... between the subject and the object 33, 50, 358
hidden 64
Serendipity 244
Shi-tai 266
Solar wind 197
Sustainability 13, 191
Stationary state 133, 135, 153, 156, 183
The subject and the object
Separation between 33, 50, 358
The fusion of 33, 49, 93, 366
Superposition, The principle of 88
Symbiotic association 198
Synchronization 145
Synectics 226
System
Closed 110, 127, 154
Isolated 110, 112, 115
Open 110, 127, 154
Tacit knowledge 223
Tai-toku 38, 54, 161, 163, 208, 228, 230
Taoism 361
Ta-riki 306, 329, 332, 337

Hongan 334, 348
Teleology 170
The theory of relativity 242
Thermodynamics, The second law
    of 14, 111, 118, 126, 127,
    132, 175, 183
Thermodynamic force 128, 139
Thomson's theorem 137
Traditional
    ..... philosophy 20, 25, 52
    ..... teachings 315, 322
    ..... view of nature 20
    ..... way to attain *Satori* 311
Transport phenomena 128, 142
Trial-and-error thinking 235
    Conscious 236
    Unconscious 236
True freedom 36, 41
The true self 219, 270
Truly real thing 20, 28, 54
Truly real idea 73, 233
Turing pattern 145
Uncertainty 92, 103, 104, 112, 115
    ..... principle 31, 100, 242
Uncompensated heat 127, 130, 153
The unconscious, collective 220
The very truth 20, 26, 95
    Arrival at 39, 49, 54, 62, 71,
    78, 252, 365, 372
    Arrival at ..... on a high level 63,
    348
    The correct idea of 314, 316
Visual image 215
Vitalism 170
Whole 48, 57, 99
Woese model 181
Yoga-gyo 304
Yui-shiki 303
Zen teachings (*Zen-shu*) 307, 344

www.ingramcontent.com/pod-product-compliance
Lightning Source LLC
Chambersburg PA
CBHW050153230526
45470CB00001B/72